THE TOXICOLOGY OF CARBON NANOTUBES

The widespread and increasing use of carbon nanotubes in scientific and engineering research and their incorporation into manufactured goods has urged an assessment of the risks and hazards associated with exposure to them. The field of nanotoxicology studies the toxicology of nanoparticles such as carbon nanotubes and has become a major growth area aimed at risk assessment of nanoparticles.

Compiled by a team of leading experts at the forefront of research, this is the first book dedicated to the toxicology of carbon nanotubes. It provides state-of-the-science information on how and why they are so potentially dangerous if breathed in, including their similarities to asbestos.

The book examines various aspects of carbon nanotubes, from their manufacture and aerodynamic behaviour to their effects at the molecular level in the lungs. It is invaluable to the many groups involved with research in this area, as well as to regulators and risk assessors.

KEN DONALDSON is the Professor of Respiratory Toxicology and Scientific Director in the ELEGI Colt Laboratory, Queen's Medical Research Institute, University of Edinburgh.

CRAIG A. POLAND is a Research Toxicologist in the Institute of Occupational Medicine, Edinburgh.

RODGER DUFFIN is an MRC-funded Senior Fellow in Respiratory Medicine at the Queen's Medical Research Institute, University of Edinburgh.

JAMES BONNER is an Associate Professor in the Department of Environmental and Molecular Toxicology at North Carolina State University.

THE TOXICOLOGY OF CARBON NANOTUBES

KEN DONALDSON
Queen's Medical Research Institute, Edinburgh, UK

CRAIG A. POLAND
Institute of Occupational Medicine, Edinburgh, UK

RODGER DUFFIN
Queen's Medical Research Institute, Edinburgh, UK

JAMES BONNER
North Carolina State University, Raleigh, NC, USA

CAMBRIDGE
UNIVERSITY PRESS

CAMBRIDGE
UNIVERSITY PRESS

University Printing House, Cambridge CB2 8BS, United Kingdom

One Liberty Plaza, 20th Floor, New York, NY 10006, USA

477 Williamstown Road, Port Melbourne, VIC 3207, Australia

314-321, 3rd Floor, Plot 3, Splendor Forum, Jasola District Centre, New Delhi - 110025, India

103 Penang Road, #05-06/07, Visioncrest Commercial, Singapore 238467

Cambridge University Press is part of the University of Cambridge.

It furthers the University's mission by disseminating knowledge in the pursuit of
education, learning and research at the highest international levels of excellence.

www.cambridge.org
Information on this title: www.cambridge.org/9781009303569

© Cambridge University Press 2012

First published 2012
First paperback edition 2022

A catalogue record for this publication is available from the British Library

ISBN 978-1-107-00837-3 Hardback
ISBN 978-1-009-30356-9 Paperback

Contents

Contents

Contributors

Rob Aitken
Institute of Occupational Medicine, Research Avenue North, Riccarton, Edinburgh EH14 4AP, UK

Catrin Albrecht
Particle Research Group, IUF – Leibniz Research Institute for Environmental Medicine, Düsseldorf, Germany

Melvin E. Andersen
The Hamner Institutes for Health Sciences, Research Triangle Park, NC 27709, USA

James C. Bonner
Department of Environmental and Molecular Toxicology, North Carolina State University, Raleigh, NC 27695, USA

Matthew Boyles
Department of Molecular Biology, Faculty of Natural Sciences, University of Salzburg, Hellbrunnerstrasse 34, 5020 Salzburg, Austria

Alison Buckley
Centre for Radiation, Chemical and Environmental Hazards, Health Protection Agency, Chilton, Oxfordshire OX11 0RQ, UK

Vincent Castranova
National Institute for Occupational Safety and Health, 1095 Willowdale Road, Morgantown, WV 26505, USA

Michael P. DeLorme
DuPont Haskell Global Centers for Health and Environmental Sciences, Newark, DE 19714, USA

Ken Donaldson
ELEGI Colt Laboratory, Centre for Inflammation Research, Queen's Medical Research Institute, Edinburgh EH16 4TJ, UK

Rodger Duffin
ELEGI Colt Laboratory, Centre for Inflammation Research, Queen's Medical Research Institute, Edinburgh EH16 4TJ, UK

Kirsten Gerloff
Particle Research Group, IUF – Leibniz Research Institute for Environmental Medicine, Düsseldorf, Germany

Helinor Johnston
School of Life Sciences, Heriot-Watt University, Edinburgh EH14 4AS, UK

Ali Kermanizadeh
School of Life Sciences, Heriot-Watt University, Edinburgh EH14 4AS, UK

Amie Lund
Lovelace Respiratory Research Institute, Albuquerque, NM 87108, USA

Laura MacCalman
Institute of Occupational Medicine, Research Avenue North, Riccarton, Edinburgh EH14 4AP, UK

Robert Maynard
Centre for Radiation, Chemical and Environmental Hazards, Health Protection Agency, Chilton, Oxfordshire OX11 0RQ, UK

Jacob D. McDonald
Lovelace Respiratory Research Institute, Albuquerque, NM 87108, USA

Robert R. Mercer
National Institute for Occupational Safety and Health, 1095 Willowdale Road, Morgantown, WV 26505, USA

Fiona A. Murphy
ELEGI Colt Laboratory, Centre for Inflammation Research, Queen's Medical Research Institute, Edinburgh EH16 4TJ, UK

Craig A. Poland
Institute of Occupational Medicine, Research Avenue North, Riccarton, Edinburgh EH14 4AP, UK

Jessica P. Ryman-Rasmussen
US Environmental Protection Agency, Washington, DC 20460, USA

Roel P. F. Schins
Particle Research Group, IUF – Leibniz Research Institute for Environmental Medicine, Düsseldorf, Germany

Charanjeet Singh
Centre for Process Innovation Ltd, NETPark, Thomas Wright Way, Sedgefield TS21 3FG, UK

Rachel Smith
Centre for Radiation, Chemical and Environmental Hazards, Health Protection Agency, Chilton, Oxfordshire OX11 0RQ, UK

Wenhui Song
Wolfson Centre for Materials Processing, Department of Mechanical Engineering, School of Engineering & Design, Brunel University, Middlesex UB8 3PH, UK

Vicki Stone
School of Life Sciences, Heriot-Watt University, Edinburgh EH14 4AS, UK

Lang Tran
Institute of Occupational Medicine, Research Avenue North, Riccarton, Edinburgh EH14 4AP, UK

Klaus Unfried
IUF – Leibniz Research Institute of Environmental Medicine, Düsseldorf, Germany

Damien van Berlo
Particle Research Group, IUF – Leibniz Research Institute for Environmental Medicine, Düsseldorf, Germany

Julia Varet
Institute of Occupational Medicine, Research Avenue South, Riccarton, Edinburgh
EH14 4AP, UK

David B. Warheit
DuPont Haskell Global Centers for Health and Environmental Sciences, Newark
DE 19714, USA

1

Carbon nanotube structure, synthesis, and applications

CHARANJEET SINGH, WENHUI SONG

1.1 Introduction

Carbon nanotubes, a member of the fullerene family, are constructed of sheets of hexagonal-shaped carbon atoms rolled up into cylinders. There are two types, single-walled carbon nanotubes (SWCNT) and multi-walled carbon nanotubes (MWCNT), depending on the number of graphite sheets. The discovery of fullerenes (C_{60} or 'buckyballs') by Kroto, Curl, Smalley and coworkers in the mid-1980s has been the basis for further discoveries of interesting, complex carbon allotropes (1). However, these carbon structures were difficult to synthesise and it took several years before the mass production of fullerenes was reported (2), although it involved only gram quantities. In 1991, Iijima reported on the formation of MWCNTs, which are elongated fullerene structures stacked concentrically (3). Such structures had been observed previously but their importance went unnoticed (4–8). Iijima recognised the scientific importance of carbon nanotubes, opening a new chapter in the science of carbon nanostructures. Two years later, SWCNTs were synthesised independently by his group at NEC and by Iijima and Ichihashi (9) and Bethune *et al.* at IBM (10). Since then, nanotubes have been experimentally proven to possess remarkable electrical and mechanical properties, which are also understood theoretically. The field of nanotubes has grown tremendously into its own discipline as a model system for nanoscience, nanomaterials, and nanotechnology. Their properties have already attracted industrial interest, with carbon nanotube commercialisation in sectors such as automotive, electrical, electronics, sporting goods, renewable energy, and drug delivery, amongst others. Challenges in both basic knowledge and industrial processes remain, with major barriers confronting the carbon nanotube field being lack of synthesis control in producing monodispersed nanotubes, a systematic understanding of the nanotube growth mechanism, and an industrial approach to processing nanotubes into devices and systems. The continued innovation in selective growth, post-synthesis purification, and sorting of carbon

The Toxicology of Carbon Nanotubes, ed. Ken Donaldson, Craig A. Poland, Rodger Duffin and James Bonner. Published by Cambridge University Press. © Cambridge University Press 2012.

nanotubes hold promise for more high-value-added products such as nanoelectronics and optoelectronics.

The synthesis of carbon nanotubes has gained tremendous momentum, from the initial production of milligrams in research laboratories to the current capability of hundreds of tonnes of MWCNTs per annum. Though large quantities and high purity MWCNTs can be produced, the material is suited only for particular applications due to its intrinsic properties and product quality. Mass production of SWCNTs is still a challenge; synthesis methods have lacked sufficient control over the tube structure, resulting in polydispersity and high impurity, with only a limited amount of material being produced currently. This has led to SWCNTs being expensive, hampering efforts to commercialise them.

The market for carbon nanotubes is predicted to grow significantly, from $215 million in 2009 to $1070 million in 2014, and potentially to surpass $9 billion by 2020 (Freedonia Group (11)). According to Mitsubishi Research Institute and Mitsui, the market for fullerenes and carbon nanotubes is forecast to expand to between $3.6 billion and $32 billion by 2020 (12).

The price for MWCNTs has dropped significantly, in some cases to as low as $100 per kilogram ($0.10 per gram), and is expected to go as low as $50 per kilogram in the near future (14, 15). SWCNTs meanwhile are still rather expensive. In 1999, SWCNTs grown by the laser ablation method were sold at $2000 per gram by Rice University. Currently the price of SWCNTs has decreased to between $80 and $1000 per gram, depending on purity, quality, and method of production (16). Recent progress in selective growth and sorting of metallic and semiconducting SWCNTs has made high-purity and nearly-single-chiral nanotubes commercially available (SouthWest Nano Technologies and NanoIntegris) though significantly more expensive, with price for metallic SWCNTs (99% purity) of $899 per milligram (17, 18).

Currently there are 21 producers having more than 5 tons per year production capacity for MWCNTs. Over 94% of the producers use chemical vapour deposition (CVD) as a method of production. Scale-up activities by nanotube manufacturers pushed production capacities from 423 tons per year in 2008 to over 2389 tons per year in 2011 (14).

1.2 Structure of carbon nanotubes

Carbon can bond in various ways to create structures with entirely different properties (19). Graphite and diamond are the two bulk solid phases of pure carbon. The mystery lies in the different hybridisation states that carbon can assume. The four valence electrons, when shared equally (sp^3 hybridised), create isotropically strong diamond. When three valence electrons are shared covalently between neighbours in a plane, while the fourth is delocalised among all atoms, the resulting material is graphite. The bonding of graphite (sp^2 type) builds a layered structure with strong in-plane bonds and weak out-of-plane van der Waals bonds. Therefore graphite is weak perpendicular to its planes and is considered a soft material due to its ability to slide along the planes. Carbon nanotubes have an architecture of sp^2-bonded carbon, and are related to fullerenes.

Based on the carbon phase diagram, graphite is the thermodynamically stable bulk phase of carbon up to very high temperatures under normal ranges of pressures. It is now known that this is not true when there are only a finite number of carbon atoms, because of the high density of dangling bonds when the graphite crystallite becomes small. At small sizes, it is energetically favourable for the structure to close onto itself, removing all the dangling bonds. Experiments done in the mid-1980s indicated that when the number of carbon atoms is smaller than a few hundred, the structure formed linear chains, rings, and closed shells (1). These closed shells are called fullerenes. Fullerenes have an even number of atoms and nominal sp^2 bonding between adjacent atoms.

In order to create curved structures such as fullerenes from a planar fragment of hexagonal graphite lattice, certain topological defects have to be introduced into the structure. To produce a convex shape, positive curvature must be introduced into the hexagonal graphite lattice, by replacing hexagons with pentagons. In addition to these pentagons, heptagons can also be found in the hexagonal network. Heptagons give negative curvature to the hexagonal graphite lattice. According to Euler's principle, exactly 12 pentagons are needed to close a hexagonal structure. Euler's theorem leads to a relation, $P = S + 12$, where P and S are the number of pentagons and heptagons incorporated into a closed hexagonal network (20). Hence, the number of pentagons should always exceed the number of heptagons by 12 in any closed surface of the hexagonal network. Therefore, C_{60} and all other fullerenes which include nanotubes (C_{2n} has n−10 hexagons) can have many hexagons but only 12 pentagons (Figure 1.1a) (21).

There are two types of carbon nanotubes depending on the number of layers present. If the nanotube is made of a single layer of a graphene sheet, this tube is called a SWCNT (Figure 1.1a) (9, 21, 22). If the nanotube consists of more than one layer, this nanotube is called a MWCNT (3). The layers in these nanotubes are concentric with a constant separation of approximately 0.34 nm, slightly larger than the separation in graphite (Figure 1.1b). If a nanotube consists of two concentric layers, it is called a double-walled carbon nanotube (DWCNT) (3, 23). DWCNTs are particularly interesting because their structure and properties are close to SWCNTs but their chemical resistance is significantly improved, especially in the case where functionalisation is required.

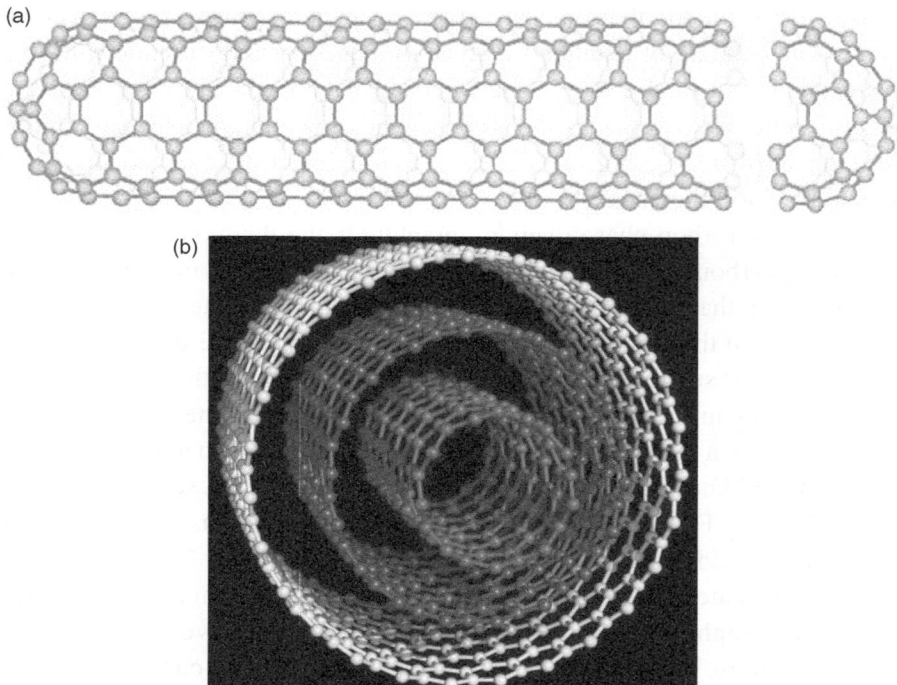

Figure 1.1 (a) Single-walled carbon nanotube (SWCNT), the basic unit of a carbon nanotube (21). (b) Multi-walled carbon nanotube (MWCNT) (courtesy of Edward Boughton).

The easiest way to imagine the structure of carbon nanotubes is to consider the conformal mapping of a finite number of two-dimensional layers of a graphite sheet into a cylinder (22). This will result in an open-ended tube, which can be closed with the addition of pentagons at the ends. The mapping can be done in many ways as long as it satisfies the criterion that the dangling bonds present at both edges are matched. A translational shift along the edges before fitting the dangling bonds will lead to a different orientation of the lattice with respect to an arbitrary tube axis. This different orientation introduces helicity into the nanotube (3, 9).

In mapping a graphene plane into a cylinder, the boundary conditions around the cylinder can be satisfied only if the Bravais lattice vectors (defined in terms of two primitive lattice vectors and a pair of integer indices (n,m)) of the graphene sheet maps to a whole circumference of the cylinder. For nanotubes, three types of folding are possible:

- $(n,0)$, known as a zigzag nanotube
- (n,n), i.e. $m = n$, known as an armchair nanotube
- all other types, known as helical nanotubes.

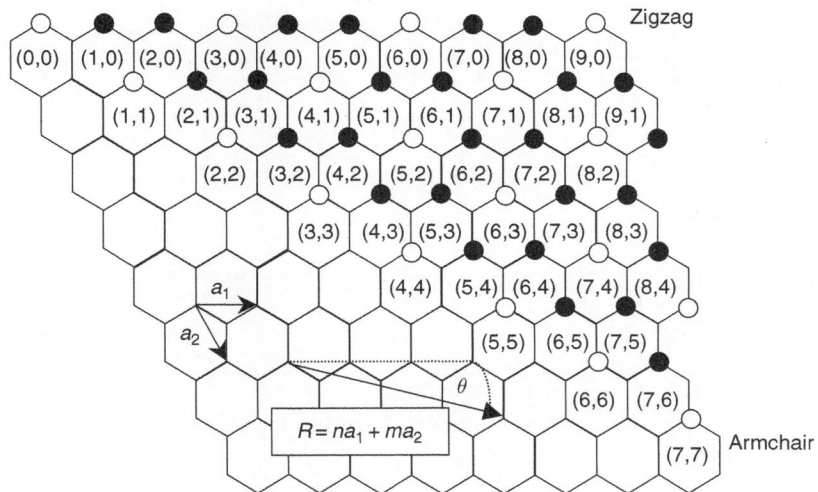

Figure 1.2 Indexing scheme showing the folding procedure to create a nanotube cylinder from a planar graphene sheet. The (n,m) naming scheme refers to the vector $R = na_1 + ma_2$, perpendicular to the tube axis, that describes how to 'roll up' the graphene sheet (a_1 and a_2 are the unit vectors of graphene in real space). Black circles indicate semiconducting behaviour, while white circles indicate metallic behaviour (19).

These foldings are shown in Figure 1.2.

Figure 1.3 illustrates the three different kinds of nanotube helicity. They are given these names for the following reason: if you take a cross section perpendicular to the nanotube axis, you will see a zigzag pattern for the $(n,0)$ nanotube, an armchair pattern for the (n,n) nanotube, and no pattern for the helical nanotubes (19). Iijima reported helicity when he first observed MWCNTs and SWCNTs (3, 9). In his paper, the helicity of the nanotubes was determined from electron diffraction patterns.

Helicity affects the electrical properties of the nanotubes. Theoretical calculations and experimental verification have shown that conductivity can be either metallic or semiconducting, depending on the nanotube diameter and helicity (21). All armchair (n,n) nanotubes are metallic, whereas zigzag and helical nanotubes are either metallic or semiconducting (24).

Some extreme nanotube structures have also been observed. The organic compound cycloparaphenylene is the shortest nanotube so far synthesised (Figure 1.4a) (25). This carbon 'nanohoop', the basic structural unit of a carbon nanotube, offers potentially more targeted chemistry to grow carbon nanotubes with a desired diameter and chirality in a controlled fashion. The thinnest freestanding SWCNT is about 3–4.3 Å in diameter (26–28). The longest carbon nanotube reported thus far has a length of 18.5 cm and was grown on a Si substrate (29). A carbon nanotube can

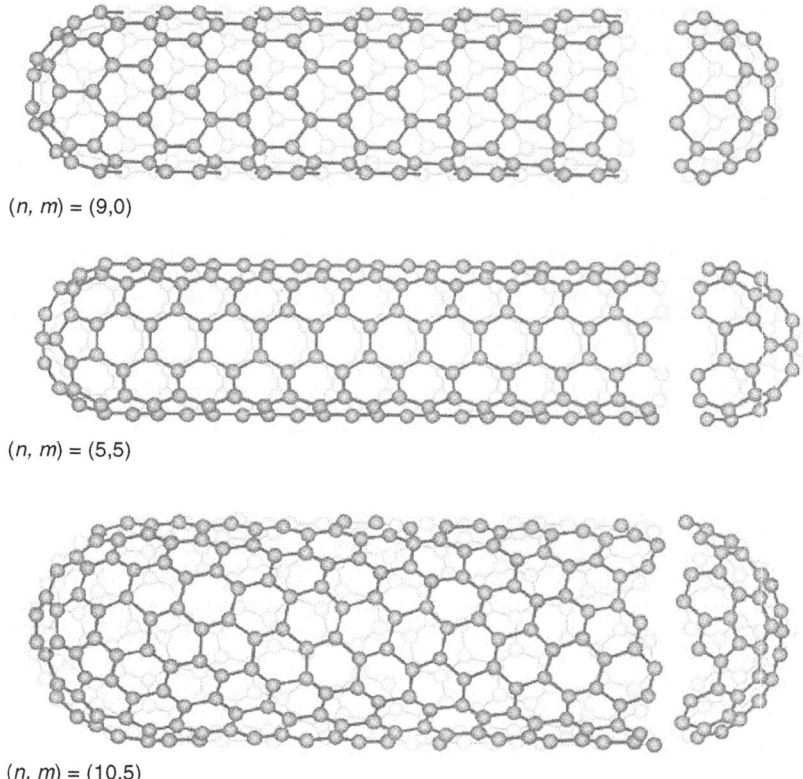

$(n, m) = (9,0)$

$(n, m) = (5,5)$

$(n, m) = (10,5)$

Figure 1.3 Examples of the three different kinds of helicity observed in single-walled nanotubes. The (9,0) zigzag nanotube, (5,5) armchair nanotube, and (10,5) helical nanotube (21).

(a)   (b)

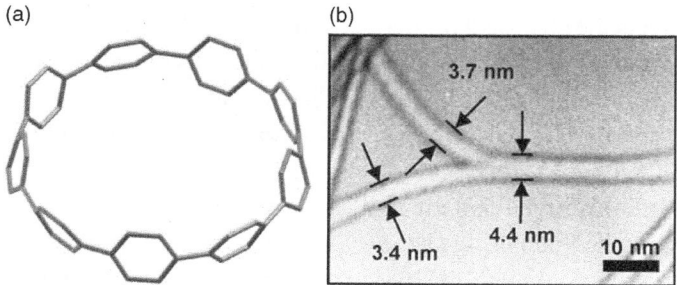

Figure 1.4 (a) Cycloparaphenylene, the shortest carbon nanotube (nanohoop) (25); (b) transmission electron micrograph (TEM) of a typical Y-junction SWCNT (34).

be bent into a 'nanotorus' (doughnut shape) which is predicted to have some unique properties, such as high magnetic moment (30). Coiled carbon nanotubes have also been predicted and observed. These CNTs are created when pentagon-heptagon atomic rings arrange themselves periodically within the hexagonal carbon network (31).

More exotic nanotube structures have also been studied. Branched carbon nanotubes are particularly intriguing because they are expected to have a profound impact on next-generation electronic devices. Branched nanotubes with T, Y, L, and more complex junctions have been observed (Figure 1.4b) (32–34). It is believed that branched nanotubes are potentially ideal nano-connect structures for nanoelectronic devices such as nanoiodides and nanotransistors. Unique carbon 'nanobuds' have been synthesised by covalently bonding spheroidal fullerenes to the outer sidewalls of carbon nanotubes (35–37). Such unique structures are believed to offer useful properties in combination with fullerenes and carbon nanotubes. The attached fullerene buds may greatly enhance the reinforcement in a nanocomposite by providing both molecular anchors and large interfaces to prevent the slipping of the nanotubes. It has been reported that such complex structures exhibit excellent field emission at a low field threshold of about 0.65 V μm^{-1} compared with SWCNTs or fullerenes alone, because of the large number of highly curved fullerene surfaces acting as emission sites on conductive SWCNTs.

1.3 Synthesis of carbon nanotubes

Rapid progress is being made in tailoring the properties of CNTs and reducing the levels of impurities and defects during production. Nevertheless, great challenges still remain in the following areas:

- low-cost, large-scale production of high-quality nanotubes, especially SWCNTs with desired structures and electronic properties
- a better understanding of the growth mechanism of carbon nanotubes
- large-scale processes for the assembly and integration, with controlled orientation and location, of carbon nanotubes into devices and systems.

Carbon nanotubes can be produced using a variety of methods. Arc discharge, laser ablation, and chemical vapour deposition (CVD) are the three main methods for both SWCNT and MWCNT production. Arc-discharge and laser ablation were among the first methods to produce nanotubes, but the equipment and energy required by either method are not cost-effective and make them difficult to scale up. A summary of these three production methods will be given, with a focus on CVD because of its efficiency, controllability, and scalability.

1.3.1 Electric arc discharge

The arc-discharge method was one of the earliest techniques for producing carbon nanotubes. It is the same apparatus used to prepare fullerenes (2). It was while investigating deposits on an electrode that MWCNTs were first observed (3). A basic schematic of the electric-arc apparatus is given in Figure 1.5 (38).

To generate nanotubes, a DC arc is usually used (with a voltage of 20 V and a current of less than 100 A) in an inert atmosphere of 500 Torr of helium or argon. The purity and yield of nanotubes is sensitive to the inert gas pressure (39). Two graphite rods are used as electrodes, one being fixed while the other is allowed to move in translation. The mobile anode electrode moves towards the cathode until the distance between them is small enough for a current to pass through the electrodes and create a plasma. The average temperature of the plasma ranges from 3500°C to 4000°C. The time-scale for the formation of carbon nanostructures is extremely small: a 5 nm MWCNT of 1000 μm length grows in 10^{-4} s (40). In order to maintain the arc between the electrodes, the anode is constantly translated to keep a constant distance. The diameter of the anode is usually smaller than that of the cathode, and both electrodes are water-cooled. By controlling the voltage and separation between the two electrodes, one can reduce fluctuations of the plasma. The arc-discharge method can produce MWCNTs or SWCNTs. For SWCNT synthesis, the anode is drilled and filled with a mixture of metal catalyst and graphite powder.

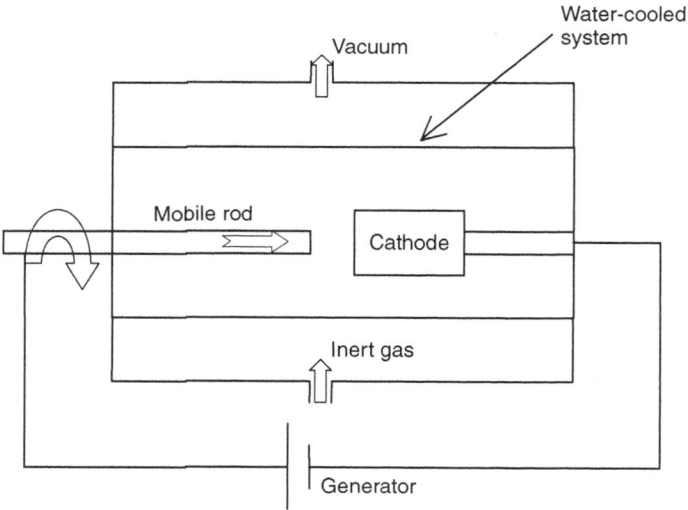

Figure 1.5 Schematic representation of the electric-arc apparatus (38).

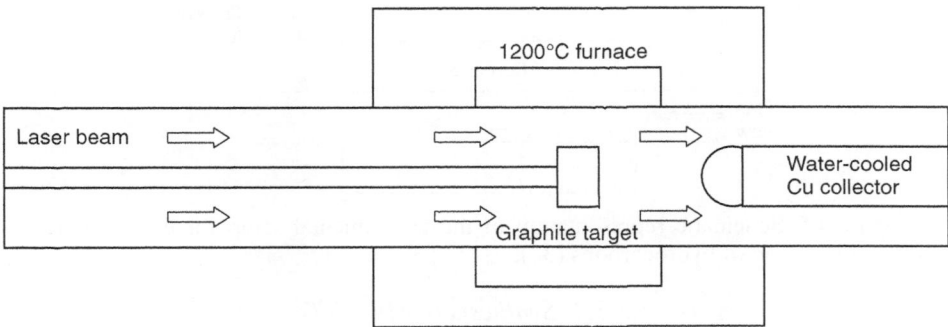

Figure 1.6 Schematic diagram of the laser-evaporation apparatus (38).

1.3.2 Laser ablation

The principle of the laser ablation method is the vaporisation of a piece of graphite by laser irradiation under an inert atmosphere (41). A graphite target is placed in the middle of a long quartz tube mounted in a temperature-controlled furnace. After the sealed tube has been evacuated, the furnace is heated to 1200°C. The tube is filled with a flowing inert gas and a scanning laser beam is focused onto the graphite target. The laser beam scans across the target surface to maintain a smooth and uniform face for vaporisation. The laser vaporisation produces carbon species, which are swept by the flowing gas from the high-temperature zone and deposited on a round water-cooled copper collector. A schematic diagram of this method is given in Figure 1.6.

Soot containing the nanotubes is collected from the water-cooled copper collector at the end of the apparatus, from the walls of the quartz tube, and from the downstream face of the graphite target. Like the arc-discharge method, the laser ablation apparatus can produce either MWCNTs or SWCNTs. SWCNTs are produced with the addition of a small amount of catalyst to the graphite target.

1.3.3 Chemical vapour deposition

Chemical vapour deposition (CVD) is a heterogeneous reaction in which both solid and volatile products are formed from a volatile precursor through chemical reaction, and the solid products are deposited on a substrate. Compared with the arc-discharge and laser-ablation methods, the main advantages of CVD are the ease in scaling up to mass production, the ability to tailor the structure and morphology of the produced nanotubes, relatively low growth temperatures, more controllable reaction parameters, low cost of production, and high yields and high purity of nanotubes. Nevertheless, CVD-grown nanotubes usually contain defects and are covered with amorphous carbon, a by-product of the thermal decomposition of hydrocarbons.

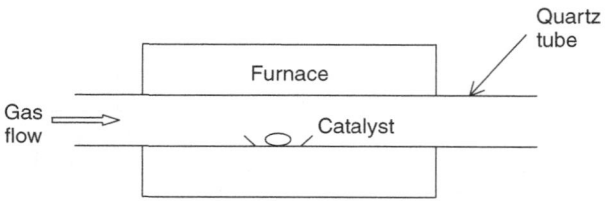

Figure 1.7 Schematic representation of the experimental setup for the catalytic decomposition of hydrocarbons (38).

1.3.3.1 Synthesis of MWCNTs

Bulk synthesis of MWCNTs by the CVD method   Joseyacaman *et al.* were the first to report the formation of carbon nanotubes by the CVD method (42). Their method involved the catalytic decomposition of acetylene over iron particles supported on graphite, at 700°C. The catalyst powder was placed in a ceramic boat, which was then placed into a quartz tube, as shown in Figure 1.7.

MWCNTs can grow in either an entangled or an aligned form. Entangled nanotubes are usually the result of synthesis using supported catalysts or the floating-catalyst method, while aligned nanotubes are formed parallel or perpendicular to a patterned or unpatterned substrate. Either process can be batch or continuous, producing large volumes of nanotubes.

MWCNT synthesis is usually carried out at high temperatures while applying a constant source of carbon diluted with a gas such as nitrogen, argon, or hydrogen (43, 44). The experimental setup consists of a high-temperature furnace in which the catalyst is placed in a crucible (Figure 1.7).

The nature and yield of nanotubes produced in the reaction are controlled by varying the catalyst and the supports, the hydrocarbon source, the gas flow, the reaction temperature, and the reaction time, among many others. By selecting the proper conditions, the structure and morphology of the MWCNTs – e.g. length, diameter, number of defects, graphitisation, shape (including straight, entangled, helical, branched, assembled), and state (including random bulks, thin films, aligned or patterned arrays) – can be tailored.

Catalysts play a crucial role in the CVD synthesis of CNTs. Various methods of catalyst preparation are used, including impregnation, ion exchange, sol-gel, co-precipitation, CVD, and mechanical grinding of the support and the metal components (43, 45–47). Catalysts can be used singly or in combination. The role of the catalyst depends on the following factors (48–51):

- reduction of the reaction barrier for decomposition of carbon feedstock vapour into carbon and its by-products
- formation of metastable carbides
- diffusion of carbon through and over the metallic particles.

Transition metals in the form of nanoparticles are considered the most effective catalysts. Among single and bimetallic catalysts that have been successfully tried are Co, Fe, Ni, Cu, Co-Fe, Co-Ni, and Fe-Ni (43, 45, 48–52). The choice of catalyst has been reported to affect the resulting nanotube deposit. The size of the catalyst particles appears to correlate with the nanotube diameter: the smaller the particle, the smaller the diameter of the nanotube formed. Furthermore, mixed catalysts generate more nanotubes with fewer defects (43, 53).

The catalyst support also plays a significant role in the production of nanotubes. The interaction between the catalyst and the support, their crystallographic orientations, and the surface roughness and porosity of the support can affect the size and distribution of metal particles and the catalytic properties (49).

Among the hydrocarbons most frequently used are acetylene, ethylene, and benzene, which are unsaturated and thus have high carbon content (42, 43, 50, 53–56). Typically, most carbon sources are diluted with a carrier gas such as argon, helium, nitrogen, or hydrogen. The optimum growth temperature varies with hydrocarbon selection. The nanotube length depends on the duration of the catalytic process, with longer durations resulting in longer nanotubes (45, 54).

Synthesis of aligned MWCNTs by the CVD method   Li *et al.* pioneered the growth of aligned MWCNTs using iron nanoparticles embedded in mesoporous silica (57). The aligned tubes grew perpendicular to the surface of the silica, with spacing between the tubes of about 100 nm. Aligned MWCNTs were also produced by the use of patterned substrates (50, 58, 59). The nanotubes were fairly uniform in diameter and of high purity. Overcrowding and the van der Waals interactions between the nanotubes were suggested as the reason for the nanotubes growing in parallel arrays. Long aligned MWCNTs of 2 mm were grown by the pyrolysis of acetylene over iron-silica substrates (60). The length of the nanotube was controlled by the duration of the reaction (50, 59, 60).

In order to produce large-scale quantities of aligned MWCNTs, organometallic catalyst precursors were used (61–66). This allowed the *in situ* formation of nanoparticles in the reaction as the precursor was sublimed. It did not require the use of pre-formed catalyst substrates. Typically, a double-stage furnace was employed, in which the catalyst precursor was placed into the first stage and nanotube growth proceeded in the second. The second-stage furnace was typically maintained between 800°C and 1100°C. Among the organometallic compounds that have been investigated are metallocenes, iron pentacorbonyl, and iron (II) phthalocyanine, while the carbon source was typically benzene or acetylene (61–66). If benzene was replaced with pyridine, carbon-nitrogen nanotubes were obtained (44). The diameter of the nanotubes was controlled by varying the relative concentration of the metallocene to carbon in the vapour phase. Satishkumar *et al.*

has proposed that the ferromagnetism of the transition-metal nanoparticle is responsible for the alignment of the nanotubes (61).

As an improvement on the organometallic sublimation setup, a syringe pump was employed to continuously feed a predetermined amount of ferrocene-xylene solution into a furnace (67). Aligned, high-yield, and pure MWCNTs were grown at 675°C, perpendicular to the surface of the quartz substrates. The conversion rate of the process was 25% of the carbon input. The low partial pressure of the solution was essential for the growth of high-purity MWCNTs. Reducing the ferrocene concentration resulted in longer and smaller-diameter nanotubes (68).

Spray pyrolysis has also been reported, using an atomiser (69, 70). The nanotube diameters produced ranged between 10 and 250 nm. The diameter, yield, alignment, and crystallinity of the nanotubes were controlled by varying the growth temperature, gas flow rate, and ferrocene concentration. The amount of encapsulated material increased with increasing ferrocene concentration.

Using photolithography, MWCNTs were grown on silica (SiO_2) while the silicon (Si) area did not result in any growth. If metal layers of nickel (Ni) were patterned onto Si, then MWCNTs were seen lifting these metal patterns during growth (71). Aligned MWCNTs have also been grown on a gold-coated substrate and subsequently used for field-emission tests (72). Vajtai *et al.* grew MWCNTs on palladium seeds using a ferrocene-xylene solution at 800°C (73).

Plasma-enhanced hot-filament CVD (PECVD) has also been used to grow aligned MWCNTs on nickel-coated glass, below 666°C (52, 74). Ammonia acted as a catalyst and dilution gas, with acetylene as the carbon source. Controllable nanotube diameters were obtained by manipulating the thickness of the nickel layer. Good alignment was achieved for nanotubes with diameters above 50 nm. Many reports on the growth of nanotubes using PECVD have appeared since this pioneering feat, with the emphasis on field-emission applications (75). With PECVD, CNTs have been reported to grow at even lower temperatures (room temperature to 100°C) (76).

1.3.3.2 Synthesis of SWCNTs

So far, focus on industrial production has been based primarily on the CVD method, including disproportionation of carbon monoxide (CO) (77), high-pressure catalytic decomposition of CO (HiPCO™) (78), the Co-Mo bimetallic catalytic process (CoMoCAT®) (79, 80), water-assisted chemical vapour deposition (Super-growth CVD) (81), and alcohol catalytic chemical vapour deposition (ACCVD) (82). More recent progress has been made toward the ultimate goal of selective growth of SWCNTs with a high percentage of metallic or semiconducting nanotubes (83) and even SWCNTs with a specific chirality distribution (79, 84–87).

Bulk synthesis of SWCNTs by the CVD method Dai *et al.* first reported the growth of SWCNTs in low yields using the disproportionation of carbon monoxide (CO) at 1200°C over nanoparticles of molybdenum, supported on alumina (77). The diameter of the produced SWCNT correlates with the size of the catalytic particle found attached to the end of the nanotube. Attention has since focused on how to obtain high yields, purity, and monodisperity of SWCNTs at large scales. Just as in the case of MWCNTs, SWCNTs have been produced by decomposition of a carbon source in the presence of various transition-metal catalysts supported on a substrate in the gas phase.

Among carbon feedstocks most frequently used are acetylene, ethylene, benzene, methane, carbon monoxide, and alcohol (77, 78, 80, 88–92). Methane is one of the most favoured carbon sources because it is a kinetically stable hydrocarbon that undergoes the least pyrolytic decomposition at high temperatures (77, 89). The duration of a run is kept short to prevent amorphous carbon formation. Similarly, CO is also used to prevent the formation of amorphous carbon overcoatings (78, 93). There is an increasing interest in alcohols such as methanol and ethanol as carbon sources in a process known as alcohol catalytic chemical vapour deposition (ACCVD). In ACCVD as developed by Maruyama *et al.*, alcohol vapour and a bimetallic Fe-Co catalyst supported on a zeolite are used to obtain SWCNTs with high purity (82, 94, 95). Alcohols may be a good carbon source for industrial-scale production of SWCNTs because of their low cost, wide availability, ease in handling and storage, and low toxicity as compared with methane and carbon monoxide.

Other carbon sources for synthesis of SWCNTs and MWCNTs without the use of any metallic catalyst have been investigated. SWCNTs with a narrow diameter distribution can grow during the decomposition of silicon carbide (SiC) at high temperature without a catalyst (96). The carbonisation of phenolic resin is also reported to produce nanotubes and other graphite polyhedral nanocrystals (97). Catalyst-free growth of SWCNTs is attractive because there is no metal residue in the products.

The size of the catalyst is of primary importance for the growth of SWCNTs. Conclusive evidence for catalyst size dependence has been reported by research groups at Duke, Stanford, and Harvard (98–100). Some reports have suggested that SWCNTs can be grown using larger catalyst particles. One hypothesis put forward is that SWCNT bundles be grown from single larger metal particles so that bundle formation would be linked to the nature of the metal surface (101). SWCNTs have also been grown from a colloid solution of ~10 nm particles, suggesting that the size of the metal nanoparticles may not need to be the same as the diameter of the SWCNT (102).

Among catalysts that have been successfully used for the synthesis of SWCNTs are Fe, Co, Ni, Cu, Pt, Pd, Mn, Mo, Cr, Sn, Au, Mg, and Al. The single-metal catalysts that have been most thoroughly investigated are Fe, Ni, Co, and Mo (77, 98–100). Mixed metal catalysts have been used to produce high yields of SWCNTs; among the most promising thus far are Fe-Mo and Co-Mo (50, 80, 90, 91, 103–107). The Co-Mo catalyst process developed by Resasco's group has made important advances in the production of large quantities of high-quality SWCNTs in a scalable fluidised-bed CVD reactor (80, 90, 91, 103). More importantly, it is reported that SWCNTs produced by their CoMoCAT® process (by SouthWest Nanotechnologies) have very narrow diameter and chirality distributions (79), commercially available in high-quality grades of both metallic and semiconducting SWCNTs.

Many types of catalyst support have also been investigated. Among the supports that have been studied are silica, silicon, alumina, magnesium oxide (MgO), zeolites, and hybrid supports (50, 89, 105, 107). Of these, silica and alumina have been thoroughly investigated.

The yield of SWCNTs varies considerably, depending on various parameters such as the metal catalyst used, metal loadings, the type of support, the reaction temperature, and the duration of the reactions.

SWCNTs can also be grown in the gas phase using organometallic precursors. Pyrolysis of iron pentacarbonyl ($Fe(CO)_5$) with excess benzene at 900°C leads to SWCNT formation (62, 108). Metallocenes can also been used as the catalyst precursor (62). The advantage of using organometallic precursors is that during decomposition they provide carbon as well as forming metal nanoparticles.

The addition of thiophene to benzene was reported to promote the synthesis of SWCNTs (88, 92). The grown nanotubes were transported out of the reaction zone by the flowing gases and were collected at the exit of the furnace. The thiophene addition was 0.5–5 wt% of the benzene, with lower weight percent resulting in purer but fewer SWCNTs. Above 5 wt% of thiophene, MWCNTs were obtained.

Rice University have developed a high-pressure gas-phase catalytic process for growing SWCNTs (78). The catalysts were formed *in situ* by the thermal decomposition of iron pentacarbonyl in a heated flow of CO, at pressures ranging from 1 to 10 atm and temperatures from 800°C to 1200°C. Although milligram quantities were obtained, such a process is continuous and has been scaled up to produce larger quantities (marketed as HiPCO™ SWCNTs by Unidym, formerly Carbon Nanotechnologies).

Another promising method that can be scaled up uses a colloidal solution of metal nanoparticles containing Co and Mo to produce SWCNTs in the gas phase (102). The colloidal solution is prepared by a reverse-micelle method and injected into a furnace at 1200°C, where the solvent (toluene) serves as the carbon source and the nanoparticles act as a catalyst. Once again, thiophene was essential for SWCNT formation. The role of sulphur was suggested for changing the surface state of the nanoparticles, blocking catalytically active surface sites by forming chemical bonds with metal atoms.

SWCNTs have also been grown by injecting a solution of ferrocene-xylene into a furnace (109). The SWCNTs bundles grow like grapevines, adhering to aligned MWCNTs. The MWCNTs act as a growth template for the directed upward growth of SWCNTs. Long strands of SWCNTs were produced using n-hexane, ferrocene, and thiophene in a vertical furnace (110).

Controlled synthesis of aligned SWCNTs by the CVD method   The synthesis of aligned arrays or designed complex networks on large substrates, and the development of patterned growth of SWCNTs with a high degree of control of size, shape, chirality, location, and orientation, are an ultimate goal for future carbon nanotube electronics. Devices made of nanotubes grown directly on substrate tend to show better performance than those using SWCNTs prepared by other methods (111). Recent progress has been made in this area despite the major challenges that remain. Hata *et al.* successfully produced millimetre-high and almost pure SWCNT arrays perpendicular to a substrate using water-assisted CVD (Super-growth CVD) (81). In this breakthrough, water was found to significantly enhance the efficiency of the catalysts by increasing their activity and lifetime during the deposition. Such a water-assisted effect has been proven to be generic for various catalysts, including Fe nanoparticles from $FeCl_3$ and sputtered metal thin films (Fe, Al-Fe, Al_2O_3-Fe, Al_2O_3-Co) on Si wafers, quartz, and metal foils. Various patterned and uniformly aligned nanotube structures have been fabricated using this method. Significant progress has also been made in the preferential growth of horizontally aligned semiconducting SWCNTs (112).

1.4 Purification of carbon nanotubes

Most carbon nanotube production methods yield an inhomogeneous raw product, consisting of nanotubes and impurities such as fullerenes, graphitised carbon structures, and amorphous carbon, as well as metal cluster impurities from the catalyst. In order to exploit nanotubes, effective purification procedures for nanotubes become necessary. To analyse the effectiveness of the purification process, scanning electron microscopy (SEM) and transmission electron microscopy (TEM) have primarily been employed. The evaluation of nanotube purity by microscopic techniques partly depends on the subjective guidelines of the experimentalist. Among techniques that have been employed for the purification of carbon nanotubes are oxidation, filtration and flocculation, chromatography, and centrifugation (113–118). Rapid progress has been made in the separation of conducting and semiconducting SWCNTs using bio-inspired sorting techniques such as density-gradient ultracentrifugation (119) and agarose gel electrophoresis (120). Systematic DNA sequences have been identified which can sort SWCNTs with single-chirality purity (121).

1.5 Properties of carbon nanotubes

Carbon nanotubes are unique because of the combination of dimension, structure, and topology that translates into a whole range of superior properties. The basic constitution of the nanotube lattice is the C–C covalent bond (as in graphite planes), which is one of the strongest structures in nature. The in-plane bond is referred to as a σ (sigma) bond. This is a strong bond that binds the atoms in the plane, and results in high stiffness and high strength of a nanotube. The remaining p-orbital is perpendicular to the plane of the σ bonds, and gives rise to π (pi) bonds, normally contributing to interlayer interactions. These π bonds, which have delocalised metallic electrons, are much weaker and less directional than the σ bonds. Here, mechanical as well as electronic and thermal properties will be reviewed.

1.5.1 Mechanical properties

The mechanical properties of carbon nanotubes have been measured experimentally, as well as calculated on the assumption of continuum elasticity. The experimental techniques are difficult to carry out because of the scale of the nanotubes; problems include placement of the nanotubes in an appropriate testing configuration and their subsequent loading for measurements.

Initial experimental measurements were made to determine the Young's modulus of a carbon nanotube. The in-plane elastic modulus of graphite is known to be 1.06 TPa (22). By measuring the amplitude of MWCNT intrinsic thermal vibrations using TEM, an exceptionally high Young's moduli of 1.8 TPa was obtained (122). The error in the measurements using such a technique can be as large as ±60%. This high Young's modulus was suggested to arise because the wrapping of the graphite sheets into a seamless cylinder gives the nanotube more deformation resistance than graphite. Using a similar technique, SWCNTs had an average modulus of about 1.25 TPa (123). Using electromechanical excitation to probe the resonant frequency of MWCNTs, a Young's modulus of 1 TPa for MWCNTs was reported (124).

Scanning probe microscopy has also been used to investigate the mechanical properties of nanotubes. Atomic force microscopes (AFM) operating in various modes have been used in this respect. For MWCNTs, a Young's modulus in the range of 0.27 to 1.26 TPa has been reported (125, 126). The MWCNT broke in the outermost layer by a 'sword-in-sheath' mechanism, indicating that the inner shells did not carry the load of the force. This breaking mechanism limits the potential use of MWCNTs for structural applications. For SWCNTs, an average Young's modulus of 1 TPa was obtained (127).

Tensile strength has also been measured. For MWCNTs, the tensile strength can vary from 1.72 to 63 GPa (126, 128). The lower value (1.72 GPa) was found in

CVD-grown MWCNTs that contained defects. The tensile strength of SWCNTs is between 13 and 55 GPa (127, 129–131). The maximum tensile strain of nanotubes has also been measured. For MWCNTs, the tensile strain at break was up to 12%, while for SWCNTs the value was 5.3% (123, 126, 127, 132).

Nanotubes are remarkably flexible and resilient. MWCNTs were repeatedly bent through large angles using the tip of an AFM (133). Based on the results obtained, local strains as large as 16% can be sustained without separating a nanotube, even when repeated bending stresses were applied. Buckling strains of ~5% and fracture strains of greater than 18% were estimated based on TEM observation of MWCNTs embedded in polymer composites (134).

Micro-Raman spectroscopy has been used to measure the compressive deformation of nanotubes embedded in an epoxy matrix (135). High Young's modulus values in the range of 2.8 to 3.6 TPa were obtained for SWCNTs, while MWCNTs had values of 1.7 to 2.4 TPa.

In theoretical predictions, the thickness of a nanotube affects the mechanical property that is being calculated. The Young's modulus of a nanotube can vary from 0.64 to 5.5 TPa (136–146). Carbon nanotube strength can be as high as 150 GPa with strains of up to 30% (138–141).

In summary, the mechanical properties of nanotubes are remarkable, with Young's modulus values in the range of ~1 TPa and extremely high strengths and strains. In spite of limited theoretical predictions and experimental measurements, the mechanical properties are size-dependent and sensitive to the concentration of defects within the tube.

1.5.2 Electronic properties

Carbon nanotubes can be metallic or semiconducting depending on their chirality and diameter. Theoretical calculations on band structure had predicted their electronic properties prior to experimental verification. All SWCNTs having armchair (n,n) chirality are metallic (147–150). Zigzag $(n,0)$ nanotubes are metallic when n is divisible by three, for example in a (9,0) nanotube. All other configuration are semiconducting. Chiral nanotubes (n,m) may be metallic or semiconducting, with metallic conduction occurring if $n-m = 3q$, where n and m are the integers which specify the nanotube structure and q is an integer. Thus all armchair nanotubes are metallic, while one-third of zigzag and chiral nanotubes will be metallic and the rest semiconducting. MWCNTs behave similarly to SWCNTs, based upon band-structure calculations (151). Interlayer coupling has been shown not to affect the electronic properties of the individual tubes.

Scanning tunnelling microscopy and spectroscopy on individual SWCNTs, from which atomically resolved images allow examination of electronic properties as a function of tube diameter and helicity, have verified earlier theoretical work (152, 153). Both metallic and semiconducting nanotubes have been observed, and their electronic properties depended sensitively on the helicity of the nanotube. For semiconducting nanotubes, the band gap scales with the inverse of the tube diameter. The band gaps of both metallic and semiconducting nanotubes were consistent with theoretical predictions.

Carbon nanotubes also behave as quantum wires. Quantum wires display ballistic transport, in which electrons pass along a conductor without experiencing any scattering from impurities or phonons, i.e. encounter no resistance. Both theoretical and experimental work has been done on ballistic transport in ideally defect-free metallic nanotubes (154).

Early experimental work concentrated on measuring the resistivity of MWCNTs. The in-plane resistivity of high-quality graphite is about 0.4 $\mu\Omega$ m (155). For MWCNTs, values ranging from 0.051 to 100 $\mu\Omega$ m have been reported (39, 155–157). The higher resistivity (100 $\mu\Omega$ m) was due to impurities and contact resistances between nanotubes (39). Results also indicate that resistance is dependent on the temperature, magnetic field, tube diameter, and the level of defects present in the nanotube (155–157). Resistance rises with falling temperature, proportional to $-\ln T$ for temperatures above 1 K, while resistance decreases with increasing tube diameter.

Resistivities of SWCNTs have also been measured. Typical resistivity values for ropes of SWCNTs are in the range of 0.3 to 1.0 $\mu\Omega$ m, while SWCNT mats have higher resistivities of about 60 $\mu\Omega$ m (158, 159).

The thermal conductivity of individual MWCNTs has also been measured. The observed thermal conductivity is more than 3000 $W\,K^{-1}\,m^{-1}$ at room temperature, greater than that of natural diamond and the basal plane of graphite (160).

1.6 Applications of carbon nanotubes

The exploitation of nanotube properties has led to a dramatic increase in the number of companies involved with nanotube synthesis and nanotube-based applications over the years (11, 161). Carbon nanotubes have become one of the most commercially relevant classes of nanomaterials, potentially having the broadest range of applications, from composites to consumer electronics, energy storage, health care, and many others. Some industrial applications are already on the market. Nanotubes were first used as polymer fillers to achieve effective anti-static charge shielding with enhanced mechanical properties. Hyperion Catalysis pioneered the mass production of MWCNTs and a range of MWCNT-polymer masterbatches for applications in the automotive and electronics industry (8). CNTs are also appealing electrode materials for improving lifetime and performance in batteries. NEC has developed a fuel cell from a type of carbon nanotube (nanohorn electrode) that has about 10 times the energy capacity of a lithium battery; if it is used for personal computers in the future, a continued usage time of several days can be expected (162). Some low-volume products, such as scanning probe microscope tips and X-ray tubes, have also been commercially available (163). Despite increasing commercial development and reports on a variety of potential applications in the research lab, carbon nanotubes have not yet fulfilled their promise. High-value-added end-user products in the global market of emerging carbon electronics, sensors, and biomedical devices have yet to appear. Here, a summary of a few industrial and other potential applications is presented.

1.6.1 Carbon nanotube composites

The lightweight hollow structure and exceptional mechanical properties of nanotubes make them ideal reinforcements at the nanoscale for polymer nanocomposites. Apart from substantially improving the mechanical properties of polymer, ceramic, and metal matrices, nanotubes can add mulifunctionality to composite systems because of their remarkable electrical and thermal conductivities and other useful physical, chemical, and biological properties (164). Therefore CNT/polymer nanocomposites can be classified as structural or functional composites. Improvements at various levels have been demonstrated in a wide range of polymer-based nanocomposites, and to a lesser extent in ceramic and metal-based composites. Nevertheless, the challenge to exploit those properties in real macroscope materials still remains. Some primary problems have been nonuniformity of material samples, dispersion, alignment, and functionalisation of nanotubes, and resulting poor interfacial interaction between nanotubes and polymer matrices.

As with many other nanocomposite systems, improvement in the mechanical properties of composites reinforced by MWCNTs does not always increase proportionally to nanotube content. In addition to the separate properties of the nanotubes and polymer, the effective surface of the nanotubes and the interface between the nanotubes and the polymer matrix are primary parameters in determining the properties of the composites (165). The small diameter (nanometre scale) with high aspect ratio (>1000) and thus extremely large surface area make CNTs different from other conventional fillers. A large surface area means a large interface area is present between the nanotubes and the matrix, even at a low loading of nanotubes. In the interfacial region, polymer chain mobility and conformation, degree of cure, and crystallinity are altered from those of the nanotubes or the polymer matrix, being attributed to interfacial interaction and nanoconfinement. Therefore properties of composites such as stiffness, strength, mechanical toughness, and conductivity can be significantly modified, even at an extremely low nanotube content, in comparison with conventional particle-filled composites. For example, an intense stirring process was used to disperse MWCNTs in epoxy, resulting in a matrix conductivity of around of 10^{-2} Sm^{-1} with filler volume fractions as low as 0.1 vol.% (166). Because of the low filler fractions, the rheological and mechanical properties of the matrix were not compromised. To obtain ideal nanocomposites, the critical challenges lie in uniformly dispersing the nanotubes, achieving desirable alignment and nanotube-matrix adhesion to provide effective stress transfer, and avoiding intratube sliding within MWCNTs and intrabundle sliding within SWCNT ropes.

Both covalent and non-covalent routes have been explored in recent decades in order to achieve uniform dispersions, alignment control of nanotubes, and high interface bonding, but with limited success (165). Covalent functionalisation of nanotube surface is an effective way of dispersing nanotubes. However, covalent sidewall chemistry often induces substantial and/or irreversible changes to the structure and properties of nanotubes. To preserve the exceptional structure and properties of CNTs or to minimise perturbations, various methods of non-covalent functionalisation have been reported. Surfactant encapsulation and polymer wrapping are typical strategies which have been investigated, in a major effort. Surfactant-assisted processing of nanotube–polymer composites improved dispersion as well as interfacial bonding of the nanotubes with poly(vinyl alcohol) hydrogel, resulting in a 133% increase in tensile strength with the addition of 1 wt% of MWCNTs (167).

Following the basic principles of advanced fibre-reinforced composites, the alignment of nanotubes into a continuous fibre or thin film is an effective way to exploit the exceptional anisotropic properties of individual CNTs and transform them for micro- or macro-scale applications. Various processing techniques have been applied to align nanotubes in the polymer matrix using external fields such as electrical or magnetic fields or shear in the flow. A self-assembling strategy, as in the liquid crystalline phase of MWCNT aqueous dispersions and SWCNTs in superacid dispersions, may provide a feasible way to facilitate large-scale alignment of nanotubes during processing (168, 169). Several production routes have also been developed to fabricate continuous CNT microfibres, including wet spinning of CNTs from polymer dispersions or acid dispersions, dry spinning from aligned CNT matrices, and direct spinning from CVD reactions (170–175). Some high-performance CNT fibres begin to match conventional carbon fibres in terms of strength and stiffness, but have exceptional flexibility (toughness), considerably higher than that of any commercial fibre. The elastic modulus of nanotube fibres has been reported between 5 and 357 GPa, the tensile strength between 0.1 and 8.8 GPa, and toughness between 2 and 121 Jg^{-1}. Nevertheless, development of continuous CNT fibre composites is hampered by current laboratory-scale production of CNT fibres, inconsistent fibre quality, and high cost.

Development of CNTs and composites in the form of continuous fibres, films, and bulks is a safe and effective way of using CNTs, and CNT/polymer functional composites have gradually entered the market. The first major commercial application of nanotubes has been fillers in conductive composites for dissipating electrostatic charge, such as automotive mirror housings that are electrostatically painted on the assembly line (176). Another early winner has been in sporting goods: CNTs have been used in tennis rackets, golf clubs, and racing bicycles (14). Applications to heat resistance, chemical sensing, electrical and thermal management, photoemission, energy storage, and electromagnetic absorbtion are under development. Industrial applications of high-performance CNT/polymer composites still have a long way to go. Potential applications include automotive and aerospace uses, where nanotube composites for structural components are expected to reduce weight and hence decrease fuel consumption.

1.6.2 Electronic devices

Scanning probe microscope tips   Dai *et al.* first demonstrated the use of carbon nanotubes for scanning probe microscope tips (177). Their structural, mechanical, and electrical properties make nanotubes ideally suited for scanning probe tips. The high aspect ratio of the nanotubes allows the probing of deep crevices that occur in microelectronic circuits, and the small effective radius of the nanotube tip improves the lateral resolution beyond what can be achieved using commercial silicon tips. Nanotubes could be used for both AFM and scanning tunneling microscopy (STM) measurements. Nanotube tips with a capability for chemical and biological discrimination have been created with acidic functionality and by coupling basic or hydrophobic functionalities or biomolecular probes to the carboxyl groups that are present at the open tip ends (178). The functionalised nanotube tips will prove especially useful for imaging self-assembled polymeric and biological materials. Individual MWCNTs have been grown using an iron catalyst deposited onto the ends of silicon tips (179). The earlier method of attaching nanotubes for tip fabrication is time-consuming and selects against the smallest nanotubes, limiting the quality of tips. SWCNTs have replaced MWCNTs as tips which allow even higher resolution, since SWCNTs have diameters of between 1 and 2 nm (180, 181).

Field-emission devices (FED)   Nanotubes have properties intrinsically suitable for field emitters, such as nanometre-scale radii of curvature, high mechanical stiffness, chemical inertness, and electrical conductivity. Rinzler *et al.* first reported the field-emission properties of carbon nanotubes (182). Subsequently many reports have emerged on this subject, using both MWCNTs and SWCNTs (183–188). Such low-cost 'cold cathodes' suggest a variety of applications, including flat-panel displays and X-ray sources, electron guns for electron microscopy, and microwave amplifiers. Various methods of fabricating carbon nanotube field emitters, including direct growth of aligned nanotubes by CVD or PECVD, screen printing, electrophoretic deposition, and spray coating, have been developed in order to achieve stable and uniform emitters with high emission current densities. In combination with photolithography, it is possible to control the density of nanotube emitters, shapes, and scalable patterns. The latest report shows stable high current densities (1.5–1.9 A/cm^2 in a pulse field) in CNT field emitters made by electrophoretic deposition (188). Saito *et al.* reported on the field emission from open and closed MWCNTs and its application to electron tubes (185). Electron beams from field emitters have distinct advantages over thermionic emitters, including small energy spread, high current density at low voltage, high coherence, and fast response time. The combination of these properties allows for finely focused electron beams with high brightness. Furthermore, nanotubes are made of carbon and so are free of precious or hazardous elements. A compact prototype imaging system based on a cold-CNT X-ray tube with fast turn-on/off is under clinical test for high-speed image-guided radiotherapy for cancer research and treatment (189). Nevertheless, short lifetime at high current density, instability under high voltage, poor emission uniformity, and pixel-to-pixel inconsistency are still major obstacles.

Transistors and optoelectronic devices   There is strong interest in the use of carbon nanotube-based electronic and photonic devices, since SWCNTs have demonstrated remarkable performance in individual field-effect transistors (FETs), photoluminescence, and electroluminiescence (190–192). The first room-temperature carbon nanotube transistor was made using a semiconducting SWCNT (190). Since then, many advances in individual CNT transistors (CNT-FETs) and nanotube network thin-film transistors (CNT-TFTs) have been made. Electron carrier mobilities of 10 000 to 100 000 cm V^{-1} s^{-1} have been reported with I_{on}/I_{off} current ratio generally at 10^5–10^7 (193). CNT-FETs with a sub-20 to 30 nm channel have demonstrated excellent performance (194). The transistors' operation characteristics can be qualitatively described by models used for traditional semiconductor devices. Because of their size and high I_{on}/I_{off} current ratio, less power is required than with silicon-based devices (195). Potential applications may be possible because of their high switching speeds and improved voltage gains at room temperature. The transistors were fabricated by lithographically applying electrodes to nanotubes that were either randomly distributed on a silicon substrate or positioned on the substrate with an AFM or by self-assembling (196). By creating junctions of metallic and semiconducting nanotubes, diodes have been developed (195). Logic circuits have been demonstrated by Bachtold *et al.* and Derycke *et al.* (197, 198). Beyond single transistor devices (AND and NOR gates), more complex structures such as ring oscillators have been fabricated and characterised (199). CNT-FET-based photonic devices such as CNT light emitters, photoswitches, and photodetectors have also been studied, showing potential in future optoelectronic devices and solar cells that cover the near-infrared region of the solar spectrum (200). The success of carbon-based electronics and photonics will depend on the production of pure single-type (diameter/chirality) semiconducting nanotubes and on how rapidly techniques for fabrication, doping, and manipulating nanotubes and interconnections can be developed.

Carbon nanotube electrodes   The conductivity of SWCNT thin films has been investigated for use in transparent, flexible, and even stretchable electrodes (201). SWCNT film has a conductivity comparable to the incumbent indium tin oxide (ITO), uniformity over large areas, and compatibility with various display technologies and fabrication processes. It also exhibits optical transmittance comparable to that of ITO in the visible spectrum, but far superior transmittance in the infrared spectral band. CNT-based films have been demonstrated to be mechanically more robust than ITO, and can be deposited using a variety of low-cost and low-impact methods. Such films are also chemically resistant, and are manufactured from carbon, one of the most abundant elements on Earth. In addition, CNT transparent conductive films (TCF) offer a 'green' solution to ITO-based film products, as hazardous chemical etchants are not used in the TCF manufacturing process. Applications for touch-screens, photovoltaics, display bus bars, and beyond are under development, with some demonstration of prototypes.

Apart from application to transparent electrodes, nanotube electrodes have shown promise for use in supercapacitors (202, 203), Li batteries (204), solar cells (200, 205), fuel cells (206, 207), actuators (208, 209), and sensors (210, 211), among other things. For example, using vertically aligned CNTs as electrode structures can lead to supercapacitors with energy densities higher than $60\,W\,kg^{-1}$ and lifetimes longer than 300 000 cycles (212). Using functionalised carbon nanotubes as the positive electrode produced a significant increase – up to 10-fold – in the amount of power delivered for a given weight of material, compared to a conventional lithium-ion battery (204). Such electrodes might find application in small portable devices, and with further research might also lead to improved batteries for larger, more power-hungry applications.

Recent rapid development of wet processes for the purification, sorting, and formulation of CNTs has made high-quality carbon nanotube ink commercially available. Although the ink is still expensive, it has been demonstrated that CNTs can be printed, with the potential for high-performing devices to be made using low-cost manufacturing process such as printing over large areas. Applications of printed nanotube electrodes may include transistors and supercapacitors which bridge the gap between batteries and capacitors, leveraging the energy density of batteries with the power density of capacitors and transistors.

1.6.3 Hydrogen storage

A safe, economical, and effective storage system is critically needed for the future utilisation of hydrogen as a pollution-free energy resource (213). Extensive experimental studies as well as theoretical calculations have been done for hydrogen storage in carbon nanotubes. Initial promising experimental results have been shown to be incorrect, and careful attention must be paid to the experimental parameters. Experimental studies can be divided into two parts: gaseous and electrochemical. The degree of H_2 adsorption by nanotubes depends critically on pressure and temperature.

H_2 storage in nanotubes was first investigated with SWCNTs (214). The adsorption of hydrogen on SWCNT soot (0.1–0.2 wt%) was probed using temperature programmed desorption (TPD) spectroscopy, in an ultrahigh-vacuum chamber. Results indicated that SWCNTs could store 5–10 wt% of H_2. H_2 uptake in herringbone-stacked graphite nanofibres was reported to be as high as 67 wt% at room temperature and 12 MPa (215). Subsequent measurements failed to reproduce the earlier results, and H_2 uptake was reported to be less than 0.1 wt% (216). Errors also occurred in the study on H_2 storage of Li-doped and K-doped MWCNTs (217). Subsequent experiments proved that these high uptake values were due to moisture contained in the H_2 cylinders (218). The purity of the H_2 gas was also found to affect the results. Values of 2.5 wt% H_2 storage for Li-doped nanotubes rather than 20 wt% were obtained.

Overall, carbon nanotubes as H_2 storage media show quite moderate performance. H_2 uptake of nanotubes achieved so far is up to ~3.0–7.0 wt% at 77 K (219, 220), still quite far from the target value set by the US Department of Energy (6 wt% at nearly ambient conditions), and makes carbon nanotubes poor candidates for hydrogen storage applications. Nevertheless, computer modelling predicts that higher H_2 uptake (total uptake amounts up to 19.0 wt% at 77K and 5.5 wt% at 300K) can be realised through 3D packing of carbon nanotubes (221).

1.7 Biomedical applications

There has been growing interest in incorporating carbon nanotubes into biological systems, including proteins, DNA, and living cells. The exploration of CNTs in biomedical applications is underway, from drug-delivery carriers to substrates for vaccines, tissue scaffolds, implants, bio-transistors, ultrafast DNA sequencing, biosensors, and other devices.

1.7.1 Drug delivery and imaging

It has been reported that individual SWCNTs and MWCNTs can act as 'nano-needles' across the plasma membrane to deliver therapeutic and diagnostic small molecules, macromolecules, or contrast agents to a variety of cells (222). Owing to their unique electrical, optical, thermal, and spectroscopic properties in biological systems, CNTs allow efficient electromagnetic stimulation and sensitive detection using various imaging techniques, offering great advances in the detection, monitoring, and treatment of diseases. The majority of *in vivo* studies using functionalised nanotubes have focused on cancer therapy as preclinical models. Nanotubes have been used to halt tumour growth in various types of treatments such as chemotherapy, hyperthermia, and gene silencing (223–225). Meanwhile, some toxicological studies have generated serious discussions regarding the overall toxicity profile of carbon nanotubes.

1.7.2 Scaffolds for tissue regeneration

The ability to functionalise the sidewalls of CNTs also leads to biomedical applications such as bone scaffolds and vascular stents, and in neuron growth and regeneration. SWCNT-reinforced polymer nanocomposite scaffolds in rabbits are found to promote a three-fold greater growth of bone and show less inflammation after 12 weeks than those implanted with polymer scaffold tissue (226). The results suggest that porous nanocomposite scaffolds are not only osteoconductive but may be bioactive, assisting osteogenesis. Carbon nanotubes are also promising materials for neural prostheses because of their fibrous nanostructure and high electrical conductivity. Carbon nanotubes have been used as substrates for promoting cell attachment and growth of neurons (227), and neurons grown on a conductive nanotube network always display more efficient signal transmission (228).

1.7.3 Bioelectronics, biosensors, and medical devices

DNA plays a vital role as the carrier of genetic information in all living species. The combination of DNA and SWCNTs offers unique properties for molecular-based electronics devices. For example, field-effect transistors fabricated from SWCNTs coated with single-strand DNA (ssDNA) show remarkable chemical sensing capability with sequence-dependent chemical reorganisation (229–231).

Owing to their small size, high electrochemical activity, excellent physical properties, low density, and biocompatibility, CNT electrodes have huge potential for implantable applications such as neurological tissue stimulation and continuous monitoring of clinically relevant analytes including glucose (with relevance to the control of diabetes), lactates, antibodies, and antigens, as well as for analysis of analytes in bioreactors, veterinary and clinical chemistry, the food industry, and environmental science. Metal neural electrodes coated with a combination of carbon nanotubes and a conducting polymer have improved the efficiency of the interface between brain cells and the surface of an implanted electrode by 1000 times or more. Such electrodes significantly enhance the quality of brain-function measurements and are compatible with existing devices (232). CNT fibre-based amperometric glucose biosensors have shown higher efficiency than traditional metal electrodes for glucose biosensing (233).

Nanotube-incorporated products already exist commercially and further products are anticipated in the coming years and decades. Some applications, such as nanomedicines and tissue-engineering scaffolds, may require long periods of development since simultaneous clinical and toxicological investigations for risk-benefit assessments will be required.

1.8 Safety and future development

Commercial application of nanomaterials and nanotechnology cannot proceed in a responsible society without an understanding of the safety implications. Standardisation of CNT products, and toxicity and hazards assessments, must take place in parallel with research and development. It is imperative that all potential industries mentioned above, and environment and health regulations, ensure safe handling in the workplace, safe products in the market, and intelligent end-of-life strategies. Once these issues are addressed, technology based on nanotubes will be available on an industrial scale.

References

1. H. W. Kroto, J. R. Heath, S. C. Obrien, R. F. Curl, and R. E. Smalley, C-60: Buckminsterfullerene. *Nature*, **318**: 6042 (1985), 162–3.
2. W. Kratschmer, L. D. Lamb, K. Fostiropoulos, and D. R. Huffman, Solid C-60: A new form of carbon. *Nature*, **347**: 6291 (1990), 354–8.
3. S. Iijima, Helical microtubules of graphitic garbon. *Nature*, **354**: 6348 (1991), 56–8.
4. J. Abrahamson, P. G. Wiles, and B. L. Rhoades, Structure of carbon fibers found on carbon arc anodes. *Carbon*, **37**: 11 (1999), 1873–4.
5. P. Ball, Roll up for the revolution. *Nature*, **414**: 6860 (2001), 142–4.
6. M. Monthioux and V. Kuznetsov, Who should be given the credit for the discovery of carbon nanotubes? *Carbon*, **44** (2006), 1621–3.
7. A. Oberlin, M. Endo, and T. Koyama, Filamentous growth of carbon through benzene decomposition. *Journal of Crystal Growth*, **32** (1976), 335–49.
8. H. G. Tennent, inventor Hyperion Catalysis Int'l, Inc. (Cambridge, MA), assignee. Fibrils, method for producing same and compositions containing same, U.S. patent 4 663 230 (A), 1987.
9. S. Iijima and T. Ichihashi, Single-shell carbon nanotubes of 1-nm diameter. *Nature*, **363**: 6430 (1993), 603–5.

10. D. S. Bethune, C. H. Kiang, M. S. Devries, *et al.*, Cobalt-catalyzed growth of carbon nanotubes with single-atomic layerwalls. *Nature*, **363**: 6430 (1993), 605–7.

11. Freedonia Group, *World Nanotubes: Industry Study With Forecasts to 2009, 2014 & 2020*. The Freedonia Group, Cleveland, OH, Study no. 2019, January 2006. http://www.freedoniagroup.com/brochure/20xx/2019smwe.pdf [cited 20 August 2011].

12. Mitsui to build carbon nanotube mass-output plant. December 2001. http://www.planetark.com/dailynewsstory.cfm/newsid/13864/story.htm [cited 28 August 2011].

14. L. M. Sherman. Carbon nanotubes: Lots of potential – If the price is right. 2007. http://www.allbusiness.com/manufacturing/chemical-manufacturing-resin-synthetic/4505811-1.html#ixzz1WBfYjVMS [cited 28 August 2011].

15. http://www.cheaptubesinc.com/carbon-nanotubes-prices.htm [cited 29 August 2011].

16. Website containing nanotube suppliers. 2002. http://www.personal.rdg.ac.uk/~scsharip/tubes.htm [cited 21 August 2011].

17. http://www.swentnano.com/index.php [cited 28 August 2011].

18. http://www.nanointegris.com/en/metallic-m [cited 26 August 2011].

19. P. M. Ajayan, Nanotubes from carbon. *Chemical Reviews*, **99**: 7 (1999), 1787–99.

20. S. Iijima, T. Ichihashi, and Y. Ando, Pentagons, heptagons and negative curvature in graphite microtubule growth. *Nature*, **356** (1992), 776–8.

21. M. S. Dresselhaus, G. Dresselhaus, and R. Saito, Physics of carbon nanotubes. *Carbon*, **33**: 7 (1995), 883–91.

22. M. S. Dresselhaus, G. Dresselhaus, and P. C. Ecklund, *Science of Fullerenes and Carbon Nanotubes* (San Diego CA: Academic Press, 1996).

23. E. Flahaut, R. R. Bacsa, A. Peigney, and C. Laurent, Gram-scale CCVD synthesis of double walled carbon nanotubes. *Chemical Communications*, **12**: 12 (2003), 1442–3.

24. M. S. Dresselhaus, Nanotechnology: New tricks with nanotubes. *Nature*, **391**: 6662 (1998), 19–20.

25. R. Jasti, J. Bhattacharjee, J. B. Neaton, and C. R. Bertozzi, Synthesis, characterization, and theory of [9]-, [12]-, and [18]cycloparaphenylene: Carbon nanohoop structures. *Journal of the American Chemical Society*, **130**: 52 (2008), 17646–7.

26. L. Guan, K. Suenaga, and S. Iijima, Smallest carbon nanotube assigned with atomic resolution accuracy. *Nano Letters*, **8**: 2 (2008), 459–62.

27. T. Hayashi, Y. A. Kim, T. Matoba, *et al.*, Smallest freestanding single-walled carbon nanotube. *Nano Letters*, **3**: 7 (2003), 887–9.

28. X. Zhao, Y. Liu, S. Inoue, *et al.*, Smallest carbon nanotube is 3 Å in diameter. *Physical Review Letters*, **92**: 12 (2004), 125502–5.

29. X. Wang, Q. Li, J. Xie, *et al.*, Fabrication of ultralong and electrically uniform single-walled carbon nanotubes on clean substrates. *Nano Letters*, **9**: 9 (2009), 3137–41.

30. L. Liu, G. Y. Guo, C. S. Jayanthi, and S. Y. Wu, Colossal paramagnetic moments in metallic carbon nanotori. *Physical Review Letters*, **88**: 21 (2002), 217206–9.

31. S. Amelinckx, X. B. Zhang, D. Bernaerts, *et al.*, A formation mechanism for catalytically grown helix-shaped graphite nanotubes. *Science*, **265**: 5172 (1994), 635–9.

32. D. Zhou and S. Seraphin, Complex branching phenomena in the growth of carbon nanotubes. *Chemical Physics Letters*, *238: 4–6* (1995), 286–9.

33. J. Li, C. Papadopoulos, and J. Xu, Nanoelectronics: Growing Y-junction carbon nanotubes. *Nature*, **402**: 6759 (1999), 253–4.

34. Y. C. Choi and W. B. Choi, Synthesis of Y-junction single-wall carbon nanotubes. *Carbon*, **43**: 13 (2005), 2737–41.

35. A. G. Nasibulin, A. S. Anisimov, P. V. Pikhitsa, *et al.*, Investigations of nanobud formation. *Chemical Physics Letters*, **446**: 1–3 (2007), 109–14.

36. A. G. Nasibulin, P. V. Pikhitsa, H. Jiang, *et al.*, A novel hybrid carbon material. *Nature Nanotechnology*, **2**: 3 (2007), 156–61.
37. J. A. Furst, J. Hashemi, T. Markussen, *et al.* Electronic transport properties of fullerene functionalized carbon nanotubes: Ab initio and tight-binding calculations. *Physical Review B*, **80**: 3 (2009), 035427–30.
38. C. Journet and P. Bernier, Production of carbon nanotubes. *Applied Physics A: Materials Science & Processing*, **67**: 1 (1998), 1–9.
39. T. W. Ebbesen and P. M. Ajayan, Large-scale synthesis of carbon nanotubes. *Nature*, **358**: 6383 (1992), 220–2.
40. E. G. Gamaly and T. W. Ebbesen, Mechanism of carbon nanotube formation in the arc-discharge. *Physical Review B*, **52**: 3 (1995), 2083–9.
41. T. Guo, P. Nikolaev, A. Thess, D. T. Colbert, and R. E. Smalley, Catalytic growth of single-walled nanotubes by laser vaporization. *Chemical Physics Letters*, **243**: 1–2 (1995), 49–54.
42. M. Joseyacaman, M. Mikiyoshida, L. Rendon, and J. G. Santiesteban, Catalytic growth of carbon microtubules with fullerene structure. *Applied Physics Letters*, **62**: 6 (1993), 657–9.
43. A. Kukovecz, Z. Konya, N. Nagaraju, *et al.*, Catalytic synthesis of carbon nanotubes over Co, Fe and Ni containing conventional and sol-gel silica-aluminas. *Physical Chemistry Chemical Physics*, **2**: 13 (2000), 3071–6.
44. M. Nath, B. C. Satishkumar, A. Govindaraj, C. P. Vinod, and C. N. R. Rao, Production of bundles of aligned carbon and carbon-nitrogen nanotubes by the pyrolysis of precursors on silica-supported iron and cobalt catalysts. *Chemical Physics Letters*, **322**: 5 (2000), 333–40.
45. A. Fonseca, K. Hernadi, P. Piedigrosso, *et al.*, Synthesis of single- and multi-wall carbon nanotubes over supported catalysts. *Applied Physics A: Materials Science & Processing*, **67**: 1 (1998), 11–22.
46. V. Ivanov, A. Fonseca, J. B. Nagy, *et al.*, Catalytic production and purification of nanotubules having fullerene-scale diameters. *Carbon*, **33**: 12 (1995), 1727–38.
47. V. Ivanov, J. B. Nagy, P. Lambin, *et al.*, The study of carbon nanotubules produced by catalytic method. *Chemical Physics Letters*, **223**: 4 (1994), 329–35.
48. M. Perez-Cabero, A. Monzon, I. Rodriguez-Ramos, and A. Guerrero-Ruiz, Syntheses of CNTs over several iron-supported catalysts: Influence of the metallic precursors. *Catalysis Today*, **93**–95 (2004), 681–7.
49. S. B. Sinnott, R. Andrews, D. Qian, *et al.*, Model of carbon nanotube growth through chemical vapor deposition. *Chemical Physics Letters*, **315**: 1–2 (1999), 25–30.
50. H. J. Dai, J. Kong, C. W. Zhou, *et al.*, Controlled chemical routes to nanotube architectures, physics, and devices. *Journal of Physical Chemistry B*, **103**: 51 (1999), 11246–55.
51. T. Y. Lee, Han, J. H., Choi, S. H., *et al.*, Comparison of source gases and catalyst metals for growth of carbon nanotubes. *Surface and Coatings Technology*, **169/170** (2003), 348–52.
52. Z. F. Ren, Z. P. Huang, J. W. Xu, *et al.*, Synthesis of large arrays of well-aligned carbon nanotubes on glass. *Science*, **282**: 5391 (1998), 1105–7.
53. L. F. Sun, J. M. Mao, Z. W. Pan, *et al.*, Growth of straight nanotubes with a cobalt-nickel catalyst by chemical vapor deposition. *Applied Physics Letters*, **74**: 5 (1999), 644–6.
54. P. Piedigrosso, Z. Konya, J. F. Colomer, *et al.*, Production of differently shaped multi-wall carbon nanotubes using various cobalt supported catalysts. *Physical Chemistry Chemical Physics*, **2**: 1 (2000), 163–70.

55. M. Endo, K. Takeuchi, S. Igarashi, *et al.*, The production and structure of pyrolytic carbon nanotubes (PCNTs). *Journal of Physics and Chemistry of Solids*, **54**: 12 (1993), 1841–8.
56. A. M. Benito, Y. Maniette, E. Munoz, and M. T. Martinez, Carbon nanotubes production by catalytic pyrolysis of benzene. *Carbon*, **36**: 5–6 (1998), 681–3.
57. W. Z. Li, S. S. Xie, L. X. Qian, *et al.*, Large-scale synthesis of aligned carbon nanotubes. *Science*, **274**: 5293 (1996), 1701–3.
58. M. Terrones, N. Grobert, J. Olivares, *et al.*, Controlled production of aligned-nanotube bundles. *Nature*, **388**: 6637 (1997), 52–5.
59. S. S. Fan, M. G. Chapline, N. R. Franklin, *et al.*, Self-oriented regular arrays of carbon nanotubes and their field emission properties. *Science*, **283**: 5401 (1999), 512–4.
60. Z. W. Pan, S. S. Xie, B. H. Chang, *et al.*, Very long carbon nanotubes. *Nature*, **394**: 6694 (1998), 631–2.
61. B. C. Satishkumar, A. Govindaraj, and C. N. R. Rao, Bundles of aligned carbon nanotubes obtained by the pyrolysis of ferrocene-hydrocarbon mixtures: Role of the metal nanoparticles produced in situ. *Chemical Physics Letters*, **307**: 3–4 (1999), 158–62.
62. C. N. R. Rao, A. Govindaraj, R. Sen, and B. C. Satishkumar, Synthesis of multi-walled and single-walled nanotubes, aligned-nanotube bundles and nanorods by employing organometallic precursors. *Materials Research Innovations*, **2**: 3 (1998), 128–41.
63. R. Sen, A. Govindaraj, and C. N. R. Rao, Carbon nanotubes by the metallocene route. *Chemical Physics Letters*, **267**: 3–4 (1997), 276–80.
64. S. M. Huang, L. M. Dai, and A. W. H. Mau, Patterned growth and contact transfer of well-aligned carbon nanotube films. *Journal of Physical Chemistry B*, **103**: 21 (1999), 4223–7.
65. Y. Y. Yang, S. M. Huang, H. Z. He, A. W. H. Mau, and L. M. Dai, Patterned growth of well-aligned carbon nanotubes: A photolithographic approach. *Journal of the American Chemical Society*, **121**: 46 (1999), 10832–3.
66. D. C. Li, L. M. Dai, S. M. Huang, A. W. H. Mau, and Z. L. Wang, Structure and growth of aligned carbon nanotube films by pyrolysis. *Chemical Physics Letters*, **316**: 5–6 (2000), 349–55.
67. R. Andrews, D. Jacques, A. M. Rao, *et al.*, Continuous production of aligned carbon nanotubes: A step closer to commercial realization. *Chemical Physics Letters*, **303**: 5–6 (1999), 467–74.
68. C. Singh, M. S. Shaffer, and A. H. Windle, Production of controlled architectures of aligned carbon nanotubes by an injection chemical vapour deposition method. *Carbon*, **41**: 2 (2003), 359–68.
69. R. Kamalakaran, M. Terrones, T. Seeger, *et al.*, Synthesis of thick and crystalline nanotube arrays by spray pyrolysis. *Applied Physics Letters*, **77**: 21 (2000), 3385–7.
70. M. Mayne, N. Grobert, M. Terrones, *et al.*, Pyrolytic production of aligned carbon nanotubes from homogeneously dispersed benzene-based aerosols. *Chemical Physics Letters*, **338**: 2–3 (2001), 101–7.
71. B. Q. Wei, Z. J. Zhang, G. Ramanath, and P. M. Ajayan, Lift-up growth of aligned carbon nanotube patterns. *Applied Physics Letters*, **77**: 19 (2000), 2985–7.
72. A. Y. Cao, L. J. Ci, D. J. Li, *et al.*, Vertical aligned carbon nanotubes grown on Au film and reduction of threshold field in field emission. *Chemical Physics Letters*, **335**: 3–4 (2001), 150–4.
73. R. Vajtai, K. Kordas, B. Q. Wei, *et al.*, Carbon nanotube network growth on palladium seeds. *Materials Science & Engineering C: Biomimetic and Supramolecular Systems*, **19**: 1–2 (2002), 271–4.
74. Z. F. Ren, Z. P. Huang, D. Z. Wang, *et al.*, Growth of a single freestanding multiwall carbon nanotube on each nananickel dot. *Applied Physics Letters*, **75**: 8 (1999), 1086–8.

75. M. Chhowalla, C. Ducati, N. L. Rupesinghe, K. B. K. Teo, and G. A. J. Amaratunga, Field emission from short and stubby vertically aligned carbon nanotubes. *Applied Physics Letters*, **79**: 13 (2001), 2079–81.

76. Y. Saito. Structure and synthesis of carbon nanotubes. In Y. Saito, ed., *Carbon Nanotube and Related Field Emitters* (Weinheim: Wiley-VCH, 2010), pp. 8–10.

77. H. J. Dai, A. G. Rinzler, P. Nikolaev, *et al.*, Single-wall nanotubes produced by metal-catalyzed disproportionation of carbon monoxide. *Chemical Physics Letters*, **260**: 3–4 (1996), 471–5.

78. P. Nikolaev, M. J. Bronikowski, R. K. Bradley, *et al.*, Gas-phase catalytic growth of single-walled carbon nanotubes from carbon monoxide. *Chemical Physics Letters*, **313**: 1–2 (1999), 91–7.

79. S. M. Bachilo, L. Balzano, J. E. Herrera, *et al.*, Narrow (n, m)-distribution of single-walled carbon nanotubes grown using a solid supported catalyst. *Journal of the American Chemical Society*, **125**: 37 (2003), 11186–7.

80. B. Kitiyanan, W. E. Alvarez, J. H. Harwell, and D. E. Resasco, Controlled production of single-wall carbon nanotubes by catalytic decomposition of CO on bimetallic Co-Mo catalysts. *Chemical Physics Letters*, **317**: 3–5 (2000), 497–503.

81. K. Hata, D. N. Futaba, K. Mizuno, *et al.*, Water-assisted highly efficient synthesis of impurity-free single-waited carbon nanotubes. *Science*, **306**: 5700 (2004), 1362–4.

82. S. Maruyama, R. Kojima, Y. Miyauchi, S. Chiashi, and M. Kohno, Low-temperature synthesis of high-purity single-walled carbon nanotubes from alcohol. *Chemical Physics Letters*, **360**: 3–4 (2002), 229–34.

83. Y. M. Li, D. Mann, M. Rolandi, *et al.*, Preferential growth of semiconducting single-walled carbon nanotubes by a plasma enhanced CVD method. *Nano Letters*, **4**: 2 (2004), 317–21.

84. X. Li, X. Tu, S. Zaric, *et al.*, Selective synthesis combined with chemical separation of single-walled carbon nanotubes for chirality selection. *Journal of the American Chemical Society*, **129**: 51 (2007), 15770–1.

85. D. Ciuparu, Y. Chen, S. Lim, G. L. Haller, and L. Pfefferle, Uniform-diameter single-walled carbon nanotubes catalytically grown in cobalt-incorporated MCM-41. *Journal of Physical Chemistry B*, **108**: 2 (2004), 503–7.

86. Y. H. Miyauchi, S. H. Chiashi, Y. Murakami, Y. Hayashida, and S. Maruyama, Fluorescence spectroscopy of single-walled carbon nanotubes synthesized from alcohol. *Chemical Physics Letters*, **387**: 1–3 (2004), 198–203.

87. Y. Chen, D. Ciuparu, S. Y. Lim, *et al.*, Synthesis of uniform diameter single-wall carbon nanotubes in Co-MCM-41: Effects of the catalyst prereduction and nanotube growth temperatures. *Journal of Catalysis*, **225**: 2 (2004), 453–65.

88. H. M. Cheng, F. Li, G. Su, *et al.*, Large-scale and low-cost synthesis of single-walled carbon nanotubes by the catalytic pyrolysis of hydrocarbons. *Applied Physics Letters*, **72**: 25 (1998), 3282–4.

89. J. Kong, A. M. Cassell, and H. J. Dai, Chemical vapor deposition of methane for single-walled carbon nanotubes. *Chemical Physics Letters*, **292**: 4–6 (1998), 567–74.

90. W. E. Alvarez, B. Kitiyanan, A. Borgna, and D. E. Resasco, Synergism of Co and Mo in the catalytic production of single-wall carbon nanotubes by decomposition of CO. *Carbon*, **39**: 4 (2001), 547–58.

91. J. E. Herrera, L. Balzano, A. Borgna, W. E. Alvarez, and D. E. Resasco, Relationship between the structure/composition of Co-Mo catalysts and their ability to produce single-walled carbon nanotubes by CO disproportionation. *Journal of Catalysis*, **204**: 1 (2001), 129–45.

92. H. M. Cheng, F. Li, X. Sun, *et al.*, Bulk morphology and diameter distribution of single-walled carbon nanotubes synthesized by catalytic decomposition of hydrocarbons. *Chemical Physics Letters*, **289**: 5–6 (1998), 602–10.

93. B. Zheng, Y. Li, and J. Liu, CVD synthesis and purification of single-walled carbon nanotubes on aerogel-supported catalyst. *Applied Physics A: Materials Science & Processing*, **74**: 3 (2002), 345–8.

94. H. E. Unalan and M. Chhowalla, Investigation of single-walled carbon nanotube growth parameters using alcohol catalytic chemical vapour deposition. *Nanotechnology*, **16**: 10 (2005), 2153–63.

95. T. Maruyama, K. Sato, Y. Mizutani, *et al.*, Low-temperature synthesis of single-walled carbon nanotubes by alcohol gas source growth in high vacuum. *Journal of Nanoscience and Nanotechnology*, **10**: 6 (2010), 4095–101.

96. V. Derycke, R. Martel, M. Radosvljevic, F. M. R Ross, and P. Avouris, Catalyst-free growth of ordered single-walled carbon nanotube networks. *Nano Letters*, **2**: 10 (2002), 1043–6.

97. Y. Gogotsi, J. A. Libera, N. Kalashnikov, and M. Yoshimura, Graphite polyhedral crystals. *Science*, **290**: 5490 (2000), 317–20.

98. C. L. Cheung, A. Kurtz, H. Park, and C. M. Lieber, Diameter-controlled synthesis of carbon nanotubes. *Journal of Physical Chemistry B*, **106**: 10 (2002), 2429–33.

99. Y. M. Li, W. Kim, Y. G. Zhang, *et al.*, Growth of single-walled carbon nanotubes from discrete catalytic nanoparticles of various sizes. *Journal of Physical Chemistry B*, **105**: 46 (2001), 11424–31.

100. Y. Li, J. Liu, Y. Q. Wang, and Z. L. Wang, Preparation of monodispersed Fe-Mo nanoparticles as the catalyst for CVD synthesis of carbon nanotubes. *Chemistry of Materials*, **13**: 3 (2001), 1008–14.

101. J. F. Colomer, G. Bister, I. Willems, *et al.*, Synthesis of single-wall carbon nanotubes by catalytic decomposition of hydrocarbons. *Chemical Communications*, **14** (1999), 1343–4.

102. H. Ago, S. Ohshima, K. Uchida, and M. Yumura, Gas-phase synthesis of single-wall carbon nanotubes from colloidal solution of metal nanoparticles. *Journal of Physical Chemistry B*, **105**: 43 (2001), 10453–6.

103. W. E. Alvarez, F. Pompeo, J. E. Herrera, L. Balzano, and D. E. Resasco, Characterization of single-walled carbon nanotubes (SWNTs) produced by CO disproportionation on Co-Mo catalysts. *Chemistry of Materials*, **14**: 4 (2002), 1853–8.

104. M. Su, B. Zheng, and J. Liu, A scalable CVD method for the synthesis of single-walled carbon nanotubes with high catalyst productivity. *Chemical Physics Letters*, **322**: 5 (2000), 321–6.

105. A. M. Cassell, J. A. Raymakers, J. Kong, and H. J. Dai, Large scale CVD synthesis of single-walled carbon nanotubes. *Journal of Physical Chemistry B*, **103**: 31 (1999), 6484–92.

106. J. H. Hafner, M. J. Bronikowski, B. R. Azamian, *et al.*, Catalytic growth of single-wall carbon nanotubes from metal particles. *Chemical Physics Letters*, **296**: 1–2 (1998), 195–202.

107. J. Kong, H. T. Soh, A. M. Cassell, C. F. Quate, and H. J. Dai, Synthesis of individual single-walled carbon nanotubes on patterned silicon wafers. *Nature*, **395**: 6705 (1998), 878–81.

108. R. Sen, A. Govindaraj, and C. N. R. Rao, Metal-filled and hollow carbon nanotubes obtained by the decomposition of metal-containing free precursor molecules. *Chemistry of Materials*, **9**: 10 (1997), 2078–81.

109. A. Y. Cao, X. F. Zhang, C. L. Xu, *et al.*, Grapevine-like growth of single walled carbon nanotubes among vertically aligned multiwalled nanotube arrays. *Applied Physics Letters*, **79**: 9 (2001), 1252–4.

110. H. W. Zhu, C. L. Xu, and D. H. Wu, Direct synthesis of long single-walled carbon nanotube strands. *Science*, **296**: 5569 (2002), 884–6.

111. A. Javey, J. Guo, Q. Wang, M. Lundstrom, and H. J. Dai, Ballistic carbon nanotube field-effect transistors. *Nature*, **424**: 6949 (2003), 654–7.

112. L. Ding, A. Tselev, J. Wang, *et al.*, Selective growth of well-aligned semiconducting single-walled carbon nanotubes. *Nano Letters*, **9**: 2 (2009), 800–5.

113. S. Bandow, S. Asaka, X. Zhao, and Y. Ando, Purification and magnetic properties of carbon nanotubes. *Applied Physics A: Materials Science & Processing*, **67**: 1 (1998), 23–7.

114. G. S. Duesberg, M. Burghard, J. Muster, G. Philipp, and S. Roth, Separation of carbon nanotubes by size exclusion chromatography. *Chemical Communications*, 3 (1998), 435–6.

115. G. S. Duesberg, J. Muster, H. J. Byrne, S. Roth, and M. Burghard, Towards processing of carbon nanotubes for technical applications. *Applied Physics A: Materials Science & Processing*, **69**: 3 (1999), 269–74.

116. T. W. Ebbesen, P. M. Ajayan, H. Hiura, and K. Tanigaki, Purification of nanotubes. *Nature*, **367**: 6463 (1994), 519.

117. H. Hiura, T. W. Ebbesen, and K. Tanigaki, Opening and purification of carbon nanotubes in high yields. *Advanced Materials*, **7**: 3 (1995), 275–6.

118. J. M. Bonard, T. Stora, J. P. Salvetat, *et al.*, Purification and size-selection of carbon nanotubes. *Advanced Materials*, **9**: 10 (1997), 827–31.

119. M. S. Arnold, A. A. Green, J. F. Hulvat, S. I. Stupp, and M. C. Hersam, Sorting carbon nanotubes by electronic structure using density differentiation. *Nature Nanotechnology*, **1**: 1 (2006), 60–5.

120. T. Tanaka, H. Jin, Y. Miyata, *et al.*, Simple and scalable gel-based separation of metallic and semiconducting carbon nanotubes. *Nano Letters*, **9**: 4 (2009), 1497–500.

121. X. Tu, S. Manohar, A. Jagota, and M. Zheng, DNA sequence motifs for structure-specific recognition and separation of carbon nanotubes. *Nature*, **460**: 7252 (2009), 250–3.

122. M. M. J. Treacy, T. W. Ebbesen, and J. M. Gibson, Exceptionally high Young's modulus observed for individual carbon nanotubes. *Nature*, **381**: 6584 (1996), 678–80.

123. A. Krishnan, E. Dujardin, T. W. Ebbesen, P. N. Yianilos, and M. M. J. Treacy, Young's modulus of single-walled nanotubes. *Physical Review B*, **58**: 20 (1998), 14013–9.

124. P. Poncharal, Z. L. Wang, D. Ugarte, and W. A. de Heer, Electrostatic deflections and electromechanical resonances of carbon nanotubes. *Science*, **283**: 5407 (1999), 1513–6.

125. E. W. Wong, P. E. Sheehan, and C. M. Lieber, Nanobeam mechanics: Elasticity, strength, and toughness of nanorods and nanotubes. *Science*, **277**: 5334 (1997), 1971–5.

126. M. F. Yu, O. Lourie, M. J. Dyer, *et al.*, Strength and breaking mechanism of multi-walled carbon nanotubes under tensile load. *Science*, **287**: 5453 (2000), 637–40.

127. M. F. Yu, B. S. Files, S. Arepalli, and R. S. Ruoff, Tensile loading of ropes of single wall carbon nanotubes and their mechanical properties. *Physical Review Letters*, **84**: 24 (2000), 5552–5.

128. Z. W. Pan, S. S. Xie, L. Lu, *et al.*, Tensile tests of ropes of very long aligned multiwall carbon nanotubes. *Applied Physics Letters*, **74**: 21 (1999), 3152–4.

129. D. A. Walters, L. M. Ericson, M. J. Casavant, *et al.*, Elastic strain of freely suspended single-wall carbon nanotube ropes. *Applied Physics Letters*, **74**: 25 (1999), 3803–5.

130. H. D. Wagner, O. Lourie, Y. Feldman, and R. Tenne, Stress-induced fragmentation of multiwall carbon nanotubes in a polymer matrix. *Applied Physics Letters*, **72**: 2 (1998), 188–90.

131. F. Li, H. M. Cheng, S. Bai, G. Su, and M. S. Dresselhaus, Tensile strength of single-walled carbon nanotubes directly measured from their macroscopic ropes. *Applied Physics Letters*, **77**: 20 (2000), 3161–3.

132. T. W. Ebbesen, Cones and tubes: Geometry in the chemistry of carbon. *Accounts of Chemical Research*, **31**: 9 (1998), 558–66.

133. M. R. Falvo, G. J. Clary, R. M. Taylor, *et al.*, Bending and buckling of carbon nanotubes under large strain. *Nature*, **389**: 6651 (1997), 582–4.

134. C. Bower, R. Rosen, L. Jin, J. Han, and O. Zhou, Deformation of carbon nanotubes in nanotube-polymer composites. *Applied Physics Letters*, **74**: 22 (1999), 3317–9.

135. O. Lourie and H. D. Wagner, Evaluation of Young's modulus of carbon nanotubes by micro-Raman spectroscopy. *Journal of Materials Research*, **13**: 9 (1998), 2418–22.

136. G. Overney, W. Zhong, and D. Tomanek, Structural rigidity and low-frequency vibrational-modes of long carbon tubules. *Zeitschrift Fur Physik D: Atoms, Molecules and Clusters*, **27**: 1 (1993), 93–6.

137. R. S. Ruoff and D. C. Lorents, Mechanical and thermal properties of carbon nanotubes. *Carbon*, **33**: 7 (1995), 925–30.

138. B. I. Yakobson, C. J. Brabec, and J. Bernholc, Nanomechanics of carbon tubes: Instabilities beyond linear response. *Physical Review Letters*, **76**: 14 (1996), 2511–4.

139. B. I. Yakobson and P. Avouris, Mechanical properties of carbon nanotubes. *Carbon Nanotubes*, **80** (2001), 287–327.

140. B. I. Yakobson and R. E. Smalley, Fullerene nanotubes: C-1000000 and beyond. *American Scientist*, **85**: 4 (1997), 324–37.

141. B. I. Yakobson, M. P. Campbell, C. J. Brabec, and J. Bernholc, High strain rate fracture and C-chain unraveling in carbon nanotubes. *Computational Materials Science*, **8**: 4 (1997), 341–8.

142. Z.-C. Ou-Yang, S. Zhao-Bin, and W. Chui-Lin, Coil formation in multishell carbon nanotubes: Competition between curvature elasticity and interlayer adhesion. *Physical Review Letters*, **78**: 21 (1997), 4055–8.

143. X. Zhou, J. J. Zhou, and Z. C. Ou-Yang, Strain energy and Young's modulus of single-wall carbon nanotubes calculated from electronic energy-band theory. *Physical Review B*, **62**: 20 (2000), 13692–6.

144. Z. C. Tu and Z. Ou-Yang, Single-walled and multiwalled carbon nanotubes viewed as elastic tubes with the effective Young's moduli dependent on layer number. *Physical Review B*, **65**: 23 (2002), 233407–10.

145. G. H. Gao, T. Cagin, and W. A. Goddard, Energetics, structure, mechanical and vibrational properties of single-walled carbon nanotubes. *Nanotechnology*, **9**: 3 (1998), 184–91.

146. E. Hernandez, C. Goze, P. Bernier, and A. Rubio, Elastic properties of single-wall nanotubes. *Applied Physics A: Materials Science & Processing*, **68**: 3 (1999), 287–92.

147. N. Hamada, S. Sawada, and A. Oshiyama, New one-dimensional conductors: Graphitic microtubules. *Physical Review Letters*, **68**: 10 (1992), 1579–81.

148. J. W. Mintmire, B. I. Dunlap, and C. T. White, Are fullerene tubules metallic? *Physical Review Letters*, **68**: 5 (1992), 631–4.

149. R. Saito, M. Fujita, G. Dresselhaus, and M. S. Dresselhaus, Electronic structure of graphene tubules based on C-60. *Physical Review B*, **46**: 3 (1992), 1804–11.

150. R. Saito, M. Fujita, G. Dresselhaus, and M. S. Dresselhaus, Electronic structure of chiral graphene tubules. *Applied Physics Letters*, **60**: 18 (1992), 2204–6.

151. R. Saito, G. Dresselhaus, and M. S. Dresselhaus, Electronic structure of double-layer graphene tubules. *Journal of Applied Physics*, **73**: 2 (1993), 494–500.

152. T. W. Odom, J. L. Huang, P. Kim, and C. M. Lieber, Atomic structure and electronic properties of single-walled carbon nanotubes. *Nature*, **391**: 6662 (1998), 62–4.

153. J. W. G. Wildoer, L. C. Venema, A. G. Rinzler, R. E. Smalley, and C. Dekker, Electronic structure of atomically resolved carbon nanotubes. *Nature*, **391**: 6662 (1998), 59–62.

154. S. Frank, P. Poncharal, Z. L. Wang, and W. A. de Heer, Carbon nanotube quantum resistors. *Science*, **280**: 5370 (1998), 1744–6.

155. H. J. Dai, E. W. Wong, and C. M. Lieber, Probing electrical transport in nanomaterials: Conductivity of individual carbon nanotubes. *Science*, **272**: 5261 (1996), 523–6.

156. L. Langer, V. Bayot, E. Grivei, *et al.*, Quantum transport in a multiwalled carbon nanotube. *Physical Review Letters*, **76**: 3 (1996), 479–82.

157. T. W. Ebbesen, H. J. Lezec, H. Hiura, *et al.*, Electrical conductivity of individual carbon nanotubes. *Nature*, **382**: 6586 (1996), 54–6.

158. A. Thess, R. Lee, P. Nikolaev, *et al.*, Crystalline ropes of metallic carbon nanotubes. *Science*, **273**: 5274 (1996), 483–7.

159. J. E. Fischer, H. Dai, A. Thess, *et al.*, Metallic resistivity in crystalline ropes of single-wall carbon nanotubes. *Physical Review B*, **55**: 8 (1997), R4921–R4.

160. P. Kim, L. Shi, A. Majumdar, and P. L. McEuen, Thermal transport measurements of individual multiwalled nanotubes. *Physical Review Letters*, **8721**: 21 (2001), 215502–6.

161. R. H. Baughman, A. A. Zakhidov, and W. A. de Heer, Carbon nanotubes: The route toward applications. *Science*, **297**: 5582 (2002), 787–92.

162. NEC, NEC uses carbon nanotubes to develop a tiny cell for mobile applications. 2001. http://www.nec.co.jp/press/en/0108/3001.html [cited 19 August 2011].

163. http://www.xintek.com/index.htm [cited 24 August 2011].

164. P. J. F. Harris, Carbon nanotube composites. *International Materials Reviews*, **49**: 1 (2004), 31–43.

165. P.-C. Ma, N. A. Siddiqui, G. Marsom, and J.-K. Kim, Dispersion and functionalization of carbon nanotubes for polymer-based nanocomposites: A review. *Composites: Part A*, **41** (2010), 1345–67.

166. J. Sandler, M. S. P. Shaffer, T. Prasse, *et al.*, Development of a dispersion process for carbon nanotubes in an epoxy matrix and the resulting electrical properties. *Polymer*, **40**: 21 (1999), 5967–71.

167. Y. Huang, Y. Zheng, W. Song, *et al.*, Poly(vinyl pyrrolidone) wrapped multi-walled carbon nanotube/poly(vinyl alcohol) composite hydrogels. *Composites: Part A*, **42** (2011), 1398–405.

168. W. H. Song, I. A. Kinloch, and A. H. Windle, Nematic liquid crystallinity of multiwall carbon nanotubes. *Science*, **302**: 5649 (2003), 1363.

169. W. Song and A. H. Windle, Size-dependence and elasticity of liquid-crystalline multi-walled carbon nanotubes. *Advanced Materials*, **20**: 16 (2008), 3149–54.

170. R. Andrews, D. Jacques, A. M. Rao, *et al.*, Nanotube composite carbon fibers. *Applied Physics Letters*, **75**: 9 (1999), 1329–31.

171. K. L. Jiang, Q. Q. Li, and S. S. Fan, Nanotechnology: Spinning continuous carbon nanotube yarns. Carbon nanotubes weave their way into a range of imaginative macroscopic applications. *Nature*, **419**: 6909 (2002), 801.

172. B. Vigolo, A. Penicaud, C. Coulon, *et al.*, Macroscopic fibers and ribbons of oriented carbon nanotubes. *Science*, **290**: 5495 (2000), 1331–4.

173. L. M. Ericson, H. Fan, H. Q. Peng, *et al.*, Macroscopic, neat, single-walled carbon nanotube fibers. *Science*, **305**: 5689 (2004), 1447–50.

174. Y. L. Li, I. A. Kinloch, and A. H. Windle, Direct spinning of carbon nanotube fibers from chemical vapor deposition synthesis. *Science*, **304**: 5668 (2004), 276–8.
175. M. Zhang, K. R. Atkinson, and R. H. Baughman, Multifunctional carbon nanotube yarns by downsizing an ancient technology. *Science*, **306**: 5700 (2004), 1358–61.
176. R. D. Leaversuch. Nanocomposites broaden roles in automotive, barrier packaging. October 2001. http://www.ptonline.com/articles/nanocomposites-broaden-roles-in-automotive-barrier-packaging [cited 23 August 2011].
177. H. J. Dai, J. H. Hafner, A. G. Rinzler, D. T. Colbert, and R. E. Smalley, Nanotubes as nanoprobes in scanning probe microscopy. *Nature*, **384**: 6605 (1996), 147–50.
178. S. S. Wong, E. Joselevich, A. T. Woolley, C. L. Cheung, and C. M. Lieber, Covalently functionalized nanotubes as nanometre-sized probes in chemistry and biology. *Nature*, **394**: 6688 (1998), 52–5.
179. J. H. Hafner, C. L. Cheung, and C. M. Lieber, Growth of nanotubes for probe microscopy tips. *Nature*, **398**: 6730 (1999), 761–2.
180. J. H. Hafner, C. L. Cheung, T. H. Oosterkamp, and C. M. Lieber, High-yield assembly of individual single-walled carbon nanotube tips for scanning probe microscopies. *Journal of Physical Chemistry B*, **105**: 4 (2001), 743–6.
181. E. Yenilmez, Q. Wang, R. J. Chen, D. W. Wang, and H. J. Dai, Wafer scale production of carbon nanotube scanning probe tips for atomic force microscopy. *Applied Physics Letters*, **80**: 12 (2002), 2225–7.
182. A. G. Rinzler, J. H. Hafner, P. Nikolaev, *et al.*, Unraveling nanotubes: Field emission from an atomic wire. *Science*, **269**: 5230 (1995), 1550–3.
183. W. A. de Heer, A. Châtelain, and U. D. Ugarte, A carbon nanotube field-emission electron source. *Science*, **270** (1995), 1179–80.
184. Y. Saito, K. Hamaguchi, K. Hata, *et al.*, Conical beams from open nanotubes. *Nature*, **389**: 6651 (1997), 554–5.
185. Y. Saito, K. Hamaguchi, S. Uemura, *et al.*, Field emission from multi-walled carbon nanotubes and its application to electron tubes. *Applied Physics A: Materials Science & Processing*, **67**: 1 (1998), 95–100.
186. W. B. Choi, D. S. Chung, J. H. Kang, *et al.*, Fully sealed, high-brightness carbon-nanotube field-emission display. *Applibed Physics Letters*, **75**: 20 (1999), 3129–31.
187. K. B. K. Teo, E. Minoux, L. Hudanski, *et al.*, Microwave devices: Carbon nanotubes as cold cathodes. *Nature*, **437**: 7061 (2005), 968.
188. X. Calderon-Colon, H. Geng, B. Gao, *et al.*, A carbon nanotube field emission cathode with high current density and long-term stability. *Nanotechnology*, **20**: 32 (2009), 325707.
189. AAPM: Researchers develop nano-based X-ray for imaging, radiotherapy. 2009. http://www.healthimaging.com/index.php?option=com_articles&view=article&id=18201:aapm-researchers-develop-nano-based-x-ray-for-imaging-radiotherapy [cited 20 August 2011].
190. S. J. Tans, A. R. M. Verschueren, and C. Dekker, Room-temperature transistor based on a single carbon nanotube. *Nature*, **393**: 6680 (1998), 49–52.
191. M. J. O'Connell, S. M. Bachilo, C. B. Huffman, *et al.*, Band gap fluorescence from individual single-walled carbon nanotubes. *Science*, **297**: 5581 (2002), 593–6.
192. J. A. Misewich, R. Martel, P. Avouris, *et al.*, Electrically induced optical emission from a carbon nanotube FET. *Science*, **300**: 5620 (2003), 783–6.
193. T. Durkop, S. A. Getty, E. Cobas, and M. S. Fuhrer, Extraordinary mobility in semi-conducting carbon nanotubes. *Nano Letters*, **4**: 1 (2004), 35–9.
194. R. V. Seidel, A. P. Graham, J. Kretz, *et al.*, Sub-20 nm short channel carbon nanotube transistors. *Nano Letters*, **5**: 1 (2005), 147–50.

195. P. G. Collins and P. Avouris, Nanotubes for electronics. *Scientific American*, **283**: 6 (2000), 62–9.
196. M. Engel, J. P. Small, M. Steiner, *et al.*, Thin film nanotube transistors based on self-assembled, aligned, semiconducting carbon nanotube arrays. *ACS Nano*, **2**: 12 (2008), 2445–52.
197. A. Bachtold, P. Hadley, T. Nakanishi, and C. Dekker, Logic circuits with carbon nanotube transistors. *Science*, **294**: 5545 (2001), 1317–20.
198. V. Derycke, R. Martel, J. Appenzeller, and P. Avouris, Carbon nanotube inter- and intramolecular logic gates. *Nano Letters*, **1**: 9 (2001), 453–6.
199. Z. H. Chen, J. Appenzeller, Y. M. Lin, *et al.*, An integrated logic circuit assembled on a single carbon nanotube. *Science*, **311**: 5768 (2006), 1735.
200. N. M. Gabor, Z. Zhong, K. Bosnick, J. Park, and P. L. McEuen, Extremely efficient multiple electron-hole pair generation in carbon nanotube photodiodes. *Science*, **325**: 5946 (2009), 1367–71.
201. Z. C. Wu, Z. H. Chen, X. Du, *et al.*, Transparent, conductive carbon nanotube films. *Science*, **305**: 5688 (2004), 1273–6.
202. K. H. An, W. S. Kim, Y. S. Park, *et al.*, Supercapacitors using single-walled carbon nanotube electrodes. *Advanced Materials*, **13**: 7 (2001), 497–500.
203. M. Hughes, M. S. P. Shaffer, A. C. Renouf, *et al.*, Electrochemical capacitance of nanocomposite films formed by coating aligned arrays of carbon nanotubes with polypyrrole. *Advanced Materials*, **14**: 5 (2002), 382–5.
204. S. W. Lee, N. Yabuuchi, B. M. Gallant, *et al.*, High-power lithium batteries from functionalized carbon-nanotube electrodes. *Nature Nanotechnology*, **5**: 7 (2010), 531–7.
205. P. Brown, K. Takechi, and P. V. Kamat, Single-walled carbon nanotube scaffolds for dye-sensitized solar cells. *Journal of Physical Chemistry C*, **112**: 12 (2008), 4776–82.
206. M. Kaempgen, M. Lebert, N. Nicoloso, and S. Roth, Multifunctional carbon nanotube networks for fuel cells. *Applied Physics Letters*, **92**: 9 (2008), 094103–5.
207. K. Gong, F. Du, Z. Xia, M. Durstock, and L. Dai, Nitrogen-doped carbon nanotube arrays with high electrocatalytic activity for oxygen reduction. *Science*, **323**: 5915 (2009), 760–4.
208. R. H. Baughman, Conducting polymer artificial muscles. *Synthetic Metals*, **78**: 3 (1996), 339–53.
209. R. H. Baughman, C. X. Cui, A. A. Zakhidov, *et al.*, Carbon nanotube actuators. *Science*, **284**: 5418 (1999), 1340–4.
210. P. G. Collins, K. Bradley, M. Ishigami, and A. Zettl, Extreme oxygen sensitivity of electronic properties of carbon nanotubes. *Science*, **287**: 5459 (2000), 1801–4.
211. J. Kong, N. R. Franklin, C. W. Zhou, *et al.*, Nanotube molecular wires as chemical sensors. *Science*, **287**: 5453 (2000), 622–5.
212. MIT carbon nanotube ultracapacitor could approach storage density of batteries. February 2006. http://www.greencarcongress.com/2006/02/mit_carbon_nano.html [cited 30 August 2011].
213. R. G. Ding, G. Q. Lu, Z. F. Yan, and M. A. Wilson, Recent advances in the preparation and utilization of carbon nanotubes for hydrogen storage. *Journal of Nanoscience and Nanotechnology*, **1**: 1 (2001), 7–29.
214. A. C. Dillon, K. M. Jones, T. A. Bekkedahl, *et al.*, Storage of hydrogen in single-walled carbon nanotubes. *Nature*, **386**: 6623 (1997), 377–9.
215. A. Chambers, C. Park, R. T. K. Baker, and N. M. Rodriguez, Hydrogen storage in graphite nanofibers. *Journal of Physical Chemistry B*, **102**: 22 (1998), 4253–6.

216. C. C. Ahn, Y. Ye, B. V. Ratnakumar, *et al.*, Hydrogen desorption and adsorption measurements on graphite nanofibers. *Applied Physics Letters*, **73**: 23 (1998), 3378–80.

217. P. Chen, X. Wu, J. Lin, and K. L. Tan, High H_2 uptake by alkali-doped carbon nanotubes under ambient pressure and moderate temperatures. *Science*, **285**: 5424 (1999), 91–3.

218. R. T. Yang, Hydrogen storage by alkali-doped carbon nanotubes: Revisited. *Carbon*, **38**: 4 (2000), 623–6.

219. G. K. Dimitrakakis, E. Tylianakis, and G. E. Froudakis, Pillared graphene: A new 3-D network nanostructure for enhanced hydrogen storage. *Nano Letters*, **8**: 10 (2008), 3166–70.

220. M. Becher, M. Haluska, M. Hirscher, *et al.*, Hydrogen storage in carbon nanotubes. *Comptes Rendus Physique*, **4**: 9 (2003), 1055–62.

221. B. Assfour, S. Leoni, G. Seifert, and I. A. Baburin, Packings of carbon nanotubes: New materials for hydrogen storage. *Advanced Materials*, **23**: 10 (2011), 1237–41.

222. K. Kostarelos, L. Lacerda, G. Pastorin, *et al.*, Cellular uptake of functionalized carbon nanotubes is independent of functional group and cell type. *Nature Nanotechnology*, **2**: 2 (2007), 108–13.

223. H. Ali-Boucetta, K. T. Al-Jamal, D. McCarthy, *et al.*, Multiwalled carbon nanotube-doxorubicin supramolecular complexes for cancer therapeutics. *Chemical Communications*, 4 (2008), 459–61.

224. C. J. Gannon, P. Cherukuri, B. I. Yakobson, *et al.*, Carbon nanotube-enhanced thermal destruction of cancer cells in a noninvasive radiofrequency field. *Cancer*, **110**: 12 (2007), 2654–65.

225. J. E. Podesta, K. T. Al-Jamal, M. A. Herrero, *et al.*, Antitumor activity and prolonged survival by carbon-nanotube-mediated therapeutic siRNA silencing in a human lung xenograft model. *Small*, **5**: 10 (2009), 1176–85.

226. B. Sitharaman, X. Shi, X. F. Walboomers, *et al.*, In vivo biocompatibility of ultra-short single-walled carbon nanotube/biodegradable polymer nanocomposites for bone tissue engineering. *Bone*, **43**: 2 (2008), 362–70.

227. M. P. Mattson, R. C. Haddon, and A. M. Rao, Molecular functionalization of carbon nanotubes and use as substrates for neuronal growth. *Journal of Molecular Neuroscience*, **14**: 3 (2000), 175–82.

228. P. Galvan-Garcia, E. W. Keefer, F. Yang, *et al.*, Robust cell migration and neuronal growth on pristine carbon nanotube sheets and yarns. *Journal of Biomaterials Science: Polymer Edition*, **18**: 10 (2007), 1245–61.

229. K. Keren, R. S. Berman, E. Buchstab, U. Sivan, and E. Braun, DNA-templated carbon nanotube field-effect transistor. *Science*, **302**: 5649 (2003), 1380–2.

230. C. Staii and A. T. Johnson, DNA-decorated carbon nanotubes for chemical sensing. *Nano Letters*, **5**: 9 (2005), 1774–8.

231. Y. Li, T. Kaneko, Y. Hirotsu, and R. Hatakeyama, Light-induced electron transfer through DNA-decorated single-walled carbon nanotubes. *Small*, **6**: 1 (2010), 27–30.

232. E. W. Keefer, B. R. Botterman, M. I. Romero, A. F. Rossi, and G. W. Gross, Carbon nanotube coating improves neuronal recordings. *Nature Nanotechnology*, **3**: 7 (2008), 434–9.

233. Z. Zhu, W. Song, K. Burugapalli, *et al.*, Nano-yarn carbon nanotube fiber based enzymatic glucose biosensor. *Nanotechnology*, **21**: 16 (2010), 165501–10.

2

The aerodynamic behaviour and pulmonary deposition of carbon nanotubes

ALISON BUCKLEY, RACHEL SMITH, ROBERT MAYNARD

2.1 Introduction

Rapid developments in the nanotechnology field have led to concerns about the possible effects on health of exposure to carbon nanotubes (CNT). It has been demonstrated that CNT have some toxicological properties in common with asbestos fibres and concerns have been voiced that exposure may lead to some of the diseases associated with exposure to asbestos. It is known that asbestos fibres are efficiently deposited in the gas exchange zone of the lung and that they can translocate to the pleural space by penetrating the visceral pleura. Recent work has suggested that such fibres are not efficiently cleared from the pleural space: they may fail to enter the stomata of the parietal pleura and this might lead to frustrated phagocytosis by macrophages moving on the surface of the parietal pleura. These ideas remain speculative, but they suggest that we should be concerned about deposition of CNT particles in the gas exchange zone of the lung. Whether aerosolised CNT particles behave in the same way as asbestos fibres in the air streams of the airways of the lung remains unknown. In this chapter we explore some of the factors that may affect their deposition. It should be noted that very little experimental work has been done in this area and that modelling the deposition of nanotubes in all the various forms in which they appear from first principles is likely to be difficult. However, the results of new work are likely to accumulate rapidly in the next few years. The ideas put forward here may provide a guide to the research that is needed; we note that these should be considered as speculative and will no doubt be refined as data accumulate.

Particles, including fibres, are deposited in the lung as a result of their leaving the air streams within the pulmonary airways and coming into contact with the walls of those airways. The main physical processes governing particle behaviour are generally understood and over the past few decades a number of theoretical models have been developed to describe the transport and deposition behaviour of particles

The Toxicology of Carbon Nanotubes, ed. Ken Donaldson, Craig A. Poland, Rodger Duffin and James Bonner. Published by Cambridge University Press. © Cambridge University Press 2012.

in the respiratory tract. In this chapter the application of these models to CNT is explored. Relevant experimental data are also considered. A key issue in relation to the deposition behaviour of carbon nanotubes is particle morphology, and this is where we start.

2.2 Carbon nanotube aerosol particle morphology

If one is asked to think of a CNT particle, it is likely that the image of a long, thin, needle-like fibre, somewhat like crocidolite asbestos, would come to mind. Electron microscopy of CNT particles sampled in real exposure situations shows, however, that the majority are bent and tangled, forming low-density agglomerates up to tens of μm in diameter. As particle form can have a significant effect on the location and efficiency of deposition in the respiratory tract one must be careful to consider realistic particle morphologies when assessing the potential health risk posed by inhalation of CNT.

There have been very few CNT exposure assessments and therefore knowledge of the form that 'real' CNT aerosol particles take is limited. The following discussion is based on what has been observed to date, although it is acknowledged that other forms of CNT aerosol particles may be seen in the future.

Airborne CNT particles have been identified in electron micrographs of samples taken in research and industrial facilities dealing with a range of CNT materials and during a range of activities. Table 2.1 summarises the key findings relating to the CNT aerosol particle properties for exposure assessments made to date. In even this limited set of studies, a wide range of particle types has been observed, varying with activity, CNT type, production method, and functionalised state. The types observed can be split into two general categories: 'fibre-like' and 'broadly spherical', the latter being more common.

'Fibre-like' particles are dispersed or lightly agglomerated individual tubes or 'ropes' of tubes, ranging in length from 1 to 6 μm with diameters typically around 50 nm (Figure 2.1). This form was only observed for multi-walled carbon nanotubes (MWCNT): in the limited observations made to date, 'fibre-like' single-walled carbon nanotube (SWCNT) aerosol particles have not been seen. Compared to the characteristics of MWCNT as manufactured, which generally have diameters between 10 and 100 nm and lengths typically of tens of μm, only short or broken MWCNT appear to be aerosolised as 'fibre-like' particles. 'Broadly spherical' particles are typically near-spherical, open agglomerated clumps of tubes, which have been observed with a wide range of diameters, from 100 nm up to hundreds of μm or more. Only one exposure assessment of SWCNT in a research facility has been reported to date (Maynard *et al.*, 2004). The release of two different types of SWCNT was investigated, both of which had a strong tendency to form large

Table 2.1 *Characteristics of CNT aerosols sampled in various exposure situations.*

Reference	Activity	CNT type	CNT observed in air samples?	Particle characteristics
Maynard *et al.*, 2004	Removing CNT from production vessel	Laser ablated SWCNT	Yes	Large, open agglomerates, 100 μm–1 mm in diameter ('broadly spherical')
		HiPCO™ SWCNT	Yes	Large, open agglomerates, a few μm in diameter ('broadly spherical')
Han *et al.*, 2008	Recovering CNT and blending composites	MWCNT	Yes	Individual and lightly agglomerated tubes, average diameter 55 nm, average length 1.5 μm ('fibre-like'), and small and larger agglomerates, 100–300 nm in diameter and a few μm in diameter ('broadly spherical'), respectively
Bello *et al.*, 2008, 2009, 2010	Machining of CNT composites	MWCNT composites	Only during drilling of composite material	Particle clusters containing CNT aggregates a few μm in diameter ('broadly spherical')
Lee *et al.*, 2010	Various (six separate facilities)	MWCNT	During CVD production with no controls	Agglomerates of fibres, diameter ~50 nm, length 3 μm ('fibre-like'), and spherical particles, diameter 100–600 nm ('broadly spherical')
Johnson *et al.*, 2010	Weighing, transferring to liquid and ultra-sonication	Raw MWCNT	Only during transferring	Loose bundles of tubes ~500 nm in diameter ('broadly spherical')
		Hydroxylated MWCNT	Only during sonication	Highly agglomerated bundles >1 μm in diameter ('broadly spherical')

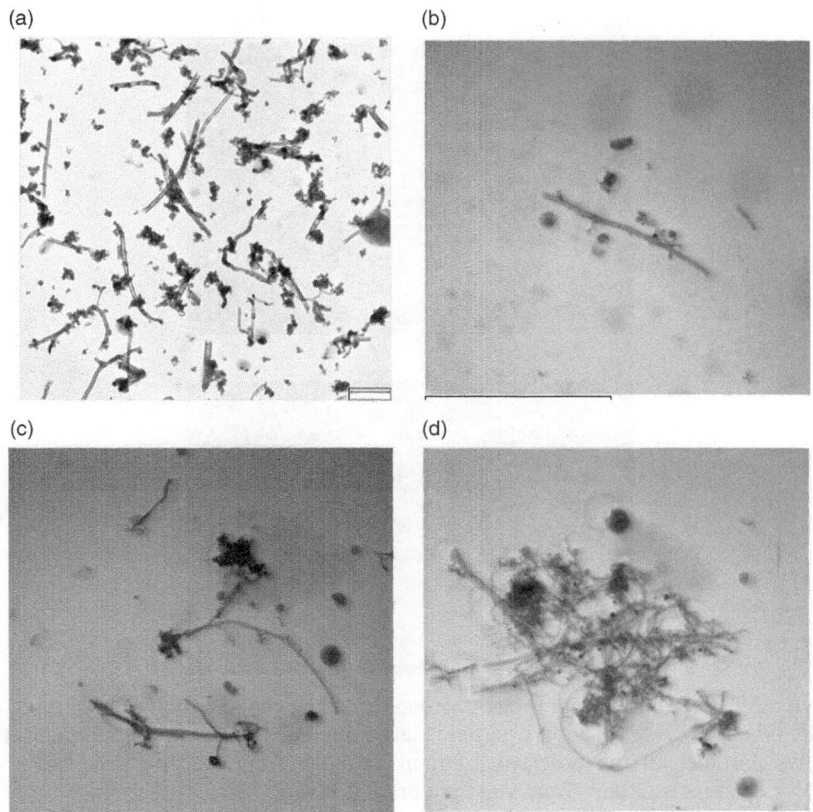

Figure 2.1 Airborne CNT sampled in a MWCNT research facility: (a) after opening of the furnace, (b) showing an individual tube structure, (c) showing a multiple-tube structure, and (d) showing a clumped-tube structure (reprinted by permission of the publisher from Han *et al.* (2008)). Scale bars in (b), (c), and (d) indicate 4 μm.

'broadly spherical' particles, ranging in size from about 1 μm up to hundreds of μm or more (Figure 2.2). 'Broadly spherical' MWCNT aerosol particles have also been seen, but, on the basis of the observations reported to date, are generally smaller than SWCNT particles. Diameters range from about 100 nm up to a few μm (Figures 2.1 and 2.3).

The 'fibre-like' CNT aerosol particles observed are smaller than typical asbestos fibres, which are usually around 500 nm in diameter and a few to tens of μm long. The aerosol dynamics and deposition pattern in the respiratory system are therefore likely to differ.

Density is an important factor in particle aerodynamics, and is of particular interest for CNT as it is likely to be significantly lower than that of most other aerosol particles considered from a potential health perspective. The density of CNT

(a)

(b)

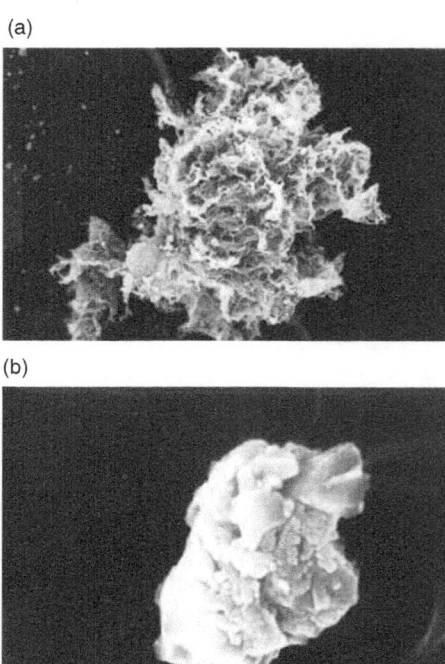

Figure 2.2 Electron micrographs of aerosol particles generated during handling of HiPCO™ SWCNT: (a) shows particles larger than about 100 μm in diameter; (b) shows particles of the order of 1 μm in diameter (reprinted by permission of Taylor & Francis Group from Maynard *et al.* (2004)).

particles is likely to be lower than that of other allotropes of carbon. Kim *et al.* (2009) measured the density of individual MWCNT grown in the gas phase on nickel catalyst particles to be $1740 \pm 160 \, \text{kg m}^{-3}$. For comparison, the density of graphite is $2220 \, \text{kg m}^{-3}$, that for carbon black has been found to range between 1840 and $2060 \, \text{kg m}^{-3}$, and airborne asbestos fibres are typically around $2500 \, \text{kg m}^{-3}$. While for 'fibre-like' CNT aerosol particles we may not expect a significant reduction in density from that for individual tubes, in 'broadly spherical' agglomerate form it has been suggested that CNT particle densities may be significantly lower, in the range $100-1000 \, \text{kg m}^{-3}$ (Hao *et al.*, 2003; Maynard *et al.*, 2004). These density measurements relate to only a few types of CNT particles, and as particle form has been shown to vary significantly with CNT production method and aerosolisation activity, densities are also expected to vary. For a full understanding of the aerodynamic and deposition behaviour of CNT particles in the respiratory tract, more measurements of the density of 'realistic' CNT aerosol particles are required.

Whilst in this review we have divided CNT particles into two general categories to simplify discussion of their transport and deposition behaviour, it should be recognised that CNT aerosols are likely to consist of a mix of such particle forms

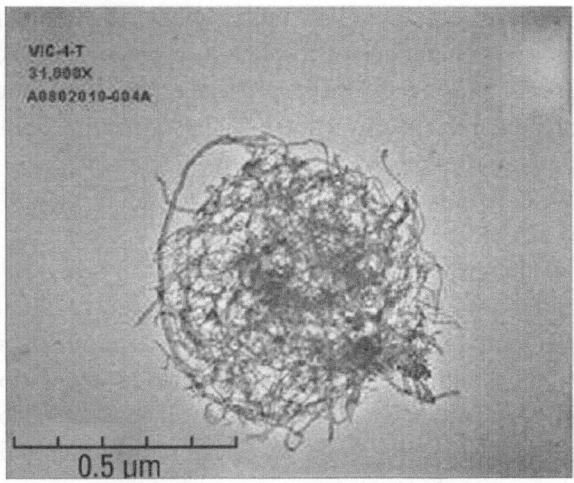

Figure 2.3 TEM image of engineered carbon nanomaterials aerosolised during the sonication of raw MWCNT in water containing 100 mg/l NOM (natural organic material) inside an unventilated enclosure (reproduced from Johnson *et al.* (2010) with permission from Environmental Health Perspectives). Scale bar indicates 0.5 μm.

and sizes, as well as 'intermediate' forms, for example, 'broadly spherical' particles with tails of single tubes or ropes protruding from the bulk, or 'fibre-like' particles with significant branches.

2.3 Mechanisms of particle deposition in the lung

Inhaled aerosol particles deposit in the lung as a result of coming into contact with the walls of the airway. There are five mechanisms which control this: inertial impaction; sedimentation; diffusion; interception; and electrostatic forces. The extent to which each mechanism influences deposition is dependent on the properties of the particle under consideration and varies with the region of the respiratory tract, depending on morphology and air-flow characteristics.

The respiratory system comprises a series of branching airways. These begin at the nose (anterior nares) and lead via the posterior nares, the naso-pharynx, the larynx, trachea, bronchi, bronchioles, and alveolar ducts to the alveoli. In the pulmonary region the trachea divides to form the right and left main bronchi. The main bronchi divide to form the lobar bronchi and further divisions lead to the small bronchi. Following the bronchi are the bronchioles; several generations are produced, ending with the terminal bronchioles. These small airways are about 0.6 mm in diameter and the diameter of the airways distal to these does not decrease by very much. In man and a number of other species, but not in laboratory rodents, the terminal bronchioles are followed by three or so generations of respiratory bronchioles. These short airways (the length-to-diameter ratios being less than 2) lead to the alveolar ducts and thence to the alveoli. The alveoli are saucer- or dome-shaped structures with a smooth inner surface; they expand on inspiration. Gas exchange occurs only across alveolar walls. There are about 30 000 terminal bronchioles and 300 000 000 alveoli in man. The total surface area of the alveoli exceeds 100 square metres.

The divisions of the airways have been described in terms of 'generations' of airways. Several numbering systems exist: if the trachea is designated as generation number 1, then the bronchi end at generation 11, the terminal bronchioles are at generation 16, and the alveoli are generally described as generation 25.

Airflow through the system is complex and is dependent on both the physical characteristics of the different regions of the respiratory tract and breathing rates, governed primarily by activity level. The upper part of the airway is characterised by a series of changes of direction of airflow. The expected flow regime is characterised by the Reynolds number. This is proportional to the air-flow rate and density and inversely proportional to the viscosity and airway radius. Laminar flow occurs at low Reynolds numbers and turbulent flow at high Reynolds numbers. In the human respiratory tract Reynolds number decreases with increasing airway generation and flow tends to be laminar in the smaller airways. Flow velocities in the larger airways can be high enough to allow turbulent flow, particularly with increased breathing rate. In the small airways the axial core of the flowing air can forge ahead during inspiration and penetrate into the alveolar ducts but when flow reverses for expiration the flow has a flatter velocity profile.

Deposition by *inertial impaction* occurs when air flow changes direction but particles continue along their original path for a short distance because of inertia. *Sedimentation* is the result of gravitational settling. Most deposition due to impaction occurs at the first airway bifurcation and to a lesser extent at subsequent bifurcations, making it of primary concern in the large airways. Sedimentation, on the other hand, is of most importance in the smaller airways where flow velocities are low and residence times are long. Deposition by both these mechanisms is governed by the *aerodynamic equivalent diameter* (often termed *aerodynamic diameter*) of a particle, with the likelihood of deposition increasing with (aerodynamic) diameter. Deposition by impaction and by sedimentation are most important for particles with aerodynamic diameters greater than 500 nm.

Diffusion is the result of collisions between air molecules and airborne particles and leads to an increased probability that particles will deposit on the airway wall. The motion is characterised by the particle *diffusion coefficient* which, for spherical particles, is inversely proportional to physical size. Diffusion is the predominant deposition mechanism for spherical particles less than about 500 nm in diameter. Deposition by diffusion is most important in the smaller airways and alveoli where distances are short and residence times relatively long.

Interception is the only deposition mechanism that is not the result of deviation from the original gas streamline and is the process by which a particle contacts the airway surface because of its physical size. For compact particles, the likelihood of interception is low, even in the smallest airways. For high-aspect-ratio particles, though, as length increases so does the probability that the ends of the particle will touch a surface, even if the particle's centre of mass is central to the airway. Interception of long, 'fibre-like' CNT particles, or 'broadly-spherical' CNT particles with tails protruding from the bulk may therefore contribute to deposition in the lung, particularly in the small airways.

The effects of electrostatic forces are often assumed to be negligible compared with other deposition mechanisms. However, CNT aerosol particles holding multiple elementary charges can be produced (Nasibulin *et al.*, 2008) so electrostatic deposition may need to be considered, particularly in the small airways where it is likely to have the greatest effect.

As the aerodynamic diameter and diffusion coefficient (or diffusion equivalent diameter) play a key role in determining the pattern of particle deposition in the respiratory tract, it is useful to consider these concepts further.

Aerodynamic diameter, d_{ae}, is defined as the diameter of a unit-density sphere ($\rho_0 = 1000 \text{ kg m}^{-3}$) with the same terminal settling velocity as the particle of interest. For a spherical particle, aerodynamic diameter is related to physical diameter, d, and particle density, ρ_p, by the expression

$$d_{ae} = d\sqrt{\frac{\rho_p}{\rho_0} \cdot \frac{C_c(d)}{C_c(d_{ae})}}, \tag{2.1}$$

where C_c is the slip correction factor, included to extend the validity of the approach to particles with diameters near the mean free path of the air molecules, and is given by the expression

$$C_c(d) = 1 + Kn\left[C_1 + C_2\exp\left(\frac{-C_3}{Kn}\right)\right], \tag{2.2}$$

where $Kn = 2\lambda/d$, λ is the mean free path of the gas, and C_1, C_2, and C_3 are constants. Based on experimental measurements for solid, spherical particles, Allen and Raabe (1985) calculated the slip constants for solid particles to be $C_1 = 1.142$, $C_2 = 0.558$, and $C_3 = 0.999$ for an assumed mean free path of 0.0673 μm for air at sea level and $23°C$ with viscosity of 183.245 μP.

Table 2.2 *Predicted aerodynamic diameters (in μm) of spherical particles with diameters and densities in the ranges observed for 'broadly spherical' CNT particles (values in brackets are the ratio d/d$_{ae}$).*

Physical diameter (μm)	Density (kg m^{-3})		
	100	1000	2000
0.1	0.012 (8.1)	0.1 (1.0)	0.17 (0.59)
2	0.58 (3.4)	2.0 (1.0)	2.9 (0.70)
100	32 (3.2)	100 (1.0)	140 (0.71)

Estimates of the aerodynamic diameters of 'broadly spherical' CNT particles with diameters and densities in the ranges observed (see earlier discussion) can be made using the above equations; see Table 2.2. These indicate a very wide range of aerodynamic diameters, from 10 nm to over 100 μm (i.e. four orders of magnitude), although it should be noted that inhalable particles are usually defined as those having aerodynamic diameters less than 10 μm. In general, information on the densities of such CNT particles (Maynard *et al.*, 2004) indicates that their aerodynamic diameter is likely to be less than or equal to their physical diameter.

The above analysis assumes that the CNT particles form perfect spheres. As seen in Figures 2.2 and 2.3, this is not the case. Particles diverge by varying degrees from spherical, with, in some cases, 'tails' protruding from the bulk. It is unclear at this stage how this will impact on their aerodynamics. A common approach to dealing with the non-spherical nature of most aerosol particles is to include a dynamic shape factor, χ, in the denominator of the square-root term on the right hand side of Equation 2.1 and to replace the physical diameter, d, by the volume equivalent diameter, d_v. The dynamic shape factor is generally greater than 1.0, meaning that the aerodynamic diameter of non-spherical particles will be less than for spherical particles. Dynamic shape factors have not been determined for carbon nanotube particles and this clearly warrants further experimental investigation.

While particle orientation is of little interest when considering spherical particles, for 'fibre-like' particles in a flow, orientation will affect their aerodynamic and diffusion behaviour. While the flow velocity gradient acts to orientate fibres, rotational Brownian motion acts to randomise orientation. For 'fibre-like' CNT aerosol particles in the human respiratory tract, Brownian spinning is expected to dominate for fibres up to a few μm long. For longer fibres, Bernstein and Shapiro (1994) have shown that as the Reynolds number increases in the laminar flow regime, fibres become aligned parallel to the direction of motion. In the human respiratory tract, Reynolds numbers are largest in the upper airways, making alignment more likely.

However, turbulence, which can also occur in the upper airways at higher Reynolds numbers, acts to randomise alignment.

For a 'fibre-like' particle, modelled as an ellipsoid whose polar axis forms an angle θ with the direction of flow, aerodynamic diameter can be shown to be related to the fibre's physical properties by the following expression:

$$d_{ae} = \sqrt{\frac{3\rho_p d_f^3 \beta}{2\rho_0} \frac{C_c(d_e)}{C_c(d_{ae})} \left(\frac{\sin^2\theta}{d_{f,\perp}} + \frac{\cos^2\theta}{d_{f,\parallel}} \right)}, \qquad (2.3)$$

where d_f, L, and $\beta(=L/d_f)$ are, respectively, fibre diameter, fibre length, and aspect ratio, and $d_{f,\perp}$ and $d_{f,\parallel}$ are the Stokes diameters for a fibre orientated perpendicular and parallel to the direction of motion (Oseen, 1927), given by the expressions

$$d_{f,\perp} = \frac{\frac{8}{3}(\beta^2-1)d_f}{\frac{2\beta^2-3}{\sqrt{\beta^2-1}} \ln\left(\beta + \sqrt{\beta^2-1}\right) + \beta}, \qquad (2.4)$$

$$d_{f,\parallel} = \frac{\frac{4}{3}(\beta^2-1)d_f}{\frac{2\beta^2-1}{\sqrt{\beta^2-1}} \ln\left(\beta + \sqrt{\beta^2-1}\right) - \beta}. \qquad (2.5)$$

There is no exact expression available for the slip correction factors for ellipsoids. Instead, the approximate method by Dahneke (1973) is applied, whereby a slip equivalent diameter, d_e, dependent on d_f, β, and θ, is calculated and used in Equation 2.2 in place of d to give the slip correction factor.

For the case of an ellipsoid orientated parallel to the direction of movement, $\theta = 0$, making Equation 2.3 dependent only on $d_{f,\parallel}$. If an average orientation due to Brownian spinning is assumed, Equation 2.3 reduces to

$$\overline{d_{ae}} = \sqrt{\frac{3\rho_p d_f^3 \beta}{2\rho_0} \frac{C_c(d_e)}{C_c(\overline{d_{ae}})} \left(\frac{2/3}{d_{f,\perp}} + \frac{1/3}{d_{f,\parallel}} \right)}. \qquad (2.6)$$

Predicted aerodynamic diameters for 'fibre-like' CNT particles for a range of diameters and lengths are presented in Table 2.3. A fibre diameter of 2 nm is taken as typical of a hypothetical individual SWCNT fibre; 10 nm, 50 nm, and

Table 2.3 *Predicted aerodynamic diameters (in nm) for
'fibre-like' CNT particles for a range of diameters and lengths.
Random orientation is assumed. Particle density is assumed to
be 1750 kg m^{-3}.[a]*

	Fibre diameter (nm)			
Fibre length (μm)	2	10	50	90
0.1	7.7	37.1	149.8	230.0
0.5	7.7	37.9	164.4	264.3
1	7.7	38.3	171.5	281.8
2	7.7	38.8	178.9	299.9
5	7.8	39.2	183.2	323.3
10	7.8	39.7	196.3	340.5
50	7.8	40.7	213.5	378.9

[a] Kim *et al.*, 2009

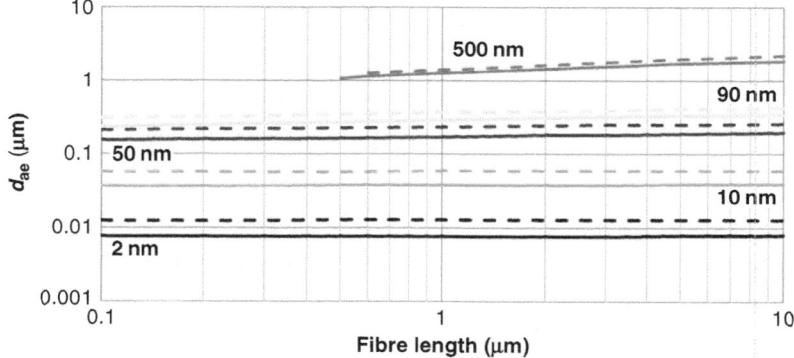

Figure 2.4 Predicted aerodynamic equivalent diameters of 'fibre-like' CNT
particles for a range of fibre diameters, for both random (solid lines) and parallel
(dashed lines) orientation to airflow, compared with that for a typical asbestos fibre
(diameter 500 nm and density 2500 kg m^{-3}).

90 nm cover the range of MWCNT diameters; and 50 nm is typical of the
'fibre-like' CNT aerosol particles observed in samples taken at research and
industrial facilities. These are also compared with the values for a typical
asbestos fibre in Figure 2.4.

From the above, it is clear that for 'fibre-like' CNT particles aerodynamic
diameter increases with both fibre diameter and length but is governed predominantly by fibre diameter. As a rule of thumb, aerodynamic diameter is 2–4 times

greater than fibre diameter; the ratio increases with fibre length, levelling out as the aspect ratio increases to around 200. Predicted aerodynamic diameters for 'fibre-like' CNT aerosol particles are in the approximate range 40–340 nm. Aerodynamic diameters are greater, by up to 50%, when fibres are assumed to be aligned parallel to the direction of motion rather than randomly orientated. This difference decreases with increasing fibre diameter.

This approach assumes particles are solid, straight ellipsoids, but exposure assessments have shown that while 'fibre-like' MWCNT aerosol particles can be straight cylinders (e.g. Figure 2.1), they are more likely to be bent and/or lightly agglomerated into parallel bundles or branching structures. While the ellipsoid model is probably reasonable for straight cylinders and parallel bundles of tubes, it may become inappropriate as particle shape deviates further from the ideal. Even for straight cylinders, the effect of ends, the hollow form of CNT, and the variation in density across the particle, for example due to metal impurities, may mean that the aerodynamic diameter of straight, 'fibre-like' CNT aerosol particles will differ significantly from that predicted above. The effect of such complex forms does not have implications simply for the theoretical formulae for deposition but also for assumptions regarding orientations. In view of these uncertainties, assessment of whether an ellipsoid model is appropriate for the full range of 'fibre-like' CNT aerosol particles possible is needed. This should form part of a larger investigation into the relationship between CNT aerosol particle morphology and transport and deposition.

The thermal movement of particles suspended in a resistant medium was expressed by Einstein (1905) as

$$\bar{x} = \sqrt{2D_t t}, \tag{2.7}$$

where $\bar{x}$ is the average particle displacement along each coordinate axis per unit time t, and D_t is the translational diffusion coefficient. For spherical particles, the translational diffusion coefficient can be described by the Stoke–Einstein equation in terms of particle diameter as

$$D_t = \frac{kTC_c(d)}{3\pi\mu d}, \tag{2.8}$$

where k is the Boltzmann constant, T is the absolute temperature, and μ is the viscosity of the fluid. For non-spherical particles a diffusion equivalent diameter, d_d, is often defined, being the diameter of a spherical particle with equivalent diffusion coefficient as the particle of interest. By replacing d with d_d in Equation 2.8, the translational diffusion coefficient for non-spherical particles can then be calculated.

Diffusion diameter is related to volume equivalent diameter, d_{ev}, by the relationship (Kasper, 1982)

$$\frac{d_d}{C_c(d_d)} = \chi \frac{d_{ev}}{C_c(d_{ev})}, \tag{2.9}$$

where χ is a dynamic shape factor. For spherical particles (with or without internal voids), $\chi = 1$ and $d_d = d_{ev} = d$. For irregular particles and aggregates, $\chi > 1$ and d_d is always greater than d_{ev} (DeCarlo *et al.*, 2004). As particle shape deviates further from spherical, the dynamic shape factor and the difference between diffusion diameter and volume equivalent diameter increases. While the relationship between d_d and d_{ev} for 'broadly spherical' CNT agglomerates has not been investigated, it has been shown previously for TiO_2, Si, and diesel exhaust agglomerates that their projected area equivalent diameter (assumed to be equivalent to volume equivalent diameter) is approximately equal to their diffusion diameter (Rogak *et al.*, 1993; DeCarlo *et al.*, 2004; Park *et al.*, 2004). For 'broadly spherical' CNT agglomerate particles, volume equivalent diameter should therefore act as a reasonable approximation to physical diameter in Equation 2.8. It should be remembered, though, that the error in this assumption will increase as particle shape deviates further from spherical.

For 'fibre-like' particles, modelled again as ellipsoids forming an angle θ with the direction of movement, their diffusion equivalent diameter (Asgharian *et al.*, 1988) can be expressed as

$$\frac{C_c(d_d)}{d_d} = C_c(d_e)\left(\frac{\sin^2\theta}{d_{f,\perp}} + \frac{\cos^2\theta}{d_{f,\parallel}}\right). \tag{2.10}$$

For the case of orientation parallel to the direction of movement, $\theta = 0$, making Equation 2.10 dependent only on $d_{f,\parallel}$. If an average orientation due to Brownian spinning is assumed, Equation 2.10 reduces to

$$\frac{C_c(\overline{d_d})}{\overline{d_d}} = C_c(d_e)\left(\frac{2/3}{d_{f,\perp}} + \frac{1/3}{d_{f,\parallel}}\right). \tag{2.11}$$

Figure 2.5 shows the relationship expressed in Equation 2.10 between diffusion equivalent diameter and length for 'fibre-like' particles with a range of diameters. The cases of parallel and random alignment with respect to the direction of movement are shown to represent the expected situations in the upper and lower regions of the lung respectively.

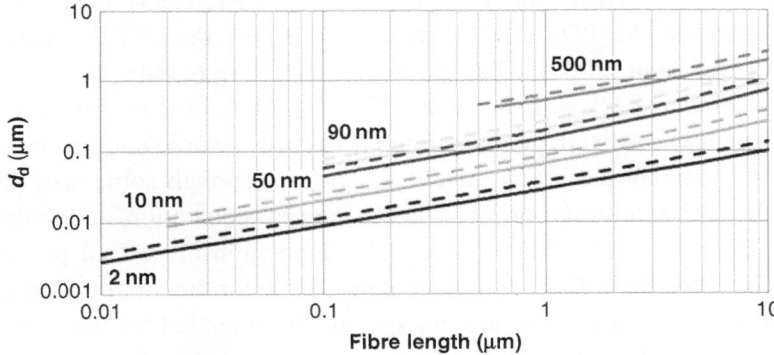

Figure 2.5 Predicted diffusion equivalent diameters of 'fibre-like' CNT particles for a range of fibre diameters, for both random (dashed lines) and parallel (solid lines) orientation to airflow, compared with that for a typical asbestos fibre (diameter 500 nm and density 2500 $kg\,m^{-3}$).

The diffusion equivalent diameter for 'fibre-like' particles increases with both fibre diameter and length and therefore a simple rule-of-thumb relationship does not exist. If SWCNT could be aerosolised as individual straight fibres, as a result of their extremely small diameters diffusion equivalent diameters would be comparatively small, ranging from a few nm up to around 100–200 nm for fibre lengths up to 10 µm. For typical 'fibre-like' MWCNT particles observed during exposure assessments, diffusion diameters range between about 50 nm and 1 µm. For asbestos fibres, diffusion diameters are likely to be in the region of a few hundred nm to a few µm.

CNT particles have a tendency to be bent and form lightly agglomerated branching structures. As for the theoretical calculation of aerodynamic diameters of 'fibre-like' CNT particles, deviations such as these from the ideal 'fibre-like' shape will result in an error in calculating diffusion diameters. Deviation from the ideal shape will also affect the particles' orientation in a flow. To address these issues an experimental investigation of the relationship between diffusion diameter and morphology of real CNT aerosol particles needs to be undertaken.

2.4 Deposition efficiency of carbon nanotubes in the respiratory tract

A large number of models have been developed over the past half-century for predicting the deposition of aerosols in the human respiratory tract and that of a small number of other animals. The best-known models are the widely available whole-lung deposition models. These are semi-empirical models derived from experimental data and theoretical considerations of the deposition mechanisms and use simplified representations of the lung structure. They include the ICRP

Human Respiratory Tract Model (HRTM) model (ICRP, 1994), the multiple path particle dosimetry (MPPD) model (Freijer *et al.*, 1999), the NCRP model (NCRP, 1997) and Hofmann (2009). The first two of these are readily available, implemented as software packages (Jarvis and Birchall, 1994; Freijer *et al.*, 1999). Such models generally predict deposition efficiencies in the whole respiratory tract and its major constituent parts (e.g. alveolar region), although some also allow the estimation of this information on a generation-by-generation or lobular basis. They are intended primarily to address the deposition of aerosol particles that are broadly spherical in form. Although some allow for the input of shape factors for non-spherical shapes, they are not specifically intended for use with fibrous particles.

Recent reviews (Kleinstreuer *et al.*, 2008; Hofmann, 2009; Rostami, 2009) have indicated that, in general, the available whole-lung models predict total and regional deposition for broadly spherical particles reasonably accurately over a wide size range. This is not entirely surprising, as many of these models include elements derived from experimental data and have been extensively validated against other available experimental data.

Considering the ICRP HRTM as one example of such models (see Figure 2.6), this predicts that deposition by aerodynamic mechanisms (impaction and sedimentation) does not occur for particles with aerodynamic diameters less than 100 nm in any region of the respiratory tract. Deposition in the alveolar region reaches about 10% as aerodynamic diameter increases to 500 nm and peaks at 2 µm, when about

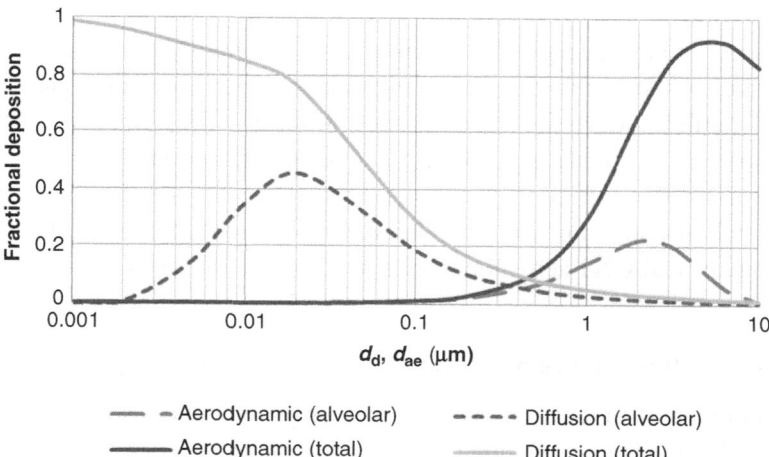

Figure 2.6 Predicted deposition in the alveolar region and the whole respiratory tract as a function of aerodynamic and diffusion equivalent diameter using the ICRP Human Respiratory Tract Model (ICRP, 1994).

20% of inhaled particles are expected to be deposited. The 'peak' alveolar deposition efficiency is still relatively low as most particles of this size will deposit in the upper airways before reaching the alveoli. Deposition by diffusion is most efficient for spherical particles with small physical diameters; diffusional deposition in the alveolar region peaks at about 45% for particles with diameters (or diffusion equivalent diameters) of 20 nm. For smaller particles, deposition in the pulmonary region decreases as more particles deposit in the upper airways before reaching the alveoli, and, at larger sizes, deposition decreases in all lung regions. By 200 nm, the expected diffusional deposition in the alveolar region has decreased to 10%, and by 10 μm to only 1%.

It is expected that such models would provide a reasonable estimate of the deposition efficiencies of 'broadly spherical' CNT aerosol particles within the respiratory tract. With aerodynamic diameters for realistic 'broadly spherical' CNT aerosol particles expected to be in the range 10 nm to 140 μm (Table 2.2) and diffusion equivalent diameters in the range 100 nm to 100 μm (Figure 2.5), the predicted deposition in the entire respiratory tract varies significantly, from a minimum of approximately 20% to over 90%. The maximum alveolar deposition efficiency is expected to be around 20% with deposition efficiencies generally between 10% and 20%, going below 10% for aerodynamic diameters >5 μm. As discussed earlier, though, some uncertainties in the estimations of diameters remain to be resolved.

As indicated above, models such as HRTM are not intended for use with fibrous materials. A number of models have been developed specifically for fibres and these are discussed below; however, an early approach to modelling in this area was to derive aerodynamic and diffusion equivalent diameters for fibres (using approaches such as those presented above) and use these in a standard particle deposition model. It is instructive therefore to consider predicted diameters for 'fibre-like' CNT particles in relation to the ICRP HRTM (Figure 2.6). With aerodynamic diameters for realistic 'fibre-like' MWCNT particles expected to be in the range 40–400 nm (Table 2.3 and Figure 2.4), the deposition efficiency of sedimentation and impaction in the alveolar region is expected to be less than a few percent. With diffusion equivalent diameters in the range 50 nm to 1 μm, diffusion would be the dominant mechanism in this region, with diffusion deposition efficiencies from around 30% at 50 nm to 5% at 400 nm. Thus total alveolar deposition efficiencies in the range from a few percent up to a maximum of 30% for the narrowest and shortest fibres would be expected. This can be contrasted with the position for asbestos fibres. These have aerodynamic diameters in the few μm range which is the size for peak alveolar deposition by aerodynamic mechanisms of 20%. With diffusion equivalent diameters in the range of a few hundred nm to a few μm the maximum deposition by this mechanism is a few percent.

Although the above gives some indication of the likely deposition efficiencies of CNT fibres, the HRTM model does not include interception as a deposition mechanism, which it is anticipated may be important for CNT fibres, and is likely to increase deposition efficiencies, particularly in the narrower airways. Scheckman and McMurry (2011) found that in a cast of a section of a human lung, deposition of agglomerates was significantly greater than that for spherical particles with equivalent mobility diameters in the range 20–300 nm. For example, the deposition efficiency of silica agglomerates with a primary particle diameter of 10 nm, mobility equivalent diameter of 100 nm, and maximum dimension of about 300 nm was 8%, compared to only 3% for oleic acid spheres of equivalent mobility diameter. Scheckman and McMurry concluded that interception was responsible for the enhanced deposition observed and hypothesised that the enhanced deposition occurs at locations where the flow undergoes transitions, for example at bifurcations and where there are irregularities in the surface of the flow pathways. Interception could therefore also enhance deposition of 'fibre-like' MWCNT particles with aspect ratios as low as 3 or 4.

There are a number of more sophisticated fibre-deposition models. Asgharian and Yu (1988; 1989) developed whole-lung models for the deposition of fibres in the human and rat respiratory tracts. The approach used was to extend their pre-existing models for spherical particles for diffusion, sedimentation, and impaction with an empirical model based on limited experimental data for interception. These single-path models allowed deposition in regions of the lung to be evaluated. Following the development of a multiple-path deposition model for spherical particles in the rat (Anjilvel and Asgharian, 1995), which allowed the estimation of deposition in each branch of the respiratory tract, Asgharian and Anjilvel (1998) extended this to address deposition of fibres. This model indicated that for a given fibre diameter deposition decreased with aspect ratio in the diffusion-dominated region and increased with aspect ratio in the impaction range. Ding *et al.* (1997) also developed Asgharian and Yu's earlier model for fibre deposition in the rat lung. They used the details of further theoretical studies on deposition in cylindrical sections combined with an empirical equation based on experimental data for combined deposition by impaction and interception in the nasal passages. The authors made a detailed comparison of model predictions with the available experimental data. They found that the model predictions were reasonable: for total and nasal deposition they tended to overestimate deposition, whilst for lower tract deposition the predictions were more accurate.

Strum and Hofmann have also developed a number of models for the deposition of fibres in the human respiratory tract (Strum and Hofmann, 2006; 2009), and the only software package available for fibres, Fibros (Strum and Hofmann, 2006). These are based upon earlier models for the deposition of particles (Koblinger and Hofmann, 1990). Aerodynamic and diffusional deposition were modelled in a similar manner to the whole-lung models discussed above, using aerodynamic and effective diffusion diameters for the fibres (derived as discussed earlier). To account for deposition by interception, correction factors for airway generations 1–12 (i.e. upper airways before bronchioles) were computed based on numerical simulations of fibre behaviour within single airway bifurcations. These were used to 'correct' the deposition by interception of spherical particles.

The results of the model for a range of fibre diameters and two aspect ratios are given in Figure 2.7. These indicate a standard 'two-peak' distribution curve for alveolar deposition, with an initial (diffusion-dominated) peak for fibre diameters in the region of 20 nm, reducing to a minimum at a few hundred nm and increasing again to a second peak in the low-μm region (deposition primarily by aerodynamic mechanisms). The peak alveolar deposition is between 10% and 20%. The study found that for fibres with diameters >0.1 μm an increase in aspect ratio shifted the peak deposition diameter to a smaller value. The effect of aspect ratio for particles with diameters <0.1 μm was less marked, with a maximum difference of a few percent between aspect ratios of 3 and 100. The model indicates that fibres with diameter <0.1 μm are chiefly deposited by diffusion.

Unlike the 'spherical' particle deposition models discussed earlier, there are limited experimental data available to validate these fibre-deposition models, and so their predictions must be considered significantly more uncertain. It is also important to recognise that these models were developed primarily to address the deposition of asbestos fibres and similarly shaped man-made fibres (e.g. glass and ceramic) as at the time these were of particular concern from a health perspective. Such fibres tend to have diameters in the range of hundreds of nm to μm and also tend to be more rigid and straight than the 'fibre-like' CNT agglomerates considered here, so there are clearly questions about their suitability for use in predicting CNT deposition without modification, in particular in relation to the treatment of interception. However, with these caveats, it is still instructive to consider the results of such models in relation to 'fibre-like' CNT particles: for a typical 'fibre-like' CNT with diameter 50 nm, Figure 2.7 indicates alveolar deposition in the region of 10% for a wide range of aspect ratios.

A more recent modelling study has focused more directly on nanofibres and developed a semi-analytical model describing the transport and deposition of fibres from nano- to micro-scale (Högberg *et al.*, 2010). This is not a whole-lung model but addresses transport and deposition of different fibre types by considering straight tubes of various dimensions and orientations to represent respiratory airways. The fibres are modelled as ellipsoidal bodies for computational ease. Results are presented

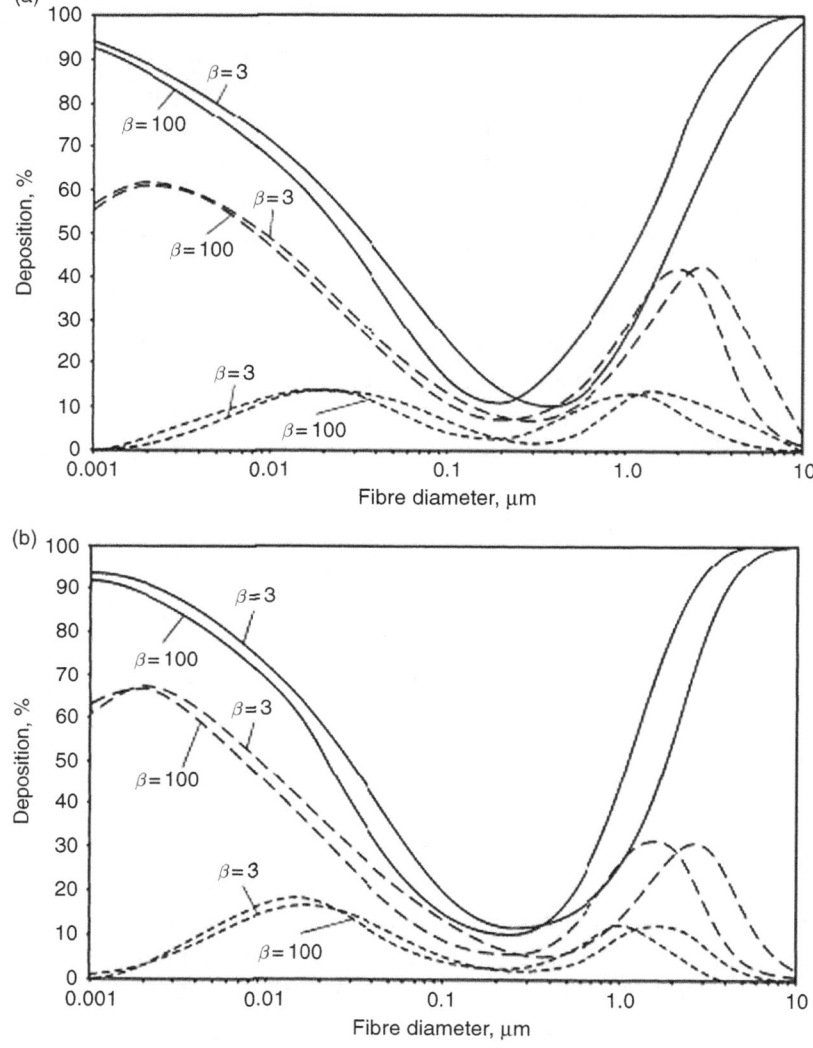

Figure 2.7 Predicted total (full lines), tracheo-bronchial (long dashed lines), and alveolar (short dashed lines) deposition for a range of fibre diameters for two aspect ratios (β), 3 and 100, for (A) sitting breathing conditions, and (B) light-work breathing conditions (reprinted from Strum and Hofmann (2009) with permission from Elsevier).

for steady laminar flow in straight horizontal and inclined airways, intended to represent lung generations to 16 (terminal bronchioles). For unit-density fibres the model indicates that, for fibres with small aspect ratio, the diameter of minimal deposition is located around 0.5 μm, which is broadly consistent with the models discussed above. For nanofibres, diffusion dominates for low aspect ratios, decreasing with length until interception becomes a significant deposition mechanism (e.g. for generation 16 for lengths >10 μm). The model predicts minimum deposition for fibres with diameters in the size range 10–100 nm and lengths of several μm. Thus the authors conclude that the probability of reaching the gas-exchange region of the lung is highest for such fibres. The key unanswered question, however, is what the deposition efficiency of such fibres would be in the alveolar region.

All the above models assume that electrostatic deposition is negligible. For compact particles, human-volunteer studies have shown that pulmonary deposition may be considerably increased by the electrostatic charge carried by an individual particle if a threshold level of charge is met (Melandri *et al.*, 1983; Prodi and Mularoni, 1985). While there have so far been no reports on the effect of charge on the deposition of CNT particles, the effect of charge on the deposition of chrysotile asbestos fibres in rats has been investigated (Davis *et al.*, 1988). Animals exposed to charged chrysotile fibre aerosols holding 13 electrons/μm and 20 electrons/μm had, respectively, 20% and 50% more material retained in their lungs compared with those exposed to neutralised aerosols (1–2 electrons/μm). The charging efficiency of CNT aerosol particles, both as produced and after charge neutralisation, has been shown to be higher than compact particles with equivalent mobility (Kulkarni *et al.*, 2009; Nasibulin *et al.*, 2008). While it is not yet known whether the charging is sufficient to noticeably increase deposition, it certainly suggests that further work on the subject is required.

As mentioned above, there is very little relevant experimental data on the deposition of CNT and other high-aspect-ratio nanofibres. Reported experimental studies on the behaviour of fibres in the respiratory tract are focused on asbestos and various man-made materials, including glass and refractory ceramic fibres. The studies fall into three general categories: human post-mortem lung tissue sample analyses; experiments using airway casts; and animal studies. *Ex vivo* studies of individuals exposed to fibres, primarily studies of the lungs of deceased asbestos workers, have provided qualitative information on the ability of various fibre types to reach different parts of the lung (e.g. Timbrell, 1982). While such studies cannot be used to derive deposition efficiencies, results have clearly revealed the ability of long fibres (longer than tens of μm) to penetrate deep into the lung. Experiments using airway casts have provided only limited information on deposition in the upper airways.

The results of a number of short-term fibre-inhalation exposure studies using rats have been usefully summarised by Ding *et al.* (1997). The majority of the fibres used in these studies were asbestos with lengths from 1 to 6 μm. In one such study, Morgan *et al.* (1975) suggested that alveolar deposition increased very little over the range considered and that a value of approximately 11% could be assumed. Unfortunately, as the majority of the studies reported used fibres with aerodynamic diameters generally in the range of 1 to a few μm, which are significantly larger than the 'fibre-like' CNT aerosol particles we are considering here, the results, although of general interest, cannot be directly applied to their deposition. However, Griffis *et al.* (1981) investigated the deposition of a glass-fibre aerosol with count median diameter 0.11 μm and count median length 8.3 μm, which are similar to the dimensions of the 'fibre-like' CNT aerosol particles considered here. Analysis of the results indicates approximate deposition in the alveolar region of a few percent, which is generally consistent with the model predictions in Figure 2.7.

2.5 Summary and research needs

The deposition efficiency of aerosolised CNT particles in the respiratory tract is a key input to any risk assessment of inhalation exposures to this material. We have attempted here to explore this using relevant experimental studies and theoretical models. The focus has been on deposition in the alveolar region as this is of particular importance in relation to considerations of the potential health effects.

It is important to note that this analysis is based on the currently limited available information on the morphology of aerosolised CNT fibres under 'realistic' exposure conditions. It is important to recognise that new manufacturing techniques and processes and applications could result in exposures to other particle forms. The information available to date indicates that SWCNT particulates have only been found as large 'broadly spherical' agglomerates. MWCNT aerosols have also frequently been observed to contain similar 'broadly spherical' forms but can also exist as single fibres or ropes of agglomerated fibres. The aerodynamic and deposition behaviour of both types of particles has been explored.

For 'broadly spherical' CNT particles it is considered that currently available 'whole-lung' models for the deposition of spherical particles (e.g. ICRP HRTM) probably provide a reasonable estimate of deposition efficiencies. The maximum alveolar deposition efficiency is expected to be around 20% with deposition efficiencies generally between 10% and 20%, going below 10% for aerodynamic diameters >5 μm. The key uncertainties relate to particle density and the effect of divergence from perfect spherical shape. Experimental studies should be undertaken to explore these factors.

The situation for 'fibre-like' CNT particles is more complex. These differ significantly from other fibrous aerosol materials, e.g. asbestos and various man-made fibres, being in general narrower, less dense, and more twisted/bent. As such, much of the theoretical modelling and experimental studies on these materials are of questionable relevance to CNT, and clearly further experimental and theoretical studies are required in this area. In particular, such studies should focus on the effects of divergence from an ideal fibre shape, the role of interception for CNT deposition, and also the potential impact of electrostatic charging on deposition. This preliminary analysis does, however, tend to suggest that aerosolised CNT fibres will deposit primarily by diffusion. The current models and very limited relevant experimental data indicate alveolar deposition in the range of a few percent to a few tens of percent depending in detail on the particle characteristics and modelling approach taken.

It is important to recognise differences between the deposition of CNT particles and of asbestos fibres. The alveolar deposition efficiency for asbestos fibres (typically tens of percent) may be similar to that for some 'fibre-like' CNT particles, but the dominant deposition mechanisms are different, with sedimentation dominating for asbestos, and diffusion (and possibly interception) for CNT. It would be interesting to explore whether the various deposition mechanisms have any implications for the potential transport of fibres from the alveolar region.

To develop understanding further on the pulmonary deposition of CNT aerosol particles, experimental and theoretical modelling work is required to investigate the following:

● The form of 'real' CNT aerosol particles, e.g. shape, size, and density
● The aerodynamic and diffusion diameters of CNT aerosol particles and the validity of the theoretical treatment presented here
● The role of interception in the deposition of CNT aerosol particles in the lungs and its inclusion in deposition models
● The deposition efficiency of CNT particles in the respiratory tract (in particular experimental results for fibre deposition model validation)
● The aerosol charge state of CNT aerosol particles and the effect of charge on their deposition in the lungs.

References

Allen, M. D. and Raabe, O. G. (1985). Slip correction measurements of spherical solid aerosol particles in an improved Millikan apparatus. *Aerosol Science and Technology*, **4** (3), 269–286.

Anjilvel, S. and Asgharian, B. (1995). A multiple-path model of particle deposition in the rat lung. *Fundamental and Applied Toxicology*, **28**, 41–50.

Asgharian, B. and Yu, C. (1988). Deposition of inhaled fibrous particles in the human lung. *Journal of Aerosol Medicine*, **1** (1), 37–50.

Asgharian, B., Yu, C. P., and Gradon, L. (1988). Diffusion of fibers in a tubular flow. *Aerosol Science and Technology*, **9**, 213–219.

Asgharian, B. and Yu, C. (1989). Deposition of fibers in the rat lung. *Journal of Aerosol Science*, **20** (3), 355–366.

Asgharian, B. and Anjilvel, S. (1998). A multiple-path model of fiber deposition in the rat lung. *Toxicology of Science*, **44**, 80–86.

Bello, D., Hart, A. J., Ahn, K., *et al.* (2008). Particle exposure levels during CVD growth and subsequent handling of vertically-aligned carbon nanotube films. *Carbon*, **46** (6), 974–977.

Bello, D., Wardle, B. L., Yamamoto, N., *et al.* (2009). Exposure to nanoscale particles and fibers during machining of hybrid advanced composites containing carbon nanotubes. *Journal of Nanoparticle Research*, **11** (1), 231–249.

Bello, D., Wardle, B. L., Jie, Z., *et al.* (2010). Characterization of exposures to nanoscale particles and fibers during solid core drilling of hybrid carbon nanotube advanced composites. *International Journal of Occupational and Environmental Health*, **16** (4), 434–450.

Bernstein, O. and Shapiro, M. (1994). Direct determination of the orientation distribution function of cylindrical particles immersed in laminar and turbulent shear flows. *Journal of Aerosol Science*, **25** (1), 13–136.

Dahneke, B. E. (1973). Slip correction factors for nonspherical bodies – III the form of the general law. *Journal of Aerosol Science*, **4** (2), 163–170.

Davis, J. M., Bolton, R. E., Douglas, A. N., Jones, A. D., and Smith, T. (1988). Effects of electrostatic charge on the pathogenicity of chrysotile asbestos. *British Journal of Industrial Medicine*, **45** (5), 292–299.

DeCarlo, P. F., Slowik, J. G., Worsnop, D. R., Davidovits, P., and Jimenez, J. L. (2004). Particle morphology and density characterization by combined mobility and aerodynamic diameter measurements. Part 1: Theory. *Aerosol Science and Technology*, **38**, 1185–1205.

Ding, J. Y., Yu, C. P., Zhang, L., and Chen, Y. K. (1997). Deposition modelling of fibrous particles in rats: Comparisons with available experimental data. *Aerosol Science and Technology*, **26**, 403–414.

Einstein, A. (1905). On the motion, required by the molecular-kinetic theory of heat, of particles suspended in a fluid at rest. *Annals of Physics*, **17**, 549–560.

Freijer, J. I., Cassee, F. R., Subramaniam, R., *et al.* (1999). Multiple Path Particle Deposition Model (MPPDep Version V1.11): A model for human and rat airway particle deposition. Bilthoven, The Netherlands, National Institute of Public Health and the Environment (RIVM). Rapport no. 650010019.

Griffis, L. C., Henderson, T. R., and Pickrell, J. A. (1981). A method for determining glass in rat lung after exposure to a glass fiber aerosol. *American Industrial Hygiene Association Journal*, **42** (8), 566–569.

Han, J. H., Lee, E. J., Lee, J. H., *et al.* (2008). Monitoring multiwalled carbon nanotube exposure in carbon nanotube research facility. *Inhalation Toxicology*, **20** (8), 741–749.

Hao, Y., Qunfeng, Z., Fei, W., Weizhong, Q., and Guohua, L. (2003). Agglomerated CNTs synthesized in a fluidized bed reactor: Agglomerate structure and formation mechanism. *Carbon*, **41** (14), 2855–2863.

Hofmann, W. (2009). Modelling particle deposition in human lungs: Modelling concepts and comparison with experimental data. *Biomarkers*, **14**(S1), 59–62.

Högberg, S. M., Åkerstedt, H. O., Lundström, T. S., and Freund, J. B. (2010). Respiratory deposition of fibers in the non-inertial regime: Development and application of a semi-analytical model. *Aerosol Science and Technology*, **44** (10), 847–860.

ICRP (1994). Human Respiratory Tract Model for Radiological Protection. *ICRP Publication 66.* Pergamon Press, Oxford.

Jarvis, N. S. and Birchall, A. (1994). LUDEP 1.0, a personal computer program to implement the ICRP respiratory tract model. *Radiation Protection Dosimetry*, 53(1–4), 191–193.

Johnson, D. R., Methner, M. M., Kennedy, A. J., and Steevens, J. A. (2010). Potential for occupational exposure to engineered carbon-based nanomaterials in environmental laboratory studies. *Environmental Health Perspectives*, **118** (1), 49–54.

Kasper, G. (1982). Dynamics and measurement of smokes. I: Size characterization of nonspherical particles. *Aerosol Science and Technology*, **1** (2), 187–199.

Kim, S. H., Mulholland, G. W., and Zachariah, M. R. (2009). Density measurement of size selected multiwalled carbon nanotubes by mobility-mass characterization. *Carbon*, **47** (5), 1297–1302.

Kleinstreuer, C., Zhang, Z., and Li, Z. (2008). Modeling airflow and particle transport/deposition in pulmonary airways. *Respiratory Physiology & Neurobiology*, **163**, 128–138.

Koblinger, L. and Hofmann, W. (1990). Monte Carlo modelling of aerosol deposition in human lungs. Part I: Simulation of particle transport in a stochastic lung structure. *Journal of Aerosol Science*, **21**, 661–674.

Kulkarni, P., Deye, G. J., and Baron, P. A. (2009). Bipolar diffusion charging characteristics of single-wall carbon nanotube aerosol particles. *Journal of Aerosol Science*, **40** (2), 164–179.

Lee, J. H., Lee, S. B., Bae, G. N., *et al.* (2010). Exposure assessment of carbon nanotube manufacturing workplaces. *Inhalation Toxicology*, **22** (5), 369–381.

Maynard, A. D., Baron, P. A., Foley, M., *et al.* (2004). Exposure to carbon nanotube material: Aerosol release during the handling of unrefined single-walled carbon nanotube material. *Journal of Toxicology and Environmental Health, Part A*, **67** (1), 87–107.

Melandri, C., Tarroni, G., Prodi, V., *et al.* (1983). Deposition of charged particles in the human airways. *Journal of Aerosol Science*, **14** (5), 657–669.

Morgan, A., Evans, J. C., and Holmes, A. (1975). Deposition and clearance of inhaled fibrous minerals in the rat: Studies using radioactive tracer techniques. *Inhaled Particles*, **4**, 259–274.

Nasibulin, A. G., Shandakov, S. D., and Anisimov, A. S. (2008). Charging of aerosol products during ferrocene vapor decomposition in N_2 and CO atmospheres. *The Journal of Physical Chemistry C*, **112** (15), 5762–5769.

NCRP (1997). *Deposition, Retention and Dosimetry of Inhaled Radioactive Substances.* NCRP report no. 125 (Bethesda MD: National Council on Radiation Protection).

Oseen, C. W. (1927). *Neuere Methoden und Ergebnisse in der Hydrodynamik.* Leipzig: Akademische Verlagsgesellschaft.

Park, K., Kittelson, D., and McMurry, P. (2004). Structural properties of diesel exhaust particles measured by transmission electron microscopy (TEM): Relationships to particle mass and mobility, *Aerosol Science and Technology*, **38** (9), 881–889.

Prodi, V. and Mularoni, A. (1985). Electrostatic lung deposition experiments with humans and animals. *Annals of Occupational Hygiene*, **29** (2), 229–240.

Rogak, S. N., Flagan, R. C., and Nguyen, H. V. (1993). The mobility and structure of aerosol agglomerates. *Aerosol Science and Technology*, **18** (1), 25–47.

Rostami, A. A. (2009). Computational modelling of aerosol deposition in respiratory tract: A review. *Inhalation Toxicology*, **21**, 262–290.

Scheckman, J. and McMurry, P. (2011). Deposition of silica agglomerates in a cast of human lung airways: Enhancement relative to spheres of equal mobility and aerodynamic diameter. *Journal of Aerosol Science*, **42**, 508–516.

Strum, R. and Hofmann, W. (2006). A computer program for the simulation of fiber deposition in the human respiratory tract. *Computers in Biology and Medicine*, **36**, 1252–1267.

Strum, R. and Hofmann, W. (2009). A theoretical approach to the deposition and clearance of fibers with variable size in the human respiratory tract. *Journal of Hazardous Materials*, **170**, 210–218.

Timbrell, V. (1982). Deposition and retention of fibres in the human lung. *Annals of Occupational Hygiene*, **26** (2), 347–369.

3

Utilising the concept of the biologically effective dose to define the particle and fibre hazards of carbon nanotubes

KEN DONALDSON, RODGER DUFFIN, FIONA A. MURPHY,
CRAIG A. POLAND

3.1 Structure, toxicity, and the biologically effective dose

Toxicology can be seen in one sense as the pursuit of a structure/toxicity relationship for any class of toxins. A complete understanding of physicochemistry and how that relates to toxicity would enable us to predict toxicity from structure – the 'philosopher's stone' of toxicology. This presumes that toxicity is a result of predictable chemical reactions that culminate in cell damage following contact between a toxin and a biological system. Such a set of rules relating chemical composition to toxicological response is called a quantitative structure–activity relationship (QSAR) in chemical toxicology and should allow prediction of the effects of any chemical based on its physicochemical structure. For particles this proposition is much more complex than for chemicals. Unlike chemicals, particles that are nominally the same (e.g. quartz, asbestos, nanosilver) may vary in multiple ways (e.g. size, surface chemistry, metal content), thus making generalisations difficult. However, if it were possible to know the physicochemistry of any particle and the impact of that physicochemistry on biological systems, it would be possible to predict the toxicological effect that this particle and similar particles would exert in a biological system. There have been efforts towards, and suggestions as to, structural characteristics of particles that could drive toxicity for nanoparticles generally using *in vitro* and *in silico* approaches (Burello and Worth, 2011; Donaldson *et al.*, 2010a; Puzyn *et al.*, 2009, 2011; Rushton *et al.*, 2010). Because of the wide variety of particle types, it is unlikely that there will be one structure-toxicity relationship (STR) that embraces all particles, e.g. asbestos versus coalmine dust. Nor is the particle STR likely to be as comprehensive as a chemical compound QSAR where individual chemical moieties have more predictable effects on biological systems. The particle STR is likely to be based on physical characteristics that are less well defined, such as size, reactivity, and soluble components. However, it may be possible to form STRs for classes of particles whose physicochemical structure are common and so might conform to the same STR. Such a

The Toxicology of Carbon Nanotubes, ed. Ken Donaldson, Craig A. Poland, Rodger Duffin and James Bonner.
Published by Cambridge University Press. © Cambridge University Press 2012.

Table 3.1 *The BED for four common pathogenic particle types.*

Particle	Biologically effective dose (BED)
Quartz	Area of reactive surface
Asbestos	Biopersistent, fibres longer than ~20 μm
PM_{10}	Organics/metals/surfaces
Low-toxicity, low-solubility particles	Surface area

STR already exists for fibres in conventional fibre toxicology (reviewed in Donaldson *et al.*, 2010a), and will be employed in the arguments set out below.

The biologically effective dose (BED) in the case of particles is the entity within the mass dose of particles that actually drives the toxicity. In any perfect STR, the BED would form the structural factor(s) that would be most closely related to toxicity. Table 3.1 shows the BED for four common pathogenic particles. The idea of the BED is easily understood for compact insoluble particles and is represented by the size and the chemical nature of the surface molecular layer. The BED for compact, low-toxicity, low-solubility particles (LTLS) is therefore the total surface area (Duffin *et al.*, 2007; Tran *et al.*, 2000). For quartz, an insoluble but highly reactive particle, the BED includes both the surface area and its associated reactivity (Duffin *et al.*, 2007) although the surface reactivity of quartz can be easily shown to be a labile and variable entity (Donaldson and Borm, 1998). For asbestos it is long, biopersistent fibres (Donaldson, *et al.*, 2010a) that form the BED, as discussed extensively below. For PM_{10} it is metal, organics, and particle surfaces that comprise the BED (Donaldson *et al.*, 2005). As can be seen it is possible for any particle to have more than one factor that contributes to the BED, including length of fibres and their biopersistence. Nonetheless, the BED is an invaluable concept since it is the quantity in any dose that translates structure into harm at the cellular level. This harm may take common toxic forms and include the ability of particles and fibres to cause oxidative stress, activating signalling pathways for a diversity of biological outcomes such as inflammation, cell proliferation, and genotoxicity. For example in relation to Table 3.1, quartz causes membrane damage and oxidative stress leading to inflammation, while long fibres cause frustrated phagocytosis and oxidative stress leading to inflammation, and PM_{10} cause oxidative stress and inflammation via its associated transition metals, organics, and surface.

3.2 Nanoparticles and carbon nanotubes

Because of the rise in nanotechnologies there is a very large number of nanoparticles and nanofibres in development and use that are untested in terms of their toxicity and

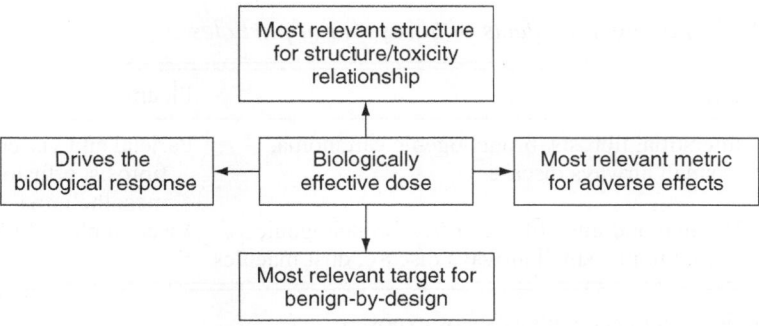

Figure 3.1 The uses of the BED.

potential risk to human health. These engineered nanoparticles vary a great deal in their composition and so our understanding is incomplete. In the last 10 years there has been a rise in the sub-speciality of nanotoxicology (Donaldson *et al.*, 2004), which seeks to address this problem. Carbon nanotubes are amongst the highest-profile of the manufactured nanoparticles and are in increasing use (Chapter 1). They also pose interesting particle toxicology problems, not least because they can exist as particles or fibres, as described in detail below. This chapter aims to set CNT in the context of their BED since this can be a very valuable exercise in the characterisation of their hazards.

3.2.1 *Biologically effective dose and structure–toxicity relationships in particle toxicology*

The BED is a key concept that unlocks a number of vital particle attributes. As shown in Figure 3.1, knowledge of the BED immediately reveals the driver for the biological response, something that can and should be plugged directly into the structure side of the structure–toxicity relationship. It is also the obvious target for designing safe particles, and is naturally the most relevant metric since it most closely approaches the harmful entity that would optimise risk management.

The BED is therefore a useful way to look at the toxicology of unknown hazards, and therefore it is vital to understand carbon nanotubes in the light of the BED paradigm. The CNT BED is the focus of the remainder of this chapter.

3.2.2 *CNT as particles or fibres*

Carbon nanotubes can present themselves in the form of particles or fibres (Figure 3.2) and so there is potential for two hazards and at least two BEDs. We do not intend here to review all of the literature on the toxicology of CNT since this

Table 3.2 *The pathogenic effects of fibres and of particles.*

	Lungs	Pleura
Fibres[a]	Interstitial fibrosis, bronchogenic carcinoma, small airways disease	Parietal and visceral pleural fibrosis, effusion, mesothelioma
Particles[b]	Nodular and interstitial fibrosis, bronchogenic carcinoma, small airways disease, dust macules	Visceral pleural fibrosis

[a] Based on the experience with asbestos exposure.
[b] Based on a range of pathogenic compact particles.

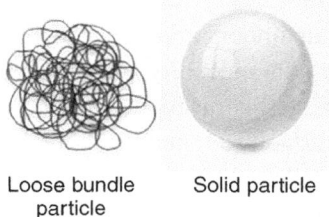

Loose bundle   Solid particle
particle

Figure 3.2 CNT as a loose partical bundle (left) and as a solid partical (right) with the same geometric diameter.

is carried out in the other chapters of this book. We focus solely on the on CNT as particles or as fibres, and the different BEDs that these might present. Identifying the BED that is appropriate for any CNT sample is important because of the different behaviour and pathogenicity of fibres versus particles (Table 3.2). This specially relates to effects in the pleura that are virtually unique to fibres including parietal pleural fibrosis, mesothelioma, and pleural effusion.

3.2.3 CNT as particles

The majority of research into the toxicology of CNT has been carried out with CNT in particulate form (indicated henceforth as CNTp). CNTp is CNT in compact form with a diameter less than about 5 μm or as very short fibres, less than 5 μm long. It is important to point out that 'fibres' shorter than 5 μm are not defined as fibres by the World Health Organization (WHO) even if they have a high aspect ratio, since short fibres behave like particles or at least do not behave like longer fibres. The bundles or tangles of CNT that have been studied are not like solid particles but are in fact

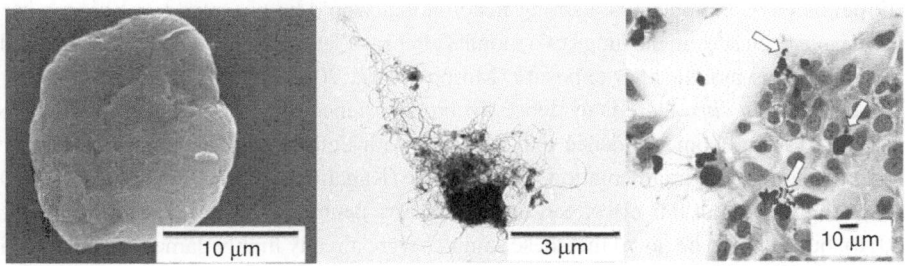

Figure 3.3 CNT as particles by SEM (left), by TEM (centre, and taken up by mesothelial cells (right, arrows).

more like a tumbleweed or a tangle of wire, in that they are composed of variable amounts of space and higher surface area than solid particles of the same diameter (Figure 3.2). Their density is correspondingly low and variable, depending on how tightly the tangle is packed, and since aerodynamic behaviour is related to density, they are likely to have different aerodynamics and deposition characteristics than conventional solid particles.

A compact particle composed of graphene does exist in the form of carbon black particles, which are composed of tightly packed, disjointed sheets of graphene. As shown in Figure 3.3, particulate bundles of CNT can appear as tight bundles (left panel) but by TEM (centre panel) they can appear more 'fluffy', which again may impart interesting aerodynamic properties. The relationship of respirable CNTp to cells is shown in Figure 3.3 (right panel), where the short tangles of of CNTp have been added to mesothelial cells in culture. The mesothelial cells can be seen to have phagocytosed some of the CNTp and these are present in the cytoplasm and on the cell surface.

Structural attributes that may represent the BED for CNTp may include the following:

- **Surface area.** For low-toxicity material, as graphene would be assumed to be, it would be predicted that the surface area might be the essential dose, as is the case with graphene in the form of carbon black (Duffin *et al.*, 2007).
- **Intrinsic free-radical activity/surface reactivity.** This could act by the CNTp transmitting oxidative stress into the cell which, depending on its nature and severity, could induce pro-inflammatory responses, DNA damage, generalised peroxidation to membranes, etc. Sources of this free radical activity are contaminating metals such as iron and nickel, which are common starting nanoparticles for CNT synthesis (Kagan *et al.*, 2006), and defects in the graphene surface crystal structure (Fenoglio *et al.*, 2008).

- **Biopersistence.** Graphene is a strong material that would be expected to be biopersistent. CNT have been seen in the lungs two months after instillation (Shvedova *et al.*, 2005) and in pleural fibrosis months after exposure (Murphy *et al.*, 2010), confirming that CNT can be biopersistent. *In vitro* there is evidence for biopersistence of MWCNT (Osmond-McLeod *et al.*, 2011) and some evidence that SWCNT with defects or surface derivatisation can undergo dissolution, fragmentation, and breakage (Kagan *et al.*, 2010; Liu *et al.*, 2010). One study suggested that this effect can be hastened by neutrophilic peroxidase (Kagan *et al.*, 2010), but it should be noted that these samples were already highly damaged and defective due to harsh acid treatment. Their fragility may therefore have made the CNT very susceptible to further oxidative damage, leading to fragmentation of the graphene crystal.
- **Zeta potential (surface charge).** Zeta potential may act as a structure in the toxicity of nanoparticles by destabilising lysosomes or phagolysosomes following uptake into these compartments. It is suggested that the particle surface charge can be revealed under acid/proteolytic conditions in the lysosomes and phagolysosomes, and that the charge interacts with the lysosomal membrane, leading to lysosomsal rupture (Nel *et al.*, 2009).

3.2.4 CNT as fibres

As shown in Figures 3.4 and 3.5, CNT can arise as fibres (hereafter designated CNTf). They can be single fibres but, because of the attractive van der Waals forces

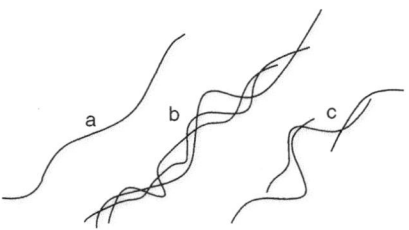

Figure 3.4 Diagrammatic representation of CNT as fibres: (a) single CNT fibre; (b) CNT rope of longer fibres; (c) CNT rope formed by short fibres.

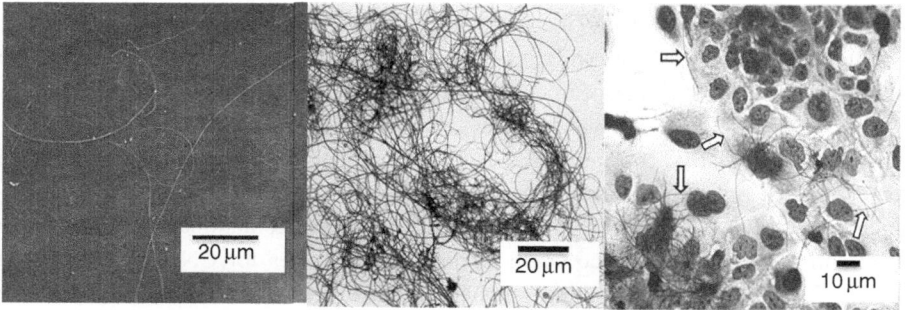

Figure 3.5 CNT as fibres by SEM (left), by TEM (centre), and partially taken up by mesothelial cells (right, arrows).

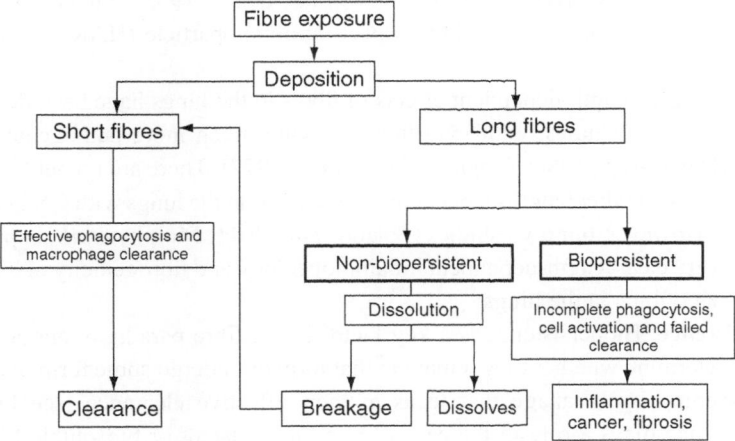

Figure 3.6 Summary of the fibre pathogenicity paradigm.

that act along the surface of CNT, they also tend to stick to each other and so can form cables or ropes, as shown in Figure 3.4(b) and (c).

The fibre pathogenicity paradigm defines the BED for any fibre. This focuses on length and biopersistence, and is summarised in Figure 3.6 in relation to the role of length in the clearance or dissolution of fibres.

The following BEDs are relevant for fibres:

- **Long fibres.**
  - **In the pleura.** The fibre pathogenicity paradigm (FPP) stresses the importance of long fibres but emphasises that non-biopersistent long fibres undergo dissolution, breakage, and shortening, leading to their effective clearance (see below). It has already been demonstrated that CNT conform to the length paradigm, i.e. they show length-dependent inflammogenicity and fibrogenicity at the peritoneal (Poland *et al.*, 2007) and pleural (Murphy *et al.*, 2010) mesothelial surfaces. The mechanism involves translocation of fibres from the lungs to the pleural space and the retention of long fibres only at stomata in the parietal pleura (Donaldson *et al.*, 2010a). There is no reason to think that other nanofibres will contravene this paradigm; data (unpublished) from our own laboratory demonstrate that other nanofibres (silver and nickel oxide nanowires) also show length-dependent pathogenicity and so conform to that part of the paradigm. CNTf can be much longer than a cell diameter and so prevent the cells from fully enclosing them. When this occurs on the surface of the airspaces and macrophages fail to fully ingest the CNTf, this is referred to as a 'frustrated phagocytosis'. This results in the macrophage being in a highly pro-inflammatory and activated state (Ye *et al.*, 1999). In addition, length-dependent genotoxicity has been demonstrated for other fibres such as asbestos and SVF fibres

(Donaldson and Golyasnya, 1995; Hesterberg *et al.*, 1985) but has not yet been shown for CNTf or any other high-aspect-ratio nanoparticle (HARN) at the time of writing.

- **In the lungs.** Length-dependent effects of fibres in the lungs have been demonstrated in a number of studies, with long fibres proving much more pathogenic than short fibres (Davis *et al.*, 1986; Wright and Kuschner, 1977). There are no published studies that address whether length-dependent effects occur in the lungs with CNT or any other HARN. However, our own studies (Poland *et al.*, 2011; Murphy *et al.*, in preparation) shows very clear length-dependent inflammogenicity and fibrogenicity of long, but not short, nanofibres in the lungs.

- **Biopersistence.** Biopersistence is a key factor in the fibre paradigm, and acts on long fibres to determine whether they remain in that form or undergo some form of dissolution and shortening by breakage that leads to their effective clearance (see Figure 3.5). Biopersistence plays a role in the case of non-biopersistent or biosoluble fibres where the environment of the lung, especially the acid milieu of the phagosomes, causes solubility, weakening, and breakage of the long fibres. Although not shown for long CNT, there is *in vitro* evidence that SWCNT with defects or surface derivatisation can undergo dissolution, fragmentation, and breakage (Kagan *et al.*, 2010; Liu *et al.*, 2010), and that neutrophilic peroxidase might hasten that effect in the case of highly oxidised SWCNT (Kagan *et al.*, 2010). Using conventional *in vitro* methods to examine the durability of MWCNT (Osmond-McLeod *et al.*, 2011), long MWCNT were found to be durable and therefore likely to be biopersistent. In addition we have found intact long MWCNT that remain for months after deposition in the pleura, suggesting that the long MWCNT are biopersistent.

- **Reactivity.** It is possible that factors associated with long fibre surfaces, or diffusible components released by them, might be a factor in their inflammogenicity and geno-toxicity. In the case of asbestos, iron in the crystal lattice is considered to play a role in catalysing free-radical generation, additionally to the long-fibre effect that retains the long fibres in the parenchyma or pleura (Kamp, 2009) and the frustrated phagocytosis effect. It is possible that CNTf might have additional factors that contribute to reactivity such as iron, nickel, or cobalt, as suggested previously (Johnston *et al.*, 2010; Kagan *et al.*, 2006).

3.3 Conclusion

We suggest that the concept of the BED can be used to good effect in characterising the hazard from CNT. Not only are there two very distinct hazards, a particle hazard and a fibre hazard, that could arise from CNT exposure, but the underlying patho-genic mechanisms arise from a multiplicity of BEDS impacting on various patho-biological processes (summarised for CNTp and CNTf in Figure 3.7). Using the BED allows us to focus on a key entity for hazard characterisation, which opens the way to a number of key relationships (Figure 3.1).

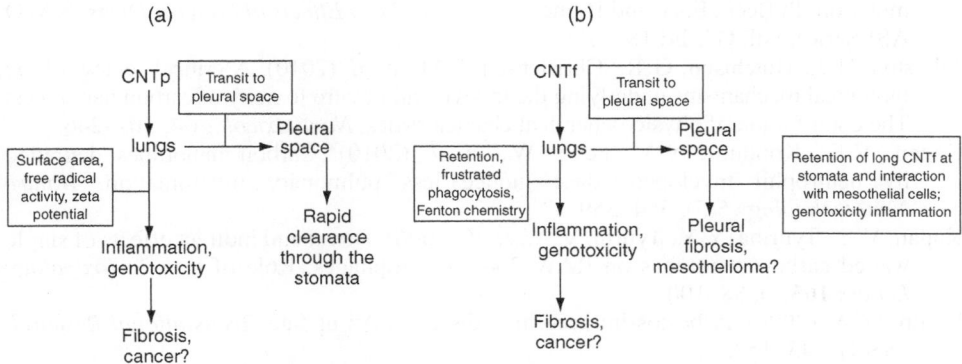

Figure 3.7 Summarised pathways for the action of CNTp (a) and CNTf (b).

Acknowledgement

KD would like to acknowledge the ongoing support of the Colt Foundation.

References

Burello, E. and Worth, A. P. (2011). A theoretical framework for predicting the oxidative stress potential of oxide nanoparticles. *Nanotoxicology* **5**(2), 228–235.

Davis, J. G., Addison, J., Bolton, R. E., *et al.* (1986). The pathogenicity of long versus short fiber samples of amosite asbestos administered to rats by inhalation and intraperitoneal injection. *British Journal of Experimental Pathology* **67**, 415–430.

Donaldson, K. and Borm, P. J. (1998). The quartz hazard: A variable entity. *The Annals of Occupational Hygiene* **42**(5), 287–294.

Donaldson, K. and Golyasnya, N. (1995). Cytogenetic and pathogenic effects of long and short amosite asbestos. *Journal of Pathology* **177**, 303–307.

Donaldson, K., Murphy, F., Duffin, R., and Poland, C. A. (2010a). Asbestos, carbon nanotubes and the pleural mesothelium: A review of the hypothesis regarding the role of long fibre retention in the parietal pleura, inflammation and mesothelioma. *Particle and Fibre Toxicology* **7**, 5.

Donaldson, K., Stone, V., Tran, C. L., Kreyling, W., and Borm, P. J. (2004). Nanotoxicology. *Occupational and Environmental Medicine* **61**(9), 727–728.

Donaldson, K., Tran, L., Jimenez, L., *et al.* (2005). Combustion-derived nanoparticles: A review of their toxicology following inhalation exposure 1. *Particle and Fibre Toxicology* **2**(1), 10.

Duffin, R., Tran, L., Brown, D., Stone, V., and Donaldson, K. (2007). Proinflammogenic effects of low-toxicity and metal nanoparticles in vivo and in vitro: Highlighting the role of particle surface area and surface reactivity 2. *InhalationToxicology* **19**(10), 849–856.

Fenoglio, I., Greco, G., Tomatis, M., *et al.* (2008). Structural defects play a major role in the acute lung toxicity of multiwall carbon nanotubes: Physicochemical aspects. *Chemical Research in Toxicology* **21**(9), 1690–1697.

Hesterberg, T., Oshimura, M., and Barrett, J. C. (1985). Transformation of mammalian cells inculture by asbestos and other mineral dusts: A mechanism involving chromosomal mutation. In Beck, E. G. and Bignon, J., eds., *In Vitro Effects of Mineral Dusts*, NATO ASI Series, vol. G3, Ed 185–196.

Johnston, H. J., Hutchison, G. R., Christensen, F. M., *et al.* (2010). A critical review of the biological mechanisms underlying the in vivo and in vitro toxicity of carbon nanotubes: The contribution of physico-chemical characteristics. *Nanotoxicology* **4**, 207–246.

Kagan, V. E., Konduru, N. V., Feng, W., *et al.* (2010). Carbon nanotubes degraded by neutrophil myeloperoxidase induce less pulmonary inflammation. *Nature Nanotechnology* **5**(5), 354–359.

Kagan, V. E., Tyurina, Y. Y., Tyurin, V. A., *et al.* (2006). Direct and indirect effects of single walled carbon nanotubes on RAW 264.7 macrophages: Role of iron 3. *Toxicology Letters* **165**(1), 88–100.

Kamp, D. W. (2009). Asbestos-induced lung diseases: An update. *Translational Research* **153**(4), 143–152.

Liu, X., Hurt, R. H., and Kane, A. B. (2010). Biodurability of single-walled carbon nanotubes depends on surface functionalization. *Carbon* **48**(7), 1961–1969.

Murphy, F., Duffin, R., Poland, C. A., *et al.* (2010). Length dependent retention of carbon nanotubes in the pleural space of mice initiates sustained inflammation and progressive fibrosis on the parietal pleura. *The American Journal of Pathology* **178**(6), 2587–2600.

Nel, A. E., Madler, L., Velegol, D., *et al.* (2009). Understanding biophysicochemical interactions at the nano-bio interface. *Nature Materials* **8**(7), 543–557.

Osmond-McLeod, M. J., Poland, C. A., Murphy, F. A., *et al.* (2011). Durability and inflammogenic impact of carbon nanotubes compared with asbestos fibres. *Particle and Fibre Toxicology* **8**, 15.

Poland, C. A, Byrne, F., Cho, W. S., *et al.* (2011). Length-dependent pathogenic effects of nickel nanowires in the lungs and the peritoneal cavity. *Nanotoxicology* (in press; Epub ahead of print)

Poland, C. A., Duffin, R., Kinloch, I., *et al.* (2007). Carbon nanotubes introduced into the abdominal cavity of mice show asbestos-like pathogenicity in a pilot study. *Nature Nanotechnology* **3**(7), 423–428.

Puzyn, T., Leszczynska, D., and Leszczynski, J. (2009). Toward the development of "nano-QSARs': Advances and challenges. *Small* **5**(22), 2494–2509.

Puzyn, T., Rasulev, B., Gajewicz, A., *et al.* (2011). Using nano-QSAR to predict the cytotoxicity of metal oxide nanoparticles. *Nature Nanotechnology* **6**(3), 175–178.

Rushton, E. K., Jiang, J., Leonard, S. S., *et al.* (2010). Concept of assessing nanoparticle hazards considering nanoparticle dosemetric and chemical/biological response metrics. *Journal of Toxicology and Environmental Health A* **73**(5), 445–461.

Shvedova, A. A., Kisin, E. R., Mercer, R., *et al.* (2005). Unusual inflammatory and fibrogenic pulmonary responses to single-walled carbon nanotubes in mice. *American Journal of Physiology: Lung Cellular and Molecular Physiology* **289**(5), L698–L708.

Tran, C. L., Buchanan, D., Cullen, R. T., *et al.* (2000). Inhalation of poorly soluble particles. II: Influence of particle surface area on inflammation and clearance. *Inhalation Toxicology* **12**(12), 1113–1126.

Wright G. and Kuschner M. (1977). THe influence of varying lengths of glass and asbestos fibers on tissue responses in the guinea pigs. In Walton W H, ed., *Inhaled Particles IV* (Oxford: Pergamon Press), pp. 455–474.

Ye, J., Shi, X., Jones, W., *et al.* (1999). Critical role of glass fiber length in TNF-alpha production and transcription factor activation in macrophages. *American Journal of Physiology* **276**(3 Pt 1), L426–L434.

4

CNT biopersistence and the fibre paradigm

DAVID B. WARHEIT, MICHAEL P. DELORME

Carbon nanotubes (CNT) have recently become the archetype for nanotechnology as they possess remarkable mechanical, optical, thermal and electrical properties which make them attractive in a variety of commercial applications. These materials are chemically defined as one of several allotropes of carbon nanoparticulates and are synthesized in the form of graphene cylindrical structures. Significant research funding has been expended on 'applications research' activities for CNTs owing both to their impressive physical properties and to their versatile potential applications in industrial, medical and diagnostic arenas (see, for example, http://www.nano.gov/initiatives).

While the many benefits of these materials for commercial engineering applications have been recognized, it is also important to note that a seemingly relevant benefit/risk paradox is operative in today's marketplace, i.e. that many of the advantageous properties associated with CNT 'applications' may be undermined by serious concerns regarding potential adverse health effects, particularly those related to inhalation exposure. It is therefore not surprising that the possibility of the 'asbestos-like toxicity' has been raised by some stakeholders and regulatory agencies with regard to forthcoming development of CNT-containing products. While asbestos and CNTs share many significant application-specific advantages such as material strength, durability, shape, a fibrous nature (i.e. increased aspect ratio) and enhanced surface area characteristics, they both also have adverse potential health impacts on the 'implications' side of the risk/benefit equation (Donaldson et al., 2011).

A number of recently published in vivo lung toxicology studies as well as in vitro cell culture reports have concluded that exposure to single-walled CNTs (SWCNTs) or multi-walled CNTs (MWCNTs; Figure 4.1) have produced significant pulmonary toxicity responses, often at relatively low airborne concentrations or in vitro doses when compared to particulates that are considered to be biologically inert. Many of the reported effects in the lung include persistent inflammation and rapidly developing pulmonary fibrosis. In a few comparative intratracheal instillation studies

The Toxicology of Carbon Nanotubes, ed. Ken Donaldson, Craig A. Poland, Rodger Duffin and James Bonner. Published by Cambridge University Press. © Cambridge University Press 2012.

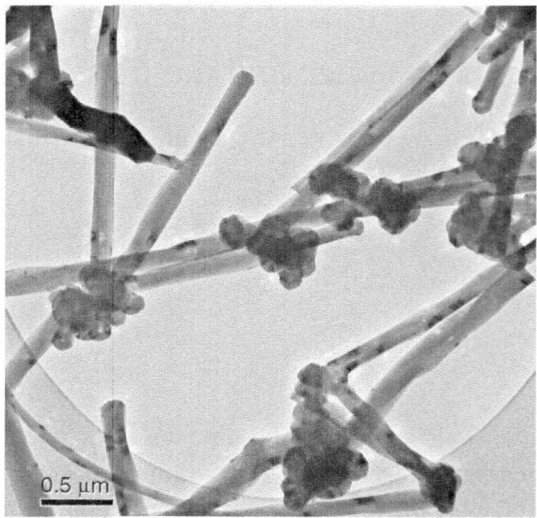

0.5 µm

Figure 4.1 Low-magnification transmission electron micrograph (TEM) of multi-walled carbon nanotubes.

with mice, CNT-induced pulmonary fibrosis developed more rapidly and at a lower mass burden when compared to either ultrafine carbon black or quartz particulates (Lam *et al.*, 2004; Shvedova *et al.*, 2005), while in other comparative studies in rats, SWCNTs did not have equivalent lung inflammatory potency when compared to quartz particulates (Warheit *et al.*, 2004; Figure 4.2). It should be noted, however, that no published epidemiological studies are currently available to assess the relevance/correlation of experimental animal data with adverse human health outcomes (NIOSH, 2010).

In addition to the potential adverse effects on the respiratory tract following inhalation exposures, it has been postulated that aerosolized CNT exposures could produce pathological impacts on the visceral- and parietal-pleura spaces in the thoracic cavity. The results of some intraperitoneal (IP) injection studies, particularly in mice, have indicated greater pleural activity for CNTs when compared to nonfibrous particulates or CNTs with shorter dimensions (Poland *et al.*, 2008). IP injection studies have traditionally been utilized as nonphysiological-type screening studies for predicting the possible hazards of mesothelial and/or pleural effects in humans. Currently, the development of mesothelial tumors in humans is considered to be uniquely identified with exposure to certain forms of asbestos fibres or erionite zeolite-type fibres. This is in contrast to findings in long-term workers exposed to other known durable fibre-types, such as refractory ceramic fibres, which have been documented to cause non-tumorous, pleural plaques and are not associated with pleural or mesothelial tumors (Donaldson *et al.*, 2006).

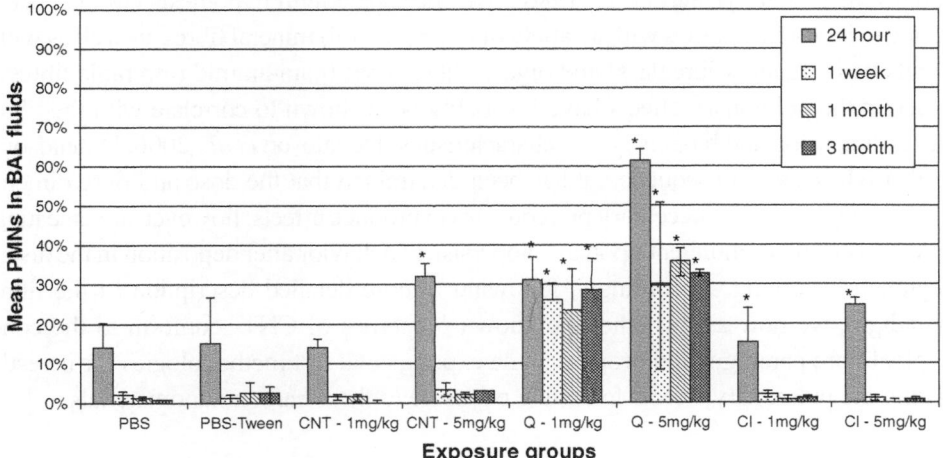

Figure 4.2 Pulmonary inflammation in particulate-exposed rats and controls as evidenced by percent neutrophils (PMN) in BAL fluids at 24 hours, 1 week, 1 month and 3 months. Instillation exposures resulted in transient inflammatory responses for nearly all groups at 24 hours pe. However, exposures to quartz particles at 1 and 5 mg/kg produced a sustained lung inflammatory response; *p < 0.05 (figure copied from Warheit *et al.*, 2004).

This chapter is designed to introduce some basic principles of fibre toxicology, and specifically to describe the concept of the 'fibre paradigm' as it may relate to assessing carbon nanotube toxicity. In addition, we will briefly review the results of two recently published subchronic (90-day) inhalation studies in rats with MWCNTs, and then propose additional experimental design protocol parameters in order to more effectively test the hypothesis that inhaled carbon nanotubes are likely to produce pathogenetic effects following inhalation exposures.

It seems clear that elucidation of mechanisms of fibre toxicity and associated fibre paradigm principles are well-established concepts which have been investigated and confirmed over many decades. The knowledge gained on fibre toxicity mechanisms has historically been based upon a long-standing and well-documented database of asbestos toxicity studies. The abundance of experimental animal studies over a period of 30–50 years has clearly demonstrated that inhalation exposures to *long, biopersistent* fibre-types were likely to produce a pathologic sequelae of impacts commencing with sustained pulmonary inflammation, leading to fibroproliferative effects such as lung fibrosis, and (with continued exposures) the development of lung cancers and possibly mesotheliomas (particularly in susceptible species like hamsters). Some of the early key studies demonstrated that the pathogenic effects of asbestos fibres were reduced or eliminated when the asbestos fibre sample was milled to produce a shorter fibre preparation in rats exposed to equal mass concentrations of the milled and unmilled long-fibre asbestos (Davis *et al.*, 1986). This fibre-paradigm hypothesis has also been tested in long-term studies with a variety of other (natural) mineral fibres, as well as with synthetic inorganic (fibreglass) and organic fibre-types (para-aramid respirable fibres), wherein the pulmonary effects have invariably been shown to correlate with the fibre dose, dimension and biopersistence characteristics (Donaldson *et al.*, 2006; Donaldson, 2009). Thus, as a consequence, it has been determined that the dose and fibre dimensions (long-thin) were necessary prerequisites to produce effects; however, adverse lung effects were also contingent upon the biopersistent behavior after deposition in the distal regions of the lung. Accordingly, following a more detailed description of the fibre paradigm, we now ask whether the known properties of CNTs conform to the conceptual fibre paradigm and propose some experimental test methodologies for investigating the pulmonary effects following exposure to these carbon nanomaterials.

4.1 Fibre toxicology primer

Hazard and risk assessments of fibrous materials generally are more complex than those of nonfibrous particulates, because potential hazards for pulmonary effects are strongly influenced not only by chemical composition but also by the aspect ratio (defined as the ratio of length to diameter) of fibres. A respirable fibre (i.e. one that can be inhaled into the alveolar/gas-exchange regions of the lung) has been defined by the World Health Organization (WHO) as a particulate with a minimum aspect ratio of 3:1, a length greater than 5 μm and a diameter less than 3 μm (WHO, 1997). Fibre dimension is a strong determinant of the dose delivered to the distal airways of the lung and greatly impacts both the lung biopersistence (i.e. length of time that fibres persist in the lung) and biological reactivity. The long-thin geometry of fibres may facilitate the inhalation and subsequent lower respiratory tract deposition of fibres as long as 100–200 μm (see Figure 4.3). However, this long-thin geometry is

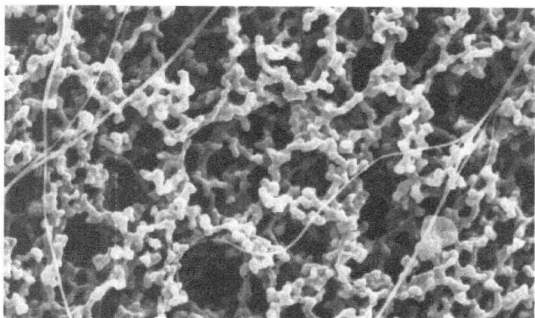

Figure 4.3 Scanning electron micrograph (SEM) of aerosolized long chrysotile asbestos fibres caught on a filter.

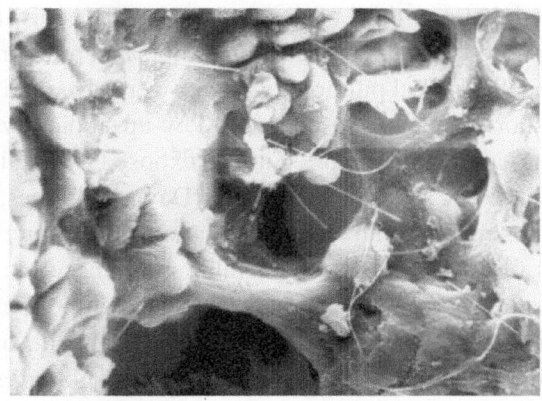

Figure 4.4 Alveolar macrophages have migrated to the sites of crocidolite asbestos fibre deposition (i.e. alveolar duct bifurcation) in an attempt to phagocytize and clear from the lung the fibrous particles following inhalation exposure in a rat model.

also problematic for lung clearance, which depends to a large extent on translocation of phagocytized particulates/fibres by alveolar macrophages to the mucociliary escalator system of the lower airways (Figure 4.4). Thus, long fibres (>20 μm in length) of very durable compositions can remain in the lung indefinitely. Longer fibres are also more biologically reactive than short fibres. When macrophages attempt to phagocytize long fibres, these long fibres conceivably could disrupt cell division, alter genetic material and/or trigger the release of inflammatory cytokines, all of which can eventually lead to the progression of lung disease (Donaldson *et al.*, 2006).

4.2 Fibre dimensions

Fibre dimensions and corresponding aspect ratio (length/diameter) play significant roles in the development or pathogenesis of fibre-related pulmonary disease. These dimensional aspects form one of the fundamental criteria that differentiate the effects of fibrous materials from the effects of nonfibrous particulates. In general, the longer the fibre, the greater the potential for causing adverse pulmonary effects, although the fibre durability or biopersistence in the lung is also a significant contributing or modifying factor of toxicity. As discussed briefly above, in a defining study published in 1986 (Davis *et al.*, 1986), two different groups of rats were exposed for 1 year by inhalation to aerosols of amosite asbestos of equivalent mass concentration that had been processed as either milled 'short' (<5 μm in length) or long (>20 μm) fibre preparations. Following the end of exposures, the results demonstrated that one-third of the rats exposed to equivalent mass concentrations of the long fibre-types developed lung tumors and pulmonary fibrosis while no significant histopathological pulmonary effects were observed in rats exposed to the preparation of short fibres. In subsequent follow-up studies by the same authors, findings of a similar nature were reported in rats exposed by inhalation for 1 year to long versus short chrysotile asbestos fibre-types (Davis and Jones, 1988). In another set of experiments, using an intratracheal instillation as the route of lung exposure, durable long glass fibres produced considerable pulmonary fibrosis in rats while instillation of the same glass fibres that were shortened at the same doses produced no adverse effects (reference in Warheit *et al.*, 1999). The results of this intratracheal instillation study further demonstrated that the impact of fibre dimension on pathological effects is not unique to asbestos fibre exposures but appears to be a general principle for assessing the mechanism of fibre-induced lung toxicity.

4.3 Lung fibre biopersistence

Lung fibre biopersistence is generally defined as the likelihood of inhaled fibres resisting changes in number, dimension, surface chemistry, chemical composition, surface area or other physicochemical characteristics when deposited in the lung. Any or all of these parameters can be altered during a fibre's residence time in the lung, and such changes affect the long-term pulmonary response to the fibre. Fibre biopersistence in the lung is determined and impacted by macrophage-mediated clearance, the dissolution rate of the fibre, along with the tendency of the fibre to undergo transverse fragmentation (i.e. shortening) in the lung. For fibres that are too long to be cleared by alveolar macrophages, the principal alternative clearance mechanisms involve translocation to other thoracic compartments, dissolution and/or transverse breakage into shorter segments. Indeed, *in vitro* fibre dissolution experiments demonstrate a broad range of dissolution rate constants for the various synthetic vitreous fibres.

In vitro fibre-dissolution studies demonstrate a correlation between the fibre's rate of leaching and its proclivity to undergo transverse fragmentation under *in vivo* conditions. It is important to note that synthetic vitreous fibres (SVF) tend to break transversely, while asbestos fibres tend to split, longitudinally. This is a significant harbinger for the pathogenetic mechanisms *in vivo* because over time in the lung the number of long asbestos fibres can actually increase as a result of splitting, while the number of long SVFs decreases because of transverse fragmentation (shortening). Generally speaking, the shortening of fibres reduces residence time in the lung by facilitating and expediting lung clearance mechanisms by macrophages, leading to enhanced and accelerated mucociliary airway clearance processes (see Warheit *et al.*, 1999).

Investigations of lung fibre biopersistence mechanisms were studied in rats for 1 year following short-term inhalation exposures either to chrysotile or crocidolite asbestos fibre-types. Following the aerosol exposures, fibres were recovered by lung digestion at various time points postexposure, and fibre dimensions were measured. The authors concluded that the mean lengths of fibres recovered from the lungs of crocidolite-exposed rats were progressively increased with recovery time; however, no significant change in mean diameters was measured over residence time in the lungs (implying clearance of shorter fibres only). In the chrysotile-exposed rats, mean fibre lengths also underwent a progressive increase; but, in contrast to the crocidolite-exposed fibres, mean diameters were significantly reduced (Davis *et al.*, 1986; Davis and Jones, 1988). The researchers concluded that long fibres were selectively retained in the lungs of both chrysotile- and crocidolite-exposed rats; however, longitudinal splitting with a corresponding decrease in mean fibre diameters occurred only with the retained chrysotile asbestos fibres. These findings were confirmed in a 2-year study of fibre clearance following intratracheal instillation. In rats instilled with crocidolite, the number of short fibres per lung declined slowly, but the number of fibres longer than 5 μm did not change significantly during 1 year of observation. In contrast, the number of chrysotile fibres per lung longer than 5 μm increased continuously over a 2-year period, which was probably a result of longitudinal splitting of fibres.

4.4 Surface chemistry

Several investigators have postulated that a chemical origin of fibre toxicity is contributory, based upon the generation of reactive oxygen species and other free radicals when some deposited fibre compositions come into contact with biological fluids. Strong oxidants are destructive to living tissues and can trigger a host of inflammatory reactions. Erionite is a fibrous-type zeolite and is one of the most potent carcinogenic minerals known. In one study, nearly 100% of rats exposed by inoculation or inhalation to erionite developed mesotheliomas, a rare form of cancer of the pleura (Wagner *et al.*, 1985). Like other zeolites, erionite has a very large internal surface area and exchanges ions with surrounding media. It is postulated that the surface characteristics of erionite contribute to its carcinogenic potential in humans and experimental animals.

4.5 Conclusions regarding the fibre pathogenicity structure: activity paradigm and the possible relevance for exposure to carbon nanotubes

As discussed earlier, the emerging and overarching fibre toxicology structure-activity paradigm can be considered in terms of dose, dimension (i.e. aspect ratio) and lung biopersistence characteristics (likely related to chemical composition). In addition it is noteworthy that erionite fibres have a high surface-area dimension (i.e. $>200 \, m^2/g$), which is considered to contribute significantly to its pathogenicity. Accordingly, the fibre toxicology paradigm identifies the geometry of fibres as a significant contributor to the toxicology of known pathogenic fibrous materials such as asbestos. The paradigm is centrally important since it defines an approach that allows for a determination of whether the various forms of carbon nanotubes fulfill the various criteria described above that are likely to be associated with pathogenic fibres. The length dimensions of carbon nanotubes cannot be easily measured; however, CNTs are known to have very small diameters, high aspect ratios and impressive surface area indices of $200–500 \, m^2/g$ and are generally considered to be very biopersistent in the lungs (Donaldson *et al.*, 2006). Therefore, from a theoretical standpoint, CNTs would appear to have many similarities to asbestos fibres and consequently would likely conform to the fibre pathogenicity structure- activity paradigm. In addition, there are emerging data which suggest that, at high aerosol concentrations (i.e. 30 mg/m^3), inhaled MWCNTs may translocate from sites of pulmonary deposition to reach subpleural tissues in mice (Ryman-Rasmussen *et al.*, 2009), and that longer MWCNTs have similar peritoneal inflammatory activity/behavior to long crocidolite asbestos fibres following intraperitoneal injection studies (Poland *et al.*, 2008). Of course crocidolite asbestos and other amphibole asbestos forms have been associated with the development of mesotheliomas in humans. Thus it will be important to gauge the pulmonary and systemic effects of CNT exposures under *in vivo* physiological conditions, such as subchronic to chronic inhalation studies.

4.6 Review of the current literature on longer-term inhalation studies with multi-walled forms of carbon nanotubes

Two well-conducted 90-day inhalation studies in rats have recently been reported which describe the effects of longer-term inhalation exposures to these carbon nanostructured materials.

The Ma-Hock *et al.* manuscript (2009) represents the first subchronic inhalation study in rodents with carbon nanotubes of either the single-walled (SWCNT) or multi-walled (MWCNT) form. In this study, male and female Wistar rats were exposed to aerosols of MWCNTs for 13 weeks at exposure concentrations of 0, 0.1, 0.5 or 2.5 mg/m^3. In addition to the study end points promulgated in the OECD 413 90-day inhalation toxicity test guideline, the lungs of MWCNT-exposed and control animals were evaluated by bronchoalveolar lavage following 13-week exposures. The test substance, Nanocyl^{TM} NC 7000, was a form of multi-walled carbon nanotube. The purity of the test sample was 90% C and 10% metal oxide, of which 9.6% was aluminum oxide with traces of iron and cobalt. The tubes have lengths of

0.1–10 μm and diameters of 5–15 nm, with specific surface area measurements of 250–300 m^2/g (Brunauer *et al.*, 1938; BET method), according to the manufacturer.

As might be expected, the major impacts of 90-day inhalation exposures to MWCNTs were measured predominantly in the respiratory tract. The adverse effects were described as hyperplastic responses in the nasal cavity and upper airways, along with elevated lung weights (versus controls), multifocal granulomatous inflammation, cellular inflammation (neutrophils and macrophages) and lipoproteinosis in alveolar regions and lung-associated lymph nodes. Lung fibrosis was not reported to occur in exposed rats. Moreover, the investigators did not observe any (extrapulmonary) systemic organ toxicity, including in major organs such as the liver, kidney or heart. This absence of systemic effects following 90-day inhalation exposures represents an important finding and suggests that the lung tissue is the primary target of toxicity upon inhalation.

A minimal granulomatous-type inflammation in the lungs and lung-associated lymph nodes was observed in rats exposed to 0.1 mg/m^3 Nanocyl^{TM} NC 7000; therefore a no-observable-adverse-effect concentration (NOAEC) could not be established. These results confirm the potency of MWCNTs as an active particulate in the lungs of exposed animals and provides the hazard criteria component for formulating risk assessment determinations. Moreover, the study results determined by an inhalation study, i.e. *physiologically relevant* route of exposure, serves to corroborate the findings of adverse pulmonary effects reported in previous studies utilizing non–physiologically relevant routes (i.e. intratracheal instillation) or short-term inhalation exposures in rats or mice.

In another study following the OECD 413 testing guideline, by Pauluhn (2010), Wistar rats were exposed nose-only to aerosols of MWCNTs (Baytubes$^{®}$) for 13 weeks at concentrations of 0, 0.1, 0.4, 1.5 and 6 mg/m^3. Animals were evaluated 1 day after the final exposure, as well as at postexposure periods of 8, 13, 17, 26 and 39 weeks. According to the author, the study was focused on respiratory tract and systemic toxicity, including analysis of MWCNT biokinetics in the lungs and lung-associated lymph nodes (LALNs). The time course and concentration dependence of pulmonary effects were assessed by bronchoalveolar lavage (BAL) and histopathology for up to 6 months postexposure.

The physicochemical composition of the Baytubes$^{®}$ was reported to be 98.6% carbon, 1.4% oxygen and 0.46% cobalt. The BET surface area was determined to be 253–257 m^2/g.

Similar to results in the Ma-Hock *et al.* (2009) study, 90-day inhalation exposures in rats to MWCNTs did not produce any systemic toxicity. Kinetic analyses demonstrated a significant retardation of MWCNT lung clearance at particle overload conditions. Translocation of MWCNTs into LALNs occurred at 1.5 and 6 mg/m^3. At the three highest exposure concentrations (i.e. 0.4, 1.5 and 6 mg/m^3), the lung and

LALN weights were significantly increased. Sustained elevations in BAL polymor-phonuclear neutrophilic leukocytes (PMNs) and soluble collagen also occurred at these concentrations, with borderline results at 0.4 mg/m^3. Histopathology revealed exposure-related lesions at 0.4 mg/m^3 and above in the upper respiratory tract and in the lower respiratory tract – described as inflammatory changes in the bronchioloal-veolar region and increased interstitial collagen staining. Granulomatous changes and a time-dependent increase of a bronchioloalveolar hyperplasia occurred at 6 mg/m^3. The NOAEC for the study was determined to be 0.1 mg/m^3.

4.7 Potential for translocation of inhaled MWCNTs

In a short-term inhalation study with mice exposed to multi-walled carbon nanoma-terials, Ryman-Rasmussen and coworkers (2009) exposed male C57BL6 mice to an aerosol of MWCNTs (1 or 30 mg/m^3) or carbon black (CB) nanoparticles (30 mg/m^3) for 6 hours and then evaluated lung tissue at 1 day, 2 weeks, 6 weeks and 14 weeks postexposure. The authors reported that inhaled MWCNTs were engulfed by macro-phages which had migrated to the subpleural region. At subsequent postexposure time periods, MWCNTs were observed in macrophages and within subpleural mesench-ymal cells and the collagen matrix of the subpleura 1 day following inhalation exposures. These investigators concluded that inhaled MWCNTs reach the subpleural region in mice after a single inhalation exposure of 30 mg/m^3 for 6 hours. MWCNTs were embedded in the subpleural wall and were also found within subpleural macro-phages. Mononuclear cell aggregates on the pleural surface increased in number and size after 1 day and macrophages containing MWCNTs were observed within these foci. Subpleural fibrosis was increased after 2 and 6 weeks following inhalation. None of these effects were seen in mice that inhaled 1 mg/m^3 MWCNTs or in animals exposed to carbon black nanoparticles.

The Ma-Hock *et al.* (2009) and Pauluhn (2010) studies represent the first 90-day subchronic inhalation studies in rodents with carbon nanotubes, either of the single-walled (SWCNT) or multi-walled (MWCNT) variety, that followed OECD testing guidelines. While these studies have set important precedents for evaluating the pulmonary effects of CNTs, future studies should be expanded to address additional relevant safety issues. Indeed, both Ma-Hock *et al.* (2009) and Pauluhn (2010) added BAL techniques in their studies to gain further insight into the pulmonary toxicity of CNTs. Some additional topics that will enhance the hazard identification and risk assessment process include:

- physicochemical characteristics as they relate to the fibre paradigm
- potential cardiovascular effects
- potential pleural effects.

These subjects will be discussed separately.

Donaldson and colleagues (2006) have raised the concern that CNTs share three properties that traditionally have been associated with the pulmonary pathological mechanisms of inhaled particles and fibres. These include the perceptions that (1) nanoparticles are more toxic to the lungs than fine-sized particles of similar composition; (2) the fibrous particulates of CNTs suggest potential similarities to asbestos fibres; and (3) the potential biopersistence of deposited CNTs during residence time in the lung could produce pathological responses in the respiratory system. Thus, it has been hypothesized that repeated exposure to SWCNTs and/or MWCNTs could result in pathological effects consistent with the fibre paradigm of concentration, dimension and biopersistence, which are known to facilitate the development of lung disease. More recently a fourth health-related concern relates to the emerging view that inhaled particulates may produce significant adverse impacts on the cardiovascular system.

4.7.1 Physicochemical characteristics

As previously discussed, it is well considered that physical properties such as length dimension and durability play important roles in the development of fibre-related pathological effects. Moreover, *in vitro* studies have demonstrated that transition metals such as bioavailable iron and nickel mediate toxic effects to lung cells. The initial synthesis of either SWCNTs or MWCNTs generally requires the contribution of metal catalysts, and these transition metals may serve to trigger metal-mediated toxicity, particularly if the metal is bioavailable following interactions with lung cells (Liu *et al.*, 2010) . The Nanocyl™ MWCNT test material in the Ma-Hock *et al.* study (2009) contains 10% metal oxides, including traces of iron and cobalt, while the Baytubes® contain traces of cobalt. Finally, the large surface area (250–300 mg/m²) of the CNTs may also play a significant role in the development of pathological pulmonary effects. Accordingly, it will be important to evaluate and compare the results of subchronic inhalation studies with CNTs of different compositions, dimensions and surface characteristics to better understand whether some formulations of CNTs are less reactive in the lungs than others.

4.7.2 Potential cardiovascular effects

Epidemiological studies have provided evidence from air pollution studies that inhaled particulate matter (PM) is associated with adverse cardiovascular events. Two distinct mechanisms of cardiovascular toxicity have been suggested: (1) oxidative stress, possibly as a component of particle-induced pulmonary inflammation (either direct or indirect effects) facilitating the development of atherothrombosis effects; and (2) transmigration of inhaled particulates from airspace to the vasculature, concomitant with direct effects on endothelial cells (Seaton *et al.*, 2009). Consequently, it would be useful to include an additional focus on assessing cardiovascular effects for future CNT inhalation studies and should be included in the experimental design regimen. In the Ma-Hock *et al.* (2009) and Pauluhn (2010) studies, no pathological cardiovascular effects were reported in the heart muscle of exposed rats. However, investigations could provide a more extensive evaluation of hemostasis and thrombosis parameters in addition to the prothrombin time (PT) and the activated partial thromboplastin time (aPTT) coagulation tests. Additional clinical pathology tests could include assessments of systemic/cardiovascular inflammation (e.g. C-reactive protein) and heart electrocardiograms.

4.7.3 Potential pleural effects

As discussed above, Donaldson and colleagues (2006; 2009) have postulated that exposure to CNTs could produce toxicological properties similar to those for asbestos fibres. Experimental results from these investigators have suggested that intraperitoneal injection exposures in mice to *long* but not *short* MWCNTs produced inflammatory and granulomatous effects in the abdominal cavity similar to amosite asbestos fibres (Poland *et al.*, 2008). Intraperitoneal injection studies in rodents have been utilized as screening assays for assessing potential mesotheliogenic activity in humans. It is also noteworthy that Ryman-Rasmussen and coworkers (2009) have reported that inhalation of high exposures of aerosolized MWCNTs may translocate from the airspace to subpleural regions. Currently, it is considered that exposure to only a few fibre-types are known to produce mesotheliomas in humans. These include amphibole asbestos fibre-types and erionite fibres. Exposure to other biopersistent fibrous particulates such as refractory ceramic fibres (RFCs) produce pleural plaques, but have not been associated with the development of mesotheliomas in humans. Therefore, it may be instructive to facilitate a more rigorous assessment of potential pleural activity in future subchronic inhalation studies with CNTs in rodents by inclusion of cell proliferation evaluations of subpleural and mesothelial regions (Warheit *et al.*, 1996), concomitant with standard histopathological evaluations.

In summary, the Ma-Hock *et al.* (2009) and Pauluhn (2010) studies represent carefully implemented and excellent starting points for studying the toxicological effects of aerosolized MWCNTs. The 90-day studies performed under OECD guidelines were well conducted and reaffirmed the hazard potential of CNT exposure in a concentration-response fashion. The results generated from the studies, in association with relevant exposure data, should form the basis for conducting reliable risk assessment determinations. It is also important to note that 13-week exposures in both studies did not produce any significant systemic effects beyond the respiratory tract. Clearly these studies have set useful precedents for assessing the health hazards related to CNT exposures. Yet there are several additional questions that should be addressed in the future testing of CNTs (Warheit, 2009). These include the sustainability of the measured responses in the respiratory tract to discriminate between sustainable, progressive effects and transient effects; the role of physicochemical characteristics including composition, size distributions, surface area determinations and transition metals; and a more intensively focused investigation into the potential cardiovascular and pleural effects. Integration of these additional parameters/evaluations into a study should provide more comprehensive information on CNT hazards and may serve to better distinguish the more potent forms of CNTs from the more benign particle types.

References

Brunauer S, Emmett PH, Teller E (1938). Adsorption of gases in multimolecular layers. *J. Am. Chem. Soc.* **60**: 309–319.

Davis JM, Jones AD (1988). Comparisons of the pathogenicity of long and short fibres of chrysotile asbestos in rats. *Br. J. Exp. Pathol.* **69**: 717–737.

Davis JM, Addison J, Bolton RE, *et al.* (1986). The pathogenicity of long versus short fibre samples of amosite asbestos administered to rats by inhalation and intraperitoneal injection. *Br. J. Exp. Pathol.* **67**: 415–430.

Donaldson K (2009). The inhalation toxicology of p-aramid fibrils. *Crit. Rev. Toxicol.* **39**: 487–500.

Donaldson K, Murphy F, Schinwald A, Duffin R, Poland CA (2011). Identifying the pulmonary hazard of high aspect ratio nanoparticles to enable their safety-by-design. *Nanomedicine (Lond).* **6**: 143–156.

Donaldson K, Aitken R, Tran L, *et al.* (2006). Carbon nanotubes: A review of their properties in relation to pulmonary toxicology and workplace safety. *Toxicol. Sci.* **92**: 5–22.

Lam CW, James JT, McCluskey R, Hunter RL (2004). Pulmonary toxicity of single-wall carbon nanotubes in mice 7 and 90 days after intratracheal instillation. *Toxicol. Sci.* **77**: 126–134.

Liu X, Hurt RH, Kane AB (2010). Biodurability of single-walled carbon nanotubes depends on surface functionalization. *Carbon N. Y.* **48**: 1961–1969.

Ma-Hock L, Treumann S, Strauss V, *et al.* (2009). Inhalation toxicity of multiwall carbon nanotubes in rats exposed for 3 months. *Toxicol. Sci.* **112**: 468–481.

NIOSH (2010). *Occupational Exposure to Carbon Nanotubes and Nanofibers.* Current Intelligence Bulletin (Washington DC: National Institute for Occupational Safety and Health).

Pauluhn J (2010). Subchronic 13-week inhalation exposure of rats to multiwalled carbon nanotubes: Toxic effects are determined by density of agglomerate structures, not fibrillar structures. *Toxicol. Sci.* **113**(1): 226–242.

Poland CA, Duffin R, Kinloch I, *et al.* (2008). Carbon nanotubes introduced into the abdominal cavity of mice show asbestos-like pathogenicity in a pilot study. *Nat. Nanotechnol.* **3**: 423–428.

Ryman-Rasmussen JP, Cesta MF, Brody AR, *et al.* (2009). Inhaled carbon nanotubes reach the subpleural tissue in mice. *Nat. Nanotechnol.* **4**: 747–751.

Seaton A, Tran L, Aitken R, *et al.* (2010). Nanoparticles, human health hazard and regulation. *J. R. Soc. Interface.* **6** (7 suppl 1), S119–129.

Shvedova AA, Kisin ER, Mercer R, *et al.* (2005). Unusual inflammatory and fibrogenic pulmonary responses to single-walled carbon nanotubes in mice. *Am. J. Physiol. Lung Cell. Mol. Physiol.* **289**:L698–L708.

Wagner JC, Skidmore JW, Hill RJ, and Griffiths DM (1985). Erionite exposure and mesotheliomas in rats. *Br. J. Cancer* **51**: 727–730.

Warheit DB, Hartsky MA, and Frame SR (1996). Pulmonary effects in rats inhaling size-separated chrysotile asbestos fibers or p-aramid fibrils: Differences in cellular proliferative responses. *Toxicol. Lett.* **88**: 287–292.

Warheit DB, Laurence BR, Reed KL, *et al.* (2004). Comparative pulmonary toxicity assessment of single-wall carbon nanotubes in rats. *Toxicol. Sci.* **77**: 117–125.

Warheit DB (2009). Long-term inhalation toxicity studies with multiwalled carbon nanotubes: Closing the gaps or initiating the debate? *Toxicol. Sci.* **112**: 273–275.

Warheit DB, Hart GA, and Hesterberg TW (1999). Fiber toxicology. In Marquardt H, Schafer SG, McClellan R, and Welsch F, eds., *Textbook of Toxicology* (San Diego: Academic Press), pp. 833–850.

WHO (1997). *Determination of Airborne Fibre Number Concentrations: A Recommended Method, by Phase-Contrast Optical Microscopy (Membrane Filter Method)* (Geneva: World Health Organization).

5

Length-dependent retention of fibres in the pleural space

CRAIG A. POLAND, FIONA A. MURPHY, KEN DONALDSON

5.1 Introduction

When considering the potential risks to human health associated with the use of carbon nanotubes (CNT) and indeed other forms of high-aspect-ratio nanoparticles (HARN), the comparison has been drawn to asbestos (1, 2) and the pandemic of disease its use has caused across the globe. Therefore, when considering the potential health effects of exposure to carbon nanotubes, it would seem sensible to look at the health effects caused by exposure to asbestos. The most common diseases associated with asbestos exposure involve the lung or the surrounding serosal tissues of the pleural and peritoneal cavities. Within the lung, the most common of these effects include lung fibrosis, or *asbestosis*, and lung cancer whilst within the pleural space these are mesothelioma, pleural plaques and pleural effusions. The formation of lung disease such as asbestosis after respiratory exposure to a toxic material is readily explicable, as the dose is presented to the lung and this leads to an adverse effect. However, disease of the pleura after respiratory exposure to pathogenic fibres is interesting as it suggests an interaction between the inhaled material and the pleurae. There are indeed causes of the pathologies mentioned above other than fibre exposure, such as trauma (pleural plaques (3)), tuberculosis (pleural plaques (3), effusions (4)) and heart failure (pleural effusions (5)). However, mesothelioma is a rare tumour of the pleura/peritoneal space which has seen a rise in its incidence in countries where asbestos mining/use has been substantial (6). This link between exposure to pathogenic fibres such as asbestos and the fibrous zeolite mineral erionite and the development of mesothelioma is now well established (7).

This raises the question of what is driving the generation of such disease pathologies and, based on the simple idea of cause and effect, one would assume that an interaction between pathogenic fibres, or some intermediate, and the pleural environment is likely. This shall be the basis for this chapter.

The Toxicology of Carbon Nanotubes, ed. Ken Donaldson, Craig A. Poland, Rodger Duffin and James Bonner.
Published by Cambridge University Press. © Cambridge University Press 2012.

5.2 The pleura structure and function

The lungs sit within the thoracic cage formed by the ribs with a narrow, fluid-filled space between the lung surface and the chest wall. The surface of this cavity is covered by a thin serous elastic membrane called the pleura, derived from the Greek *pleuron* for side of the body or rib (8). The main roles of the pleura are to reduce the friction created by the constantly moving lungs against the chest wall and to form a coupling between the lung and chest to allow breathing. This serous membrane is crucial to the normal action of the vital organs contained within both body cavities and this is evident as this lubricating surface is found enveloping both the abdominal and cardiac compartments as well as that of the pleural space.

The pleura lines the external surface of the lung (visceral pleura), lining all interlobular fissures and collecting at the hilar region, folding back on itself to form the parietal pleura covering the chest wall/ diaphragm. The parietal pleura can be further subdivided based on the area of the thoracic cavity it covers, i.e. the costal pleura (ribs and intercostal muscles), the diaphragmatic pleura (the diaphragm), the cervical pleura (summit of the lung/neck) and the mediastinal pleura (9). The pleura encapsulates each lung, and by doing this the pleura divides the thoracic cavity into three separate chambers (10) enclosing the left lung, right lung and mediastinum, thus helping to prevent interference between the organs during normal activity, and the spread of infection and traumatic injury, e.g. pneumothorax.

5.2.1 The pleural space

The pleural surface is a smooth surface composed of five layers, consisting of a single layer of mesothelial cells (I), a thin sub-mesothelial connective tissue layer (II), a thin superficial elastic layer (III), a loose connective tissue layer (IV) and a deep fibroelastic layer (V; absent in the visceral pleura of small mammals) (9, 11). This surface layer of mesothelial cells makes up the mesothelium, which is also found covering the abdominal and cardiac compartments.

The pleural cavity formed between the parietal and visceral pleura is filled with a fluid layer approximately 10 μm thick forming a continuous barrier between the visceral and parietal pleura, preventing their contact (11). Pleural fluid has several roles in the pleural cavity which are essential to health, including being a lubricating layer and also providing a liquid coupling between the chest wall and the lung surface, enabling breathing to occur. Because of elastic recoil of the lung and surface tension created within the alveoli, there is pressure drawing the visceral pleura away from the parietal pleura of the chest wall. It is only the thin layer of pleural fluid under constant negative pressure that anchors the pleurae together. As the chest expands and the diaphragm lowers against the recoil of the lung during inhalation, the pressure of the pleural fluid drops (12) but maintains this coupling and thus draws the lungs open, pulling in air. As such, any conditions which alter the dynamics of this pleural space can be potentially life-threatening as this alters this transpulmonary pressure, potentially leading to collapsing or compressing of the lung.

Pleural fluid originates from blood vessels supplying the parietal pleura (11) and is reabsorbed via lymphatic drainage points on the surface of the parietal pleura. In a healthy state, the flow of fluid between the lung and the pleural space is restricted but in cases of certain conditions, such as adult respiratory distress syndrome (ARDS) or congestive heart failure, lung permeability increases and the pleural space forms the main route of exit and reabsorption for lung oedema.

The generation of pleural fluid from sub-mesothelial blood vessels by hydrostatic pressure requires a constant outflow of fluid in order to maintain the tight coupling between the pleurae. This outflow of fluid does not occur via passive diffusion through the mesothelium but rather through a specialised network of sub-mesothelial lymphatic channels which open up into the pleura cavity as stomata. The placement of these stomatal drainage points is not even across the parietal pleura. Instead they are most abundant at points where fluid build-up is greatest because of gravity and pressure, for example at the diaphragmatic pleura or the costadiaphragmatic recess (13). The sub-mesothelial lymphatic capillaries which drain the parietal space terminate in dilations called lacunae, as described by Recklinghausen in 1862 (14). These lacunae are covered with a loose layer of connective tissue called the macula cribriformis, forming a sieve-like structure which supports covering mesothelial cells (15). Rather than the common squamous appearance of mesothelial cells, the cells covering the lacunae are more cuboidal in shape, and line these channels forming distinctive structures along the surface of the mesothelium (Figure 5.1).

At points between these cells, 2–8 µm pore-like stomata (14) can be seen, which allow outflow of fluid, cells such as resident macrophages or lymphocytes, and particles out of the pleural space (16) (see Figure 5.2). It is worth noting that, in the study by Li (16), pleural stomata were only noted on the diaphragmatic pleura and not on the thoracic or visceral wall, possibly indicating that the main outflow of fluid was generated because of gravity and pleural pressures.

The lymphatic channels which drain from these stomatal openings contain valves (17) to prevent retrograde flow of fluid from the lymphatic channel. At inspiration the chest wall expands, opening the stomata and drawing in fluid, cells and particles to the lymphatic channels. Upon expiration the chest reduces in volume, closing the stomata and forcing the fluid along the lymphatic channel (9).

The mesothelium   The mesothelium was first described in 1827 by Bichat (18), although the term 'mesothelium' was coined later (19). As mentioned, the mesothelium is the layer which makes up the surface of the serosal membrane that covers the

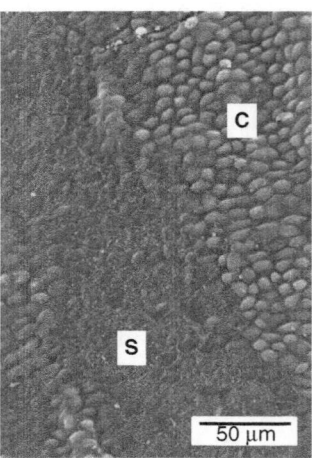

Figure 5.1 Scanning electron micrograph (SEM) of the mesothelium of a female C57/BL6 mouse. The flat squamous cells forming a smooth surface can be seen (marked 'S'), as well as microvilli around the periphery of each cell. The cuboidal cells covering lacunae regions are marked 'C'.

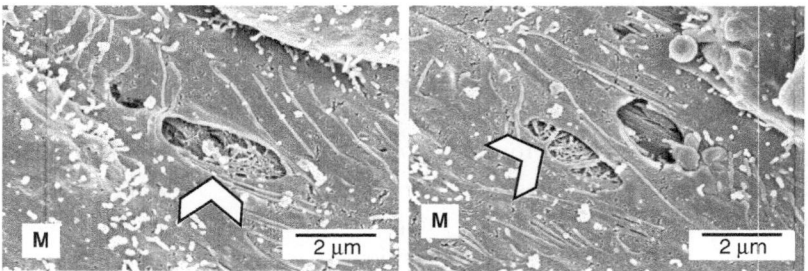

Figure 5.2 Mesothelial structures. Scanning electron micrographs of the parietal pleural mesothelium of a C57BL/6 mouse. Shown is a stoma (marked with chevron) surrounded by mesothelial cells (M) and occasional microvilli.

surface of the pleural, peritoneal and pericardial cavity as well as the lining of the testes (20). The mesothelium in each site has the common role of reducing friction and allowing the organ(s) it encapsulates to move and function without hindrance. For some time there has remained a concept that the mesothelium is simply a barrier cell performing the mundane, although crucial, task of lubricating organ movement. However, in the last few decades the dynamic nature of the mesothelium has become evident, and this is most eloquently shown in the 1982 review by Whitaker *et al.* (18) and more recently by Mutsaers (21).

The surface of the mesothelial cell can be covered with microvilli, although these can vary immensely in location and density. The microvilli serve to increase the surface area of the cells and trap hyaluronic acid–rich glycoproteins, forming a negatively charged glycocalyx which serves to lubricate the cells' surface and repel cells and particles. This has raised the interesting analogy by Antony (22) that the negative charge caused by these anionic sialomucins allows the surface of the mesothelium to operate as 'Teflon' as opposed to 'Velcro', which may aid the rapid clearance of foreign material from the pleural space.

Mesothelial cells are known to play a prominent role in the initiation, perpetuation and resolution of inflammation in all the serosal cavities. Such a role is suggested as mesothelial cells are known to secrete a whole host of mediators, including cytokines such as interleukins (IL) 1 and 6 (23) as well as a range of chemokines (e.g. IL-8 (24), MCP-1 (25)). Apical secretion of chemokines such as IL-8 (26) leads to leukocyte recruitment from the surrounding circulation and transmigration into the mesothelial cavity via mesothelial expression of cell-adhesion molecules such as intracellular adhesion molecule 1 (ICAM-1) (27) and vascular-cellular adhesion molecule 1 (VCAM-1) (27, 28). Indeed, in a rabbit model of pleurisy, it has been shown that treatment with anti-IL-8 antibodies prevents the normal influx of neutrophils into the pleural space (29).

During injury, the mesothelium has a regenerative capacity and can secrete a wealth of growth factors (e.g. PDGF (30, 31), TGF-β (32), VEGF (33)) as well as extracellular matrix proteins such as collagen, elastin and fibronectin. At a site of injury, studies have shown that mesothelial cells at the edge of the injury and also on the opposing surface begin proliferating. It has been hypothesised that mesothelial cells may repopulate areas of damage by migration from opposing surfaces using fibrin deposits as a scaffold, and thereby facilitate repair (34). The pathogenesis of pleural fibrosis and tissue regeneration is well described by the 2004 review by Mutsaers *et al.* (35).

5.2.2 Fibre-associated pleural disease

Occupational exposure to pathogenic fibres is associated with a range of diseases. Of the diseases associated with asbestos exposure, lung fibrosis was one of the first to be noted and subsequently tackled with dust control measures (36). Pathologically, asbestosis is characterised by an overexpression of extracellular matrix (ECM) beginning in and around the alveolar ducts and respiratory bronchioles, the site of fibre deposition (37), which impairs the transfer of oxygen into the blood and reduces the elastic nature of the lung, causing restrictive lung disease.

Lung cancer is also associated with exposure to pathogenic fibres, and estimates place asbestos as being related to 2–3% of all lung cancer deaths between 1980 and 2000 in the UK alone (38), and approximately 5–7% of all lung cancers worldwide (39). The casual link between lung cancer and asbestos exposure was shown by Doll in 1955. In his study into the mortality of asbestos workers, Doll found that of a cohort of 105 asbestos workers, 18 developed lung cancer, of which 15 were associated with asbestosis (40). This association between fibrosis and lung cancer is also evident in the literature where rats were exposed to asbestos and other mineral fibres by inhalation. It was noted that animals with pulmonary tumours had double the amount of fibrosis of animals that did not, and where it could be identified, the origin of the tumours was fibrotic areas (42).

Pleural plaques   Pleural plaques are white dense areas of hyaline fibrosis which form a discreet smooth lump which may become larger and calcified over time without further asbestos exposure, and which have been likened to icing on a cake (43). Like mesothelioma, they arise on and are confined to the parietal pleura, likely as a result of parietal deposition of asbestos fibres. Again like mesothelioma, pleural plaques are far more common in asbestos-exposed individuals than in non-exposed populations (44) and can have a long latency between exposure and plaque development of up to 30 years (43, 45). Despite these similarities with mesothelioma, pleural plaques are not a pre-malignant lesion and remain benign and self-limiting (43). They rarely cause morbidity, often being identified as a result of the investigation of other problems.

Pleural effusion   Pleural effusion occurs when there is a build-up of fluid within the pleural space, typically in excess of 15–20 ml, as a result of an imbalance between the influx and efflux of pleural fluid (46). Pleural effusions can be symptomatic or asymptomatic, with symptoms typically consisting of chest pain and dyspnoea. They can occur as benign pleural effusions, which is sometimes bloody and/or eosinophilic (47) and may resolve spontaneously or be recurring (43). Malignant pleural effusions can also occur as a complication of mesothelioma as well as other forms of cancer such as lung and breast cancer, as well as of other metastatic cancers, and can contribute to a decreased quality of life (48). Therefore pleural effusions are not restricted to asbestos exposure and can be seen as a result of other causes such as pulmonary embolism or complications from certain drug therapies (47).

Malignant mesothelioma   Malignant mesothelioma is an aggressive and uniformly fatal tumour (49) originating from the mesothelial lining of the pleura and, in fewer cases, from the peritoneum. In the UK alone, by 2003, 50 000 people had died from mesothelioma, with around 1700 cases a year (50). The main risk factor for malignant mesothelioma is asbestos exposure which can have occurred 20–60 years previously. However, in rare instances mesothelioma can also occur in

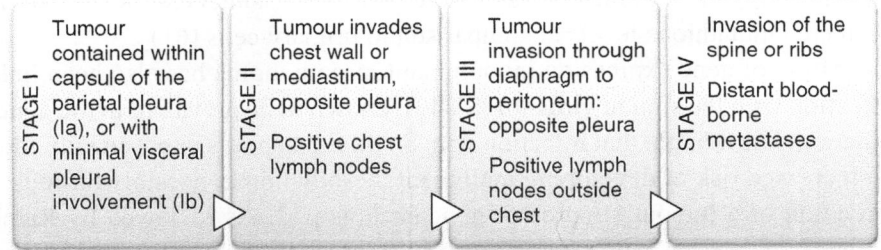

STAGE I	STAGE II	STAGE III	STAGE IV
Tumour contained within capsule of the parietal pleura (Ia), or with minimal visceral pleural involvement (Ib)	Tumour invades chest wall or mediastinum, opposite pleura Positive chest lymph nodes	Tumour invasion through diaphragm to peritoneum: opposite pleura Positive lymph nodes outside chest	Invasion of the spine or ribs Distant blood-borne metastases

Figure 5.3 The Butchart staging criteria for malignant mesothelioma (53).

patients who have undergone radiation therapy, chronic pleural inflammation or exposure to certain chemical carcinogens (51). There are several treatment options for mesothelioma but these are limited in their effectiveness and, as such, median survival time from initial presentation to death is only 9–17 months (52).

Reflecting the site of initiation, mesothelioma develops first at the parietal pleura and then spreads to the opposing visceral pleura, followed by invasion of the chest wall; it is only at the end-stage of the disease that there is blood-borne metastasis.

The initial site of development of mesothelioma is shown in the clinical staging of mesothelioma, as summarised in the Butchart staging criteria (53) in Figure 5.3, and as also suggested by others including the International Mesothelioma Interest Group, with early stage (T1a) involving only the parietal pleura with no involvement of the visceral pleura (54).

Whilst the link between asbestos exposure and the development of mesothelioma is clear, the mechanisms behind it are not. One hypothesis is that there is generation of reactive oxygen species (ROS), leading to direct interaction and mutations within the cellular DNA of the mesothelium (55). The generation of ROS leading to oxidative DNA damage has been shown in several systems, including both cell-free production of ROS by asbestos fibres and phagocyte-derived oxidative stress as a result of uptake by cells (56). The result of a pro-oxidant environment can be both genotoxic and pro-inflammatory/ fibrotic (57) due to the elicitation of various oxidant-sensitive transcription pathways such as that of nuclear factor κ-B (NFκB) (58) and activator protein-1 (AP-1) (59), leading to activation and secretion of various pro-inflammatory cytokines. It has been proposed that the heightened sensitivity of mesothelial cells to asbestos damage is due to their greater susceptibility than epithelial cells of the respiratory tract to DNA damage (60), which, if coupled with increased mesothelial cell survival and/ or reduced apoptosis, may lead to a transformed mesothelial cell. This, in turn has been suggested to be due possibly to lower levels of glutathione in mesothelial cells than in alveolar epithelial cells (3.5-fold), although these studies were performed using cell lines which may have altered glutathione levels in comparison to primary cells (61).

Alteration of gene expression in malignant mesothelioma has also been linked to infection with the simian virus-40 (SV40), known to be prevalent in the general population. It is thought that infection with SV40 predisposes exposed individuals to an increased risk of developing malignant mesothelioma, possibly through the inactivation of a tumour suppressor gene such as p53. As reviewed by Ramos-Nino *et al.* (62), it has been shown that SV40 infection alone was not associated with the development of malignant mesothelioma, but asbestos was. However, synergistically, the risk of developing malignant mesothelioma with SV40 infection was 27-fold higher than with asbestos exposure alone (63).

The exact mechanism behind the development of malignant mesothelioma is likely to remain unclear. However, it does seem that a pro-inflammatory and pro-oxidant environment created by the presence of long biopersistent fibres in the pleural space, and specifically at the parietal pleura close to a uniquely sensitive mesothelium, is important. This therefore poses the question: how do fibres gain entry to the pleura?

5.3 Particle translocation

The clearance of particles from the airways is not limited to removal along the airways to the mucociliaray escalator. Particles and fibres may also move from the airways into the interstitium and out of the airways via a number of possible routes (see Miserocchi *et al.* (64) for a comprehensive review). Because of their small size, the potential for nanoparticles to move more readily from the site of deposition has caused concern that nanoparticles may have unforeseen effects in organs not immediately associated with inhalation exposure, such as the brain (41). As discussed, when considering fibres the primary area in which disease occurs is the lungs and the pleura. This would mean that fibres may interact with these structures causing disease, although it has been suggested that fibres and particles may cause other systemic effects which may not require direct interaction, such as cardiovascular disease (65). However, whilst important, this is not the focus of this chapter. The potential for interaction between deposited fibres and the airspace surface is obvious, but far less obvious is the interaction between fibres and the pleura.

5.3.1 Particle transit to the pleura

The idea of particle transit to the pleura is not a new one. In fact through the use of post-mortem studies there is now evidence that a proportion of all lung-deposited particles reach the pleura and pass through the pleural space, from which they exit via the stomata on the parietal pleura. At these points of egress, particles may become interstitialised owing to the leaky nature of the parietal mesothelium (9) and lymphatic endothelium, forming characteristic 'black spots' on the parietal pleura. In the study by Mitchev *et al.* (66), 150 consecutive necropsies of normal urban dwellers were examined in Belgium, during which they commonly found black spots on the parietal pleura. This transit of particles from the lung into the pleural space is also seen in dusty occupations such as coal mining, with low-grade inflammatory reaction and fibrosis of black spots being very pronounced (67). During this study into the black spots of miners, Muller and colleagues examined the black spots histologically and found them extremely well-demarcated, following lines of lymph flow across and through the parietal pleura. Within these black spots, mineral particles were clearly visible.

The elongated nature of fibres and subsequent drag means that passage through to the pleura may be slower, which may partially account for the large lag time between asbestos fibre exposure and the development of pleural disease. There has been some controversy over the lengths of asbestos fibres that can translocate from the lung into the pleural space, with a number of studies showing a predilection for short-fibre translocation. The steric hindrance caused by long fibres may account for the fact that often mostly short fibres are found at the parietal pleura, as these fibres have clearly more likelihood of reaching the pleural compartment (64). However, a study carried out by Boutin *et al.* (68) specifically examined the fibre burden in areas rich in black spots and found as much as 22% of fibres measured to be greater than 5 μm in length. Heterogeneity of fibre deposition in the pleural space and differences in sampling methods may account for the lack of a consensus regarding the size of fibres that will translocate into the pleural space.

The movement of asbestos fibres into the pleural cavity after inhalation exposure in rats, and the subsequent pathogenic effects, were examined by Bernstein *et al.* (69), who reported the presence of amosite asbestos fibres on the diaphragm 7 days after the end of the inhalation exposure, with accumulations of macrophages associated with the fibres clearly seen by 14 days. The interface between the visceral and parietal pleura was examined using a non-invasive confocal microscopy technique to minimize the potential for contamination between tissue compartments, and the numbers and lengths of fibres at the visceral surface were quantified (70). The number of fibres counted at the visceral surface decreased over the course of the experiment from 181 to 365 days post-exposure; however, at each time point the sizes of the fibres counted ranged from 1 to 12 μm in length (70). The decrease in the numbers of fibres at this site over time would suggest that the fibres are progressively moving away from the visceral surface and into the pleural space, but the consistent range of fibre lengths measured at each time point indicate that there is no preference for the translocation of a particular fibre length within the range of 1 to 12 μm measured here. Unfortunately this study did not report the numbers of lengths of fibres detected along the parietal pleural of the diaphragm.

The movement of airborne particulates including asbestos fibres from the lung into the pleural space described here suggests that a proportion of all particles and fibres that can deposit in the distal regions of the lung will translocate to the pleural space. This also holds true for HARN, as has been demonstrated recently in a number of studies examining the pleural fate of CNT. In a study published in *Nature Nanotechnology*, Ryman-Rasmussen *et al.* (71) reported the deposition of CNT in the distal regions of the lung, directly adjacent to the visceral pleura, after a single inhalation exposure. This in itself demonstrates the ability of carbon nanotubes both to reach the distal lung and to become interstitialised within the sub-pleural tissue, suggesting that pleural transfer is indeed possible. A later study by Mercer *et al.* (72), using a lung instillation model rather than inhalation, quantified the number of CNT penetrating into the pleural space overtime. At 1 day post-exposure 0.6% of the lung fibre burden was detected in the sub-pleural regions of the lung, with a number of fibres penetrating into the pleural space. The number of fibres penetrating into the pleural space decreased by 7 days post-exposure but returned to a steady-state level for the remainder of the study, up to 56 days (72), with the authors suggesting the lung was acting as a reservoir of fibres.

5.4 Particle effects: retention in the pleural cavity

The well-established movement of distally deposited particles and fibres into the pleural space suggests that the more pertinent question when addressing the 'fibre-like' toxicity of new forms of manufactured fibres, including HARN, is not whether the fibre will reach the pleural space but how they are dealt with once there.

This formed the basis of the question addressed in our recent publication (73). In their study, the asbestos-like effects of CNT in the pleural space were examined using a method of directly injecting CNT into the pleural space and examining the inflammatory and fibrotic responses up to 6 months post-exposure. The study found that only those samples which contained long fibres – long fibre amosite asbestos and two long CNT samples – could induce a response in the pleural space, whereas samples containing short fibres or small particles caused no effect, suggesting a lack of toxicity or removal from the pleural space. The latter hypothesis was then further investigated by injecting quartz or coal-dust particles, which are known to cause effects in the lung, into the pleural space. These particles were small enough to negotiate the stomata and did not cause inflammation in the pleural space, suggesting they passed out of the pleural space without causing inflammation.

The length-dependent fibre response seen with long-fibre asbestos and long CNT was characterised by an early influx of inflammatory cells followed by progressive fibrosis and lesion formation along the parietal pleura: the initial site of mesothelioma development (73). Aggregates of the long fibres were associated with the formation of granulomatous lesions, which led to the hypothesis that the length-dependent response to fibres may be due to the size-restricted clearance of particles from the pleural space.

As discussed, clearance of fluid and therefore any particles in the pleural space is via drainage into the lymphatic capillaries through stomata or pores of defined size. These stomata, which are approximately 2–8μm in diameter, are abundantly present along the parietal pleural of the chest wall and diaphragm for clearance from the pleural space, and present on the posterior side of the diaphragm for clearance from the peritoneal cavity. A study by Kane *et al.* (74) noted the accumulation of long fibres injected into the peritoneal cavity around the stomata on the diaphragm, which led them to propose that retention of long fibres at the diaphragmatic mesothelial surface initiated inflammation, proliferation and granuloma formation. The concept that the size restriction imposed on the clearance of particles or fibres from the pleural space is responsible for long-fibre pathogenicity was tested by Murphy *et al.* (73) in a study in which two samples of polystyrene beads of identical composition but different size were injected into the pleural cavity and the inflammatory response subsequently examined. As expected, the larger beads, which were approaching the maximum diameter of the stomata, caused an inflammatory response, whereas the smaller beads, which would have been easily cleared, did not. These findings reflected those of a previous study using intrapleural injection of different-sized aminopolystyrene particles labelled with [111]In (75). In this study, the authors noted that there was a ~3–4-fold greater transfer of 2.18 μm particles than the larger 11.2 μm particles to the mediastinal lymph nodes.

Using a similar labelling approach as that of Liu, Murphy and colleagues (73) labelled short CNT with ^{111}In and injected them. The animals were then placed in a single-photon emission computed tomography scanner and observed over time. Using this approach, Murphy noted that initially the CNT remained in the pleural space, showing up as a diffuse glow across the chest region which rapidly became more focal in a region corresponding to mediastinal lymph nodes of the animals, and within 24 hours the signal from the labelled carbon nanotubes was almost exclusively found in these lymph nodes. The fibre burden of the lymph nodes to which the pleural fluid initially drains – the mediastinal lymph nodes – was also examined by the authors, who found, as expected, a significantly larger number of fibres in the lymph nodes of mice injected with short fibres, which supports the hypothesis that the long fibres are retained in the pleural space. This study clearly showed the importance of length in the pathogenic response to fibres or HARN in the pleural space; however, the exact length cut-off above which there will be a response is unknown and needs to be elucidated (76). This hypothesis, based on the transfer of CNT and other particles from the lung into the pleural space, as shown by Mercer *et al.* (72), and size-selective retention of long fibres at the parietal pleural surface at points of fluid egress (i.e. stomata), as shown by Murphy *et al.* (73), is shown diagrammatically in Figure 5.4.

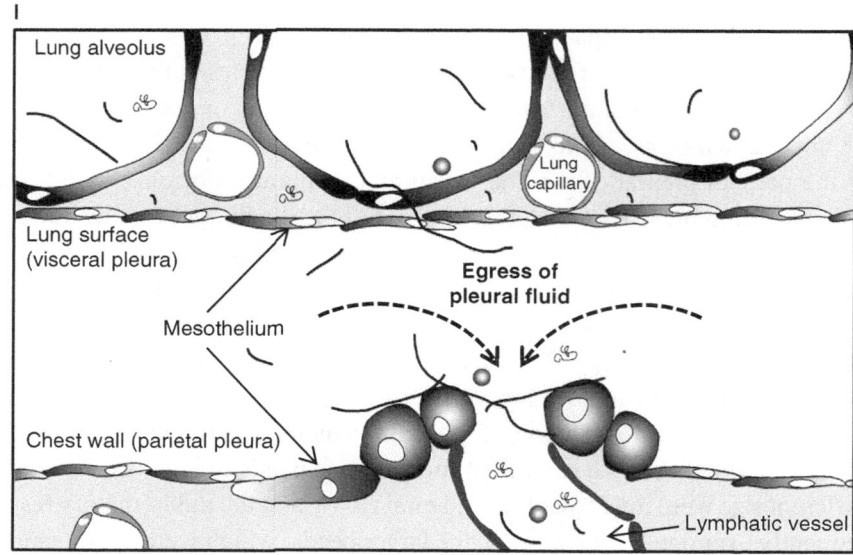

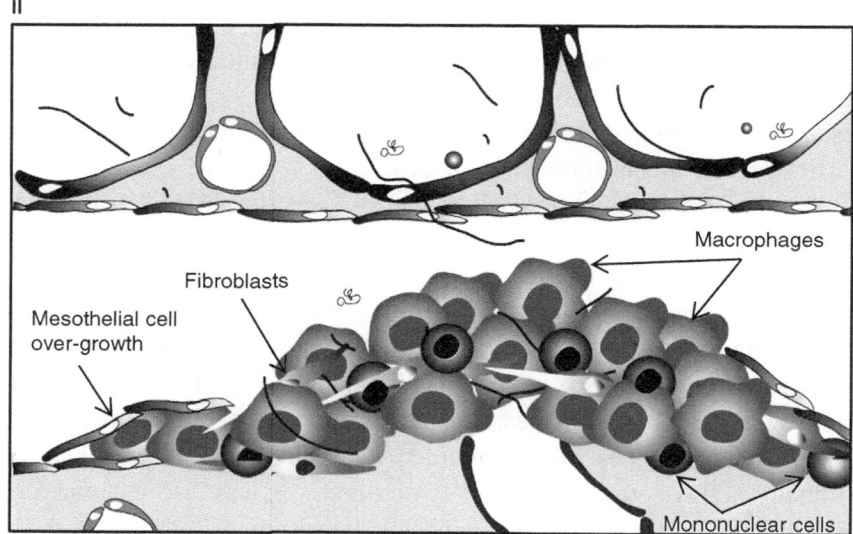

Figure 5.4 Particle transit and long-fibre retention at the parietal pleura. The upper panel (I) shows some deposited particles and CNT transiting through into the sub-pleural tissue and out into the pleural space. From here they move in the flow of pleural fluid out of the pleural space, via stomata on the parietal pleura, into the underlying lymphatics and out to the mediastinal lymph nodes. Long fibres cannot negotiate the stomata and so are retained at the parietal pleural surface. The lower panel (II) shows the resultant effect of particle retention at the pleural surface, with accumulation of inflammatory cells associated with retained fibres, denuding of the mesothelial surface and fibroblast proliferation, leading to the formation of a fibrosed granuloma. With time, the granuloma can become overgrown with mesothelial cells in the restoration of the mesothelial surface.

As a concept, the issue of retention and length is not new, but the evidence that it is also relevant to nanoparticles is. During the 1980s there were several publications which advanced the theory of length-dependent retention of fibres in the pleural space, such as that of Kane *et al.* (74), who used the peritoneal cavity to examine the inflammatory and fibrotic effects of asbestos. This emphasises the contention that particle/fibre transfer to the pleural space has long been known. Of the studies published, perhaps the best-known is that of Stanton and colleagues (77), who over the course of 72 experiments implanted a range of mineral fibres into the pleural space of rats and monitored them over a period of a year or more for signs of tumour formation. They recorded the frequency of malignant mesenchymal neoplasms and compared the resultant data to physicochemical attributes of the implanted fibres. They concluded that the probability of developing such pleural pathology correlated best with fibres that were 8 μm in length and 0.25 μm in diameter. The particular strength of this study was the large array of particles tested, which allowed them to conclude that the carcinogenicity of fibres depended on its dimensions and durability.

5.5 Conclusions

The link between exposure to harmful fibres such as certain forms of asbestos and erionite and the development of pleural pathologies and disease is unequivocal, and this has been linked to several physicochemical factors, with length being perhaps the most important.

The pleural space is a dynamic space that has well-developed mechanisms for the homeostatic maintenance of the coupling between the chest wall and lungs to enable that most crucial activity, breathing. We now know that there is a (small) flow of particulate matter from the lungs into the pleural space and that, for the most part, this is not retained. This is due to efficient clearance mechanisms operating within the pleural cavity, involving the outflow of pleural liquid and uptake of particles by pleural macrophages, causing the rapid transit through the cavity to the outlying lymph nodes. However, just as long fibres can hinder effective clearance in the lung, they also cause difficulty for more passive clearance mechanisms based around the outflow of pleural fluid from the cavity via parietal stomata. The resultant inflammation and pathology after injection of pathogenic fibres such as asbestos into pleural/peritoneal (mesothelial lined) cavities is well described (74, 78) and has been suggested for high-aspect-ratio nanoparticles including carbon nanotubes (79, 80). Increasingly, evidence both for transfer of carbon nanotubes to the pleural space (71, 72) and for the negative effects of long but not short high-aspect-ratio nanoparticles (73, 81) suggest that, if sufficient exposure were to occur to long-fibre HARN, the development of pleural effects is a possibility. This emphasises the need for proper control measures to limit exposure of workers and the public to long, biopersistent high-aspect-ratio nanoparticles to prevent the occurrence of avoidable disease.

References

1. Service RF. Chemistry: Nanotubes – The next asbestos? *Science* **281**(5379):941 (1998).
2. *Nanoscience and Nanotechnologies: Opportunities and Uncertainties.* Royal Society Policy Document 19/04 (London: The Royal Society and The Royal Academy of Engineering, 2004).
3. Clarke CC, Mowat FS, Kelsh MA, and Roberts MA. Pleural plaques: A review of diagnostic issues and possible nonasbestos factors. *Arch Environ Occup Health* **61**(4):183–92 (2006).
4. McGrath EE, Warriner D, and Anderson PB. Pleural fluid characteristics of tuberculous pleural effusions. *Heart Lung* **39**(6):540–3 (2010).
5. Devroey D and Van Casteren V. Symptoms and clinical signs associated with hospital admission and mortality for heart failure. *Cent Eur J Public Health* **18**(4):209–14 (2010).
6. Raja S, Murthy SC and Mason DP. Malignant pleural mesothelioma. *Curr Oncol Rep* **13**(4):259–64 (2011).
7. Pedata P, Feola D, Laieta MT, and Garzillo EM. Peritoneal mesothelioma: Description of a case and review of literature. *Int J Immunopathol Pharmacol* **24**(suppl. 1):85S-8S (2011).
8. Thompson DF, Fowler HW, and Fowler FG. *The Concise Oxford Dictionary of Current English* (Oxford: Clarendon Press, 1995).
9. Bouros D. *Pleural Disease* (New York: Marcel Dekker, 2004).
10. Marieb EN. *Human Anatomy and Physiology* (Menlo Park CA: Benjamin/Cummings, 1998).
11. Agostoni E and Zocchi L. Pleural liquid and its exchanges. *Respir Physiol Neurobiol* **159**(3):311–23 (2007).
12. Loring SHBJ. Chest wall mechanics. In Hlastala MPRH, ed., *Complexity in Structure and Function of the Lung* (New York: Marcel Dekker, 1998), pp. 151–79.
13. Moore KL and Agur AMR. *Essential Clinical Anatomy* (Baltimore, London: Lippincott Williams & Wilkins, 1996).
14. Tsilibary EC and Wissig SL. Absorption from the peritoneal cavity: SEM study of the mesothelium covering the peritoneal surface of the muscular portion of the diaphragm. *Am J Anat* **149**(1):127–33 (1977).
15. Miura T, Shimada T, Tanaka K, Chujo M, and Uchida Y. Lymphatic drainage of carbon particles injected into the pleural cavity of the monkey, as studied by video-assisted thoracoscopy and electron microscopy. *J Thorac Cardiovasc Surg* **120**(3):437–47 (2000).
16. Li J. Ultrastructural study on the pleural stomata in human. *Funct Dev Morphol* **3**(4):277–80 (1993).
17. Wang NS. The preformed stomas connecting the pleural cavity and the lymphatics in the parietal pleura. *Am Rev Respir Dis* **111**(1):12–20 (1975).
18. Whitaker D, Papadimitriou JM, and Walters MN. The mesothelium and its reactions: A review. *Crit Rev Toxicol* **10**(2):81–144 (1982).
19. Minot CS. The mesoderm and the coelom of vertebrates. *Am Nat* **24**(286):877 (1890).

20. Mutsaers SE. Mesothelial cells: Their structure, function and role in serosal repair. *Respirology* **7**(3):171–91 (2002).
21. Mutsaers SE. The mesothelial cell. *Int J Biochem Cell Biol* **36**(1):9–16 (2004).
22. Antony VB. Pathophysiology of the pleura. In Bouros D, ed., *Pleural Disease* (New York: Marcel Dekker, 2004), pp. 50–75.
23. Lanfrancone L, Boraschi D, Ghiara P, *et al.* Human peritoneal mesothelial cells produce many cytokines (granulocyte colony-stimulating factor [CSF], granulocyte-monocyte-CSF, macrophage-CSF, interleukin-1 [IL-1], and IL-6) and are activated and stimulated to grow by IL-1. *Blood* **80**(11):2835–42 (1992).
24. Zhang XY, Guckian M, Nasiri N, *et al.* Normal and SV40 transfected human peritoneal mesothelial cells produce IL-6 and IL-8: Implication for gynaecological disease. *Clin Exp Immunol* **129**(2):288–96 (2002).
25. Tanaka S, Choe N, Iwagaki A, Hemenway DR, and Kagan E. Asbestos exposure induces MCP-1 secretion by pleural mesothelial cells. *Exp Lung Res* **26**(4):241–55 (2000).
26. Nasreen N, Mohammed KA, Hardwick J, *et al.* Polar production of interleukin-8 by mesothelial cells promotes the transmesothelial migration of neutrophils: Role of intercellular adhesion molecule-1. *J Infect Dis* **183**(11):1638–45 (2001).
27. Jonjic N, Peri G, Bernasconi S, *et al.* Expression of adhesion molecules and chemotactic cytokines in cultured human mesothelial cells. *J Exp Med* **176**(4):1165–74 (1992).
28. Sikkink CJ, Reijnen MM, Duffhues BA, *et al.* Intercellular adhesion molecule-1 and gelatinase expression in human peritoneal mesothelial cells during propagation in culture. *Transl Res* **153**(5):240–8 (2009).
29. Broaddus VC, Boylan AM, Hoeffel JM, *et al.* Neutralization of IL-8 inhibits neutrophil influx in a rabbit model of endotoxin-induced pleurisy. *J Immunol* **152**(6):2960–7 (1994).
30. Versnel MA, Hagemeijer A, Bouts MJ, van der Kwast TH, and Hoogsteden HC. Expression of c-sis (PDGF B-chain) and PDGF A-chain genes in ten human malignant mesothelioma cell lines derived from primary and metastatic tumors. *Oncogene* **2**(6):601–5 (1988).
31. Walker C, Bermudez E, Stewart W, *et al.* Characterization of platelet-derived growth factor and platelet-derived growth factor receptor expression in asbestos-induced rat mesothelioma. *Cancer Res* **52**(2):301–6 (1992).
32. Decologne N, Kolb M, Margetts PJ, *et al.* TGF-beta1 induces progressive pleural scarring and subpleural fibrosis. *J Immunol* **179**(9):6043–51 (2007).
33. Ohta K, Yamashita N, Tajima M, *et al.* Diesel exhaust particulate induces airway hyperresponsiveness in a murine model: Essential role of GM-CSF. *J Allergy Clin Immunol* **104**(5):1024–30 (1999).
34. Mutsaers SE, Whitaker D, and Papadimitriou JM. Mesothelial regeneration is not dependent on subserosal cells. *J Pathol* **190**(1):86–92 (2000).
35. Mutsaers SE, Prele CM, Brody AR, and Idell S. Pathogenesis of pleural fibrosis. *Respirology* **9**(4):428–40 (2004).
36. Cooke WE. Fibrosis of the lungs due to the inhalation of asbestos dust. *Br Med J* **2**(3317):147 (1924).
37. Brody AR, Warheit DB, Chang LY, *et al.* Initial deposition pattern of inhaled minerals and consequent pathogenic events at the alveolar level. *Ann N Y Acad Sci* **428**:108–20 (1984).
38. Darnton AJ, McElvenny DM, and Hodgson JT. Estimating the number of asbestos-related lung cancer deaths in Great Britain from 1980 to 2000. *Ann Occup Hyg* **50**(1):29–38 (2006).

39. LaDou J. The asbestos cancer epidemic. *Environ Health Perspect* **112**(3):285–90 (2004).
40. Doll R. Mortality from lung cancer in asbestos workers. *Br J Ind Med* **12**(2):81–6 (1955).
41. Oberdorster G, Sharp Z, Elder AP, *et al*. Translocation of inhaled ultrafine particles to the brain. *Inhal Toxicol* **16** 437–45 (2004).
42. Davis JM and Cowie HA. The relationship between fibrosis and cancer in experimental animals exposed to asbestos and other fibers. *Environ Health Perspect* **88**:305–9 (1990).
43. Peacock C, Copley SJ, and Hansell DM. Asbestos-related benign pleural disease. *Clin Radiol* **55**(6):422–32 (2000).
44. Greillier L and Astoul P. Mesothelioma and asbestos-related pleural diseases. *Respiration* **76**(1):1–15 (2008).
45. Chapman SJ, Cookson WO, Musk AW, and Lee YC. Benign asbestos pleural diseases. *Curr Opin Pulm Med* **9**(4):266–71 (2003).
46. Owens MW and Milligan SA. Pleuritis and pleural effusions. *Curr Opin Pulm Med* **1**(4):318–23 (1995).
47. Kalomenidis I and Light RW. Eosinophilic pleural effusions. *Curr Opin Pulm Med* **9**(4):254–60 (2003).
48. Neragi-Miandoab S. Malignant pleural effusion, current and evolving approaches for its diagnosis and management. *Lung Cancer* **54**(1):1–9 (2006).
49. Tomasetti M, Amati M, Santarelli L, Alleva R, and Neuzil J. Malignant mesothelioma: Biology, diagnosis and therapeutic approaches. *Curr Mol Pharmacol* **2**(2):190–206 (2009).
50. Watterson A, Gorman T, Malcolm C, Robinson M, and Beck M. The economic costs of health service treatments for asbestos-related mesothelioma deaths. *Ann N Y Acad Sci* **1076**:871–81 (2006).
51. Cugell DW. Asbestos and the pleura: A review. *Chest* **125**(3):1103–17 (2004).
52. Heeschen C, Dimmeler S, Hamm CW, *et al*. Serum level of the anti-inflammatory cytokine interleukin-10 is an important prognostic determinant in patients with acute coronary syndromes. *Circulation* **107**(16):2109–14 (2003).
53. Butchart EG, Ashcroft T, Barnsley WC, and Holden MP. Pleuropneumonectomy in the management of diffuse malignant mesothelioma of the pleura: Experience with 29 patients. *Thorax* **31**(1):15–24 (1976).
54. Rusch VW. A proposed new international TNM staging system for malignant pleural mesothelioma (from the International Mesothelioma Interest Group). *Chest* **108** (4):1122–8 (1995).
55. Kamp DW and Weitzman SA. The molecular basis of asbestos induced lung injury. *Thorax* **54**(7):638–52 (1999).
56. Liu W, Ernst JD, and Broaddus VC. Phagocytosis of crocidolite asbestos induces oxidative stress, DNA damage, and apoptosis in mesothelial cells. *Am J Respir Cell Mol Biol* **23**(3):371–8 (2000).
57. Pociask DA, Sime PJ, and Brody AR. Asbestos-derived reactive oxygen species activate TGF-beta1. *Lab Invest* **84**(8):1013–23 (2004).
58. Brown DM, Beswick PH, and Donaldson K. Induction of nuclear translocation of NF-kappaB in epithelial cells by respirable mineral fibres. *J Pathol* **189**(2):258–64 (1999).
59. Mossman BT, Faux S, Janssen Y, *et al*. Cell signaling pathways elicited by asbestos. *Environ Health Perspect* **105**:1121–5 (1997).
60. Nygren J, Suhonen S, Norppa H, and Linnainmaa K. DNA damage in bronchial epithelial and mesothelial cells with and without associated crocidolite asbestos fibers. *Environ Mol Mutagen* **44**(5):477–82 (2004).

61. Puhakka A, Ollikainen T, Soini Y, *et al.* Modulation of DNA single-strand breaks by intracellular glutathione in human lung cells exposed to asbestos fibers. *Mutat Res* **514** (1–2):7–17 (2002).
62. Ramos-Nino ME, Testa JR, Altomare DA, *et al.* Cellular and molecular parameters of mesothelioma. *J Cell Biochem* **98**(4):723–34 (2006).
63. Porta C, Ardizzoni A, Gaudino G, *et al.* Malignant mesothelioma in 2004: How advanced technology and new drugs are changing the perspectives of mesothelioma patients (Highlights from the VIIth meeting of the International Mesothelioma Interest Group). *Med Lav* **96**(4):360–9 (2005).
64. Miserocchi G, Sancini G, Mantegazza F, and Chiappino G. Translocation pathways for inhaled asbestos fibers. *Environ Health* **7**:4 (2008).
65. Fukagawa NK, Li M, Sabo-Attwood T, *et al.* Inhaled asbestos exacerbates atherosclerosis in apolipoprotein E-deficient mice via CD4+ T cells. *Environ Health Perspect* **116**(9):1218–25 (2008).
66. Mitchev K, Dumortier P, and De VP. 'Black spots' and hyaline pleural plaques on the parietal pleura of 150 urban necropsy cases. *Am J Surg Pathol* **26**(9):1198–206 (2002).
67. Muller KM, Schmitz I, and Konstantinidis K. Black spots of the parietal pleura: Morphology and formal pathogenesis. *Respiration* **69**(3):261–7 (2002).
68. Boutin C, Dumortier P, Rey F, Viallat JR, and De VP. Black spots concentrate oncogenic asbestos fibers in the parietal pleura: Thoracoscopic and mineralogic study. *Am J Respir Crit Care Med* **153**(1):444–9 (1996).
69. Bernstein DM, Rogers RA, Sepulveda R, *et al.* The pathological response and fate in the lung and pleura of chrysotile in combination with fine particles compared to amosite asbestos following short-term inhalation exposure: Interim results. *Inhal Toxicol* **22**(11):937–62 (2010).
70. Bernstein DM, Rogers RA, Sepulveda R, *et al.* Quantification of the pathological response and fate in the lung and pleura of chrysotile in combination with fine particles compared to amosite-asbestos following short-term inhalation exposure. *Inhal Toxicol* **23**(7):372–91 (2011).
71. Ryman-Rasmussen JP, Cesta MF, Brody AR, *et al.* Inhaled carbon nanotubes reach the subpleural tissue in mice. *Nat Nanotechnol* **4**(11):747–51 (2009).
72. Mercer RR, Hubbs AF, Scabilloni JF, *et al.* Distribution and persistence of pleural penetrations by multi-walled carbon nanotubes. *Part Fibre Toxicol* **7**:28 (2010).
73. Murphy FA, Poland CA, Duffin R, *et al.* Length-dependent retention of carbon nanotubes in the pleural space of mice initiates sustained inflammation and progressive fibrosis on the parietal pleura. *Am J Pathol* **178**(6):2587–600 (2011).
74. Kane AB, Macdonald JL, and Moalli PA. Acute injury and regeneration of mesothelial cells produced by crocidolite asbestos fibers. *Am Rev Respir Dis* **133**:A198 (1986).
75. Liu J, Scollard DA, Reilly RM, Wu XY, and Johnston MR. Effect of particle size on the lymphatic distribution of 111 indium-aminopolystyrene through intrapleural administration. *Lymphology* **41**(4):153–60 (2008).
76. Donaldson K, Murphy F, Schinwald A, Duffin R, and Poland CA. Identifying the pulmonary hazard of high aspect ratio nanoparticles to enable their safety-by-design. *Nanomedicine (Lond)* **6**(1):143–56 (2011).
77. Stanton MF, Layard M, Tegeris A, *et al.* Relation of particle dimension to carcinogenicity in amphibole asbestoses and other fibrous minerals. *J Natl Cancer Inst* **67**(5):965–75 (1981).
78. Moalli PA, Macdonald JL, Goodglick LA, and Kane AB. Acute injury and regeneration of the mesothelium in response to asbestos fibers. *Am J Pathol* **128**:426–45 (1987).

79. Donaldson K, Murphy F, Schinwald A, Duffin R, and Poland CA. Identifying the pulmonary hazard of high aspect ratio nanoparticles to enable their safety-by-design. *Nanomedicine (Lond)* **6**(1):143–56 (2011).
80. Donaldson K, Murphy FA, Duffin R, and Poland CA. Asbestos, carbon nanotubes and the pleural mesothelium: A review and the hypothesis regarding the role of long fibre retention in the parietal pleura, inflammation and mesothelioma. *Part Fibre Toxicol* **7**(1):5 (2010).
81. Poland CA, Duffin R, Kinloch I, *et al*. Carbon nanotubes introduced into the abdominal cavity of mice show asbestos-like pathogenicity in a pilot study. *Nat Nanotechnol* **3**(7):423–8 (2008).

6

Experimental carcinogenicity of carbon nanotubes in the context of other fibres

KLAUS UNFRIED

Besides lung inflammation and fibrosis, the induction of lung tumours and mesothelioma is the most prominent adverse effect of inhaled natural and mineral fibres. Particle dimensions (length and diameter) as well as biopersistence are considered as the determining factors of fibre-induced carcinogenesis (Stanton *et al.*, 1981; WHO, 1985). As some carbon nanotubes seem to have dimensions matching the so-called fibre paradigm, and appear to be highly biopersistent, a very pressing question is whether these materials also bear a carcinogenic potential when they are inhaled by humans.

6.1 Possible mechanisms of fibre carcinogenicity

The key question for hazard identification and risk assessment with regard to respirable fibres, including nanotubes, is whether there is a common carcinogenic mechanism associated with the fibrous shape of these particles which may be also triggered by carbon nanotubes.

For asbestos fibres, the question of whether this material is tumour-inducing, or rather tumour-promoting, has been addressed in many experimental and epidemiological studies. Early investigations on the combinatory effects of chrysotile asbestos and chemical carcinogens indicated a tumour-promoting effect (Topping and Nettesheim, 1980). These experimental data seemed to reflect the fact that lung tumour rates in man induced by tobacco smoking are multiplied by asbestos exposure. Meta-analyses of epidemiological studies, however, suggested that besides a tumour promoting effect, which appears to enhance chemical carcinogenicity, asbestos fibres contribute to tumour initiation and can therefore be considered as genotoxic *in vivo* (Saracci, 1977; Nelson and Kelsey, 2002). For most of the synthetic vitreous fibres, which have been designed as substitutes for the natural fibres classified as carcinogens, such epidemiological data are not available, and likewise for the relatively new carbon nanotubes.

The Toxicology of Carbon Nanotubes, ed. Ken Donaldson, Craig A. Poland, Rodger Duffin and James Bonner.
Published by Cambridge University Press. © Cambridge University Press 2012.

Besides the question of whether a fibrous material is carcinogenic or not, it is important to understand the mechanisms by which the fibres trigger carcinogenicity. As a prerequisite for the design of studies investigating the carcinogenic potential, whether these kinds of xenobiotics are primary or secondary carcinogens must be evaluated. Primary carcinogens directly interact with target cells and elicit DNA damage by their intrinsic chemical or physicochemical reactivity. Because of the stochastic nature of DNA damage and mutagenicity, each single molecule or particle can initiate tumorigenesis. Secondary carcinogens are considered to act via cellular mechanisms like xenobiotic metabolising enzymes, but also by activating inflammatory cells which may contribute to genotoxicity by the release of reactive oxygen species. The possible relevance of such a mechanism for genotoxicity induced by non-fibrous inhalable particles has been impressively demonstrated in a system investigating the genotoxic potential of model carbon particles *in vivo* (Driscoll *et al.*, 1997). Neutrophilic granulocytes isolated from lungs of particle-exposed animals induced HPRT (hypoxanthine phosphoribosyl transferase) mutations in lung epithelial cells by the release of oxygen radicals. Identifying fibrous particles as secondary carcinogens would permit the definition of thresholds in the sense of no-effect levels for exposure regulation.

As a method to detect tumour-initiating events, some researchers applied *in vivo* genotoxicity assays, which have a very high sensitivity compared to carcinogenicity assays. These assays make use of transgenic animals bearing, in each genome, several copies of a bacterial reporter gene (e.g. *lacI*), which after isolation can be screened *ex vivo* for possible mutations (Dycaico *et al.*, 1994). With this experimental setting more than 100 000 individual events per animal could be analysed, while in the carcinogenicity studies the number of events was limited to the animal number per group which was manageable for the respective research groups. For asbestos fibres, this method has been successfully adopted in the experimental systems of inhalation, intraperitoneal injection, and intratracheal instillation (Rhin *et al.*, 2000; Unfried *et al.*, 2002; Topinka *et al.*, 2004).

The exposure of *lacI*-transgenic mice by inhalation with 5.75 mg/m^3 crocidolite (6 hours/day, 5 days) led to a significant increase in mutations in the lungs 4 weeks after the exposure (Rihn *et al.*, 2000). The molecular analysis of the nature of the mutations revealed no significant differences in the mutation spectrum between asbestos-induced and spontaneous mutations. This observation may be indicative of a proliferative effect of the asbestos fibres in the lung tissue. An accumulation of mutations that naturally occur during replication is probably an indication of increased cell proliferation. This effect, however, was not observed when crocidolite fibres were injected into the peritoneum of *lacI*-transgenic rats (Unfried *et al.*, 2002). In this study, increases in mutation frequencies in the *omentum majus* were observed 4 and 12 weeks after a single instillation of 2 mg or 5 mg crocidolite asbestos. Interestingly, the spectrum of crocidolite-induced mutations was significantly different from that of spontaneous mutations. Moreover, in the exposed animals an increase of G to T transversions was detectable. As this kind of mutation has been associated with oxygen-derived, pre-mutagenic DNA lesions, 8-hydroxydeoxyguanosine (8-OHdG) was investigated and found to have increased after asbestos exposure in these tissues. In a later study, the induction of 8-OHdG was correlated with the activation of inflammatory cells by crocidolite asbestos in the same system (Schürkes *et al.*, 2004). From these data it was concluded that crocidolite asbestos can induce mutations by a reactive oxygen species-dependent mechanism. The third study on using this system investigated mutations in the lungs of *lacI*-transgenic rats after instillation of 2 mg or 4 times 2 mg amosite asbestos (Topinka *et al.*, 2004). Again a dose-dependent increase in mutations was observed. The authors did not investigate mutations on the molecular level but were able to describe an exposure-dependent, persistent lung inflammation that correlated with the mutation frequencies. Taken together, these studies indicate at least an involvement of cellular mechanisms like proliferation and inflammation in the genotoxicity and carcinogenicity of asbestos fibres. Although for carbon nanotubes such *in vivo* investigations have not so far been described, several studies have addressed the question of whether these materials are pro-inflammatory in action. However, controversial results, which may be explained by the different nature (multi-walled and single-walled carbon nanotubes) and the different ways of preparing the test samples, have been observed (Shvedova *et al.*, 2005; Poland *et al.*, 2008; Pauluhn, 2010).

In the context of the induction of oxidative stress, fibre surface as well as chemical composition and the reactivity depending on these two parameters have to be considered. Chemical reactions triggered by surface charge or by ions leaching from the material may be directly responsible either for oxidative stress or for the induction of inflammation. Both events are supposed to be responsible for pre-cancerous DNA lesions. Iron as a contaminant frequently present in asbestos fibres can trigger Fenton-like reactions resulting in the generation of reactive oxygen species. For natural and, recently, also for synthetic nanosized asbestos, the generation of reactive oxygen species has been demonstrated to depend on the iron content of the materials (Turci *et al.*, 2011). Elemental analyses of carbon nanotubes showed that these materials can also contain considerable amounts of iron contamination (Shvedova *et al.*, 2008). The latter study clearly demonstrates the induction of oxidative stress in the lung tissue of exposed mice as well as indications for increased mutagenesis in the K-ras oncogene. However, the contribution of the iron contamination in the induction of these events is not clear. Earlier studies comparing iron-containing asbestos fibres and iron-free synthetic vitreous fibres with regard to 8-OHdG induction demonstrated that this parameter was correlated with the activation of inflammation rather than with the iron content (Schürkes *et al.*, 2004). With regard to surface reactivity, a recent study demonstrated that cytotoxicity *in vitro* but also inflammatory markers *in vivo* induced by multi-walled carbon nanotubes can be modulated by coating the fibres with polymers (Tabet *et al.*, 2011). While acid-based polymers increased oxidative stress and inflammation, polystyrene-based polymers reduced the examined parameters.

Although there are several indications that carbon nanotubes may trigger cellular and molecular reactions comparable to natural and man-made mineral fibres, hazard identification or risk assessment using these well-investigated materials as benchmarks appears problematic, as long as molecular mechanisms of the inter-action of carbon nanotubes with target cells are not well understood. These mechanistic data can be considered to be helpful in gaining insight into the potential mechanisms of the carcinogenicity of carbon nanotubes. However, as long as there are no epidemiological data, a clear correlation between exposure to carbon nanotubes and the occurrence of tumours must be demonstrated in a suitable animal system.

6.2 Carcinogenicity testing of fibrous materials

Carcinogenicity studies are usually designed to mimic real-life exposure in an animal system which appears representative of the human organism. Real-life doses as well as exposure routes are aimed at being reflective of the situation in humans. Comparison of animal species, however, reveals species-specific differences, at least with regard to fibre deposition and induction of pulmonary inflammation (Gelzleichter *et al.*, 1999). Evaluation of the can-cerogenic potential of respirable fibres just by inhalation studies is therefore questionable.

Moreover, it appears difficult to select dosages that can be easily compared to human exposure, which can happen over time intervals which cannot be covered by the lifespan of experimental animals. In the case of poorly soluble materials like mineral fibres and carbon nanotubes, this is of importance since these materials may remain in the body for years and may deploy their adverse effects, as with asbestos, decades after the exposure period (Acheson *et al.*, 1982). Increasing the dose in order to get experimental designs comparable to human exposure is also proble-matic, as natural mechanisms contributing to fibre clearance but also possibly to fibre accumulation in certain body compartments, may be impaired by an overload of the involved organ systems (Oberdörster, 1995).

The classical design of a carcinogenicity study, moreover, has some more practical deficiencies. Usually, after a defined exposure period or a single applica-tion of the fibres, animals are kept until visible signs of tumour occur or until the animals die. Depending on the chosen organism, the experiment may take more than a year. On the other hand, the sensitivity of the assay is strongly limited by the number of animals which can be treated and kept in such experiments. This means that fibres with a low carcinogenic potential may not be identified if the number of tumours does not appear to be statistically significant above the control groups.

Extrapolating high-dose effects to the low-dose range is highly problematic as overload effects cannot be excluded as responsible for false-positive results. Summarising these considerations, it appears plausible that, in addition to carcinogenicity studies, mechanistic experiments must be performed to indicate whether respirable fibres are hazardous with regard to carcinogenicity. Experimental approaches studying pre-cancerous events like *in vivo* DNA damage and *in vivo* mutagenicity as bioassays may help to clarify the cellular and molecular mechanisms underlying fibre carcinogenesis, as well as to identify the carcinogenic potential in the low-dose range.

A well-established but controversial method, the intraperitoneal injection of fibres (mostly in rats), has, however, been used and discussed over decades. Carcinogenic fibres like asbestos injected into body cavities which are lined by a mesothelial cell layer have been demonstrated to induce mesothelioma (Pott *et al.*, 1974; Churg *et al.*, 1978; Van der Meeren *et al.*, 1992). As possible clearance mechanisms or dispersion throughout the body are bypassed with this approach, fibre numbers and fibre dispersion may significantly differ from inhalation studies. The discrepancy between the tumour frequencies and tumour types was obvious in a comparative study performed by Davis *et al.* (1986). While inhaled chrysotile asbestos in rats induced lung carcinoma in 25% of the animals, the intraperitoneal injection of these fibres resulted in mesothelioma in 90% of the animals. These differences have been considered by some groups as an advantage of the intraperitoneal assay as it appears to be more sensitive than inhalation assays (Pott *et al.*, 1994). Others, however, describe the induction of mesothelioma after direct injection, which cannot be reproduced in inhalation studies, as false positives (Bernstein, 2007). For synthetic mineral fibres, however, it is accepted that within a defined dose range of fibres well characterised according to length and diameter, both assays may be used as bioassays to predict carcinogenic potential (Bernstein *et al.*, 2001a, 2001b). The reliability of the method of intraperitoneal injection is indicated by a study investigating the carcinogenic potential of biosoluble vitreous fibres in comparison to crocidolite asbestos (Grimm *et al.*, 2002). While the positive control (asbestos) induced high tumour rates after a single application of 5×10^5 or 5×10^6 cancerogenic fibres, only three of the vitreous fibres induced mesothelioma, with a very low frequency after application of a fibre number which was two orders of magnitude higher (Figure 6.1).

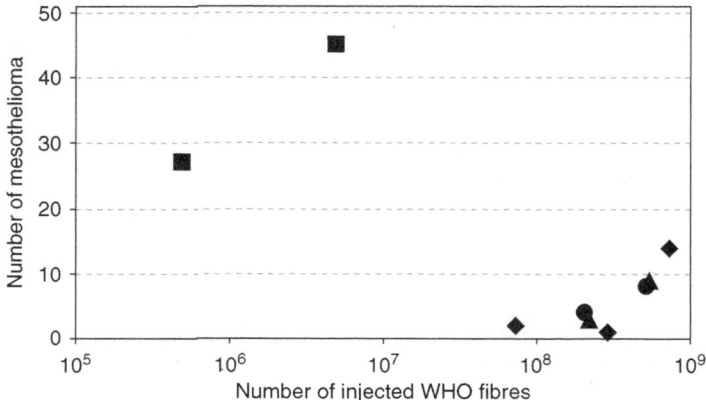

Figure 6.1 Number of mesothelioma in a long-term intraperitoneal injection study with biosoluble synthetic mineral fibers in relation to the number of injected WHO fibres (Grimm *et al.*, 2002). ■ Positive control: crocidolite asbestos. The other symbols respectively indicate three different kinds of synthetic mineral fibers with low biopersistence. Saline controls as well as two additional fibre samples induced no tumours at all.

6.3 Current testing strategy for fibrous particles

The possible mechanisms of carcinogenicity of fibrous particles have been considered in a working group report of the ILSI Risk Science Institute (Bernstein *et al.*, 2005). This paper suggests a tiered testing strategy for non-vitreous respirable fibres with the main focus on short-term assays allowing quick and reliable hazard identification and risk assessment of new fibrous materials (Figure 6.2). The authors suggest subchronic inhalation studies with rats in which physicochemical characteristics of the inhaled material as well as toxicological endpoints are determined. Biopersistence and pleural fibre burden as well as inflammatory markers in bronchoalveolar lavage (BAL), proliferation and fibrosis are the key endpoints which, if enhanced as a result of exposure, indicate the need for long-term inhalation studies to demonstrate carcinogenic effects. *In vitro* assays on durability, (surface) reactivity, and cell stress such as oxidative stress or genotoxicity are considered as optional, additional assays which may be considered as additional concerns indicating the necessity of long-term studies. The unambiguous identification, preparation, and characterisation of the used fibre sample is considered as an essential prerequisite of any testing strategy, with the aim of identifying hazards or risk assessment.

According to this strategy, several serious concerns regarding the carcinogenic potential of carbon nanotubes come from the studies described above as mechanistically relevant. Although not performed as inhalation studies, some of these studies may be considered as subchronic studies which have investigated

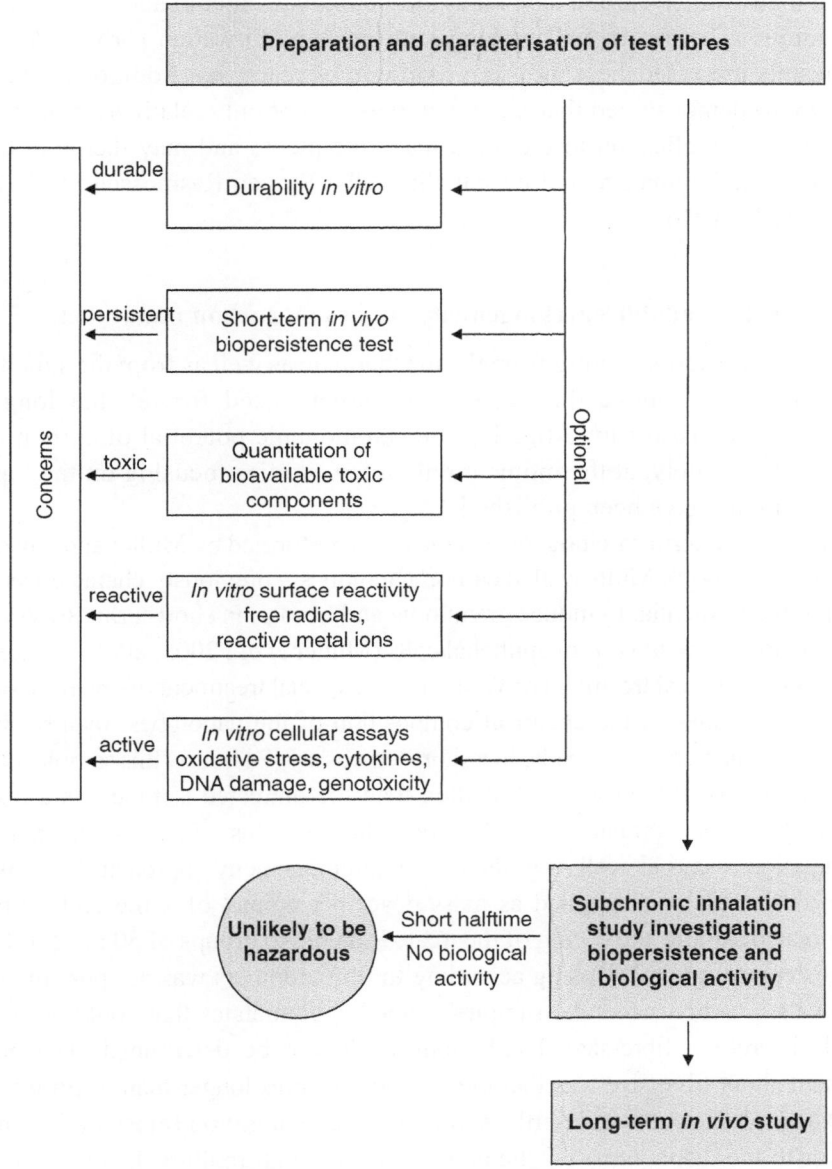

Figure 6.2 Testing strategy for fibrous particles which are not classified as synthetic vitreous fibres, recommended by the ILSI working group (Bernstein *et al.*, 2005).

biopersistence and biological reactivity. A very impressive study in this context compared asbestos fibres with carcinogenic and non-carcinogenic dimensions and multi-walled carbon nanotubes of various dimensions and shapes (tangled versus linear) (Poland *et al.*, 2008). While short fibrous asbestos, particulate carbon, and tangled multi-walled carbon nanotubes did not induce inflammation or granuloma, long fibrous asbestos as well as long and linear multi-walled carbon nanotubes significantly increased these measures of *in vivo* bioreactivity. Additional investigations *in vivo* demonstrated that these materials are not only relatively biopersistent but also, after application to the lung, reach the pleura and may therefore trigger reactions of inflammatory and mesothelial cells (Ryman-Rasmussen *et al.*, 2009; Mercer *et al.*, 2010).

6.4 Available carcinogenicity studies on carbon nanotubes

The serious concerns coming from the mechanistic as well as from the subchronic studies described above demonstrate the urgent need for reliable long-term carcinogenicity assays investigating the carcinogenic potential of carbon nanotubes. Unfortunately, at the moment only a few studies focusing on the topic of carcinogenicity have been published.

A classic long-term carcinogenicity study was performed by Muller and coworkers (Muller *et al.*, 2009). Multi-walled carbon nanotubes – previously characterised with regard to their potential to induce granuloma and fibrosis in short-term assays and to induce genotoxicity *in vitro* in epithelial cells (Muller *et al.*, 2005, 2008) – were used for the intraperitoneal treatment of Wistar rats. A special treatment of one fibre sample resulted in changes of the chemical composition of the nanotubes, mainly characterised by the depletion of metals. In a short-term assay (24 hours) this sample induced significantly less peritoneal inflammation than the untreated sample. As a positive control, 2 mg per animal of UICC crocidolite asbestos was used in short-term inflammation assays as well as in the 2-year carcinogenicity approach. Multi-walled carbon nanotubes were applied as mass doses per animal of 2 mg and 20 mg of the untreated sample and 20 mg of the treated sample to groups of 50 rats (26 for the asbestos control group). Dosing according to fibre numbers was not possible as the multi-walled carbon nanotube samples formed agglomerates that could not be dispersed. Therefore, fibre-size distributions could not be determined. The general statement about fibre size was that the number of fibres longer than 5 μm was very low. Although the authors describe some tissue reorganisation (mostly adherence of the liver to the diaphragm) for the animals treated with multi-walled carbon nanotubes, no significant increase in mesothelioma or other tumours was observed in these groups, while asbestos treatment resulted in a tumour rate of about 30%. This lack of tumorigenicity stands in contrast to the observed histological changes. Therefore, the authors carefully discuss the possibility of false-negative results. The tumour rate for 2 mg crocidolite asbestos is below the results of others who applied the same total dose of fibres as a number of smaller single doses (Roller *et al.*, 1996). Owing to the application of a single bolus of nanotubes, deposits may have been encapsulated and the inflammatory response may have subsided. In conclusion, further mechanistic and carcinogenicity studies were requested.

A second study describing no carcinogenic effects made use of the uncommon method of implanting 10 mg of gelatin-encapsulated multi-walled or single-walled carbon nanotubes into the peritoneum of F344 rats (Varga and Szendi, 2010). Twelve months after the implantation, granulomas, but no tumours, were observed in the peritoneal cavity of the animals. Although the study gives some information on the generation of granulomas when fibres are concentrated at a single site on the mesothelial cell layer, it unfortunately does not match the standards of a carcinogenicity study. With six animals per group and an observation period of 12 months, only a very strong carcinogenic effect would have been observable.

In contrast to the two reports of a lack of genotoxicity stand two reports in which the induction of mesothelioma by multi-walled carbon nanotubes has been observed. The first made use of an experimental system which has been described as sensitive to asbestos-induced mesothelioma (Takagi *et al.*, 2008). The system of p53$^{+/-}$ heterozygous mice was used in order to shorten carcinogenicity studies significantly, because of the accelerated occurrence of mesothelioma in this system (Marsella *et al.*, 1997; Vaslet *et al.*, 2002). The authors applied 3 mg per animal of multi-walled carbon nanotubes with a 27.5% proportion of fibres matching the WHO criteria. Within 180 days 90% of the nanotube-treated mice (19 animals per group) developed mesothelioma. As a positive control, another group was treated with an equal mass dose of crocidolite asbestos. Only 50% of these animals developed mesothelioma within this time frame, while in the control group (3 mg fullerene) no tumours were observed. The data are considered by the authors mainly as a first contribution to hazard identification of multi-walled carbon nanotubes. However, the work has been discussed controversially because of the calculation of particle numbers, the distribution of the particles, the (relatively high) dose and in particular the experimental system of heterogeneous knock-out animals (Donaldson *et al.*, 2008; Ichihara *et al.*, 2008). The tumour-suppressor gene p53 has only been associated with mesothelioma carcinogenesis after asbestos exposure in mice (Cistrulli *et al.*, 1996). In rats as well as in humans, fibre-induced carcinogenesis appears not to depend on mutations in this gene (Unfried *et al.*, 1997; Ni *et al.*, 2000; Kitamura *et al.*, 2002). The lack of one allele of a tumour-suppressor gene increases the chance of complete loss of gene function by a single mutation which may occur spontaneously as a result of an increased proliferation rate. Moreover, substances with very weak genotoxic potential in such a system may be overestimated.

To address the problem of an experimental system which may provide false-positive results, another small-scale study using the same carbon nanotube sample was performed (Sakamoto *et al.*, 2009). In this approach, a comparably lower dose (0.24 mg per animal) was injected in the scrotum of seven Fischer 344 rats. Again, crocidolite asbestos (0.47 mg per animal) was used as positive control in 10 animals. Within 1 year, six of the seven nanotube-treated animals developed mesothelioma while all asbestos-treated animals survived the observation period with no mesothelioma detected. As the scrotum contains a mesothelial cell layer, i.e. the tunica testis, mesothelioma may arise in this organ. However, it is not clear why the authors chose this form of application and not the intraperitoneal injection. It is not clear how multi-walled carbon nanotubes or asbestos fibres are dispersed within the animal body after application in the scrotum. The authors observed particle deposits in the peritoneal cavity. The tumours, described as metastatic spreading, appeared mainly in the peritoneum and less frequently in the pleura. From this description it is at least possible that because of the high concentration of nanotubes in the small body compartment of the scrotum, primary tumours developed there and spread to the other tissues. Unfortunately, owing to the small size of the study and the non-standardised method of application, this study is of limited value with regard to hazard identification and risk assessment of multi-walled carbon nanotubes.

6.5 Conclusion

The data on carcinogenicity of carbon nanotubes so far available in the open literature are either critically discussed by the authors themselves (Muller *et al.*, 2009) or do not fit the criteria of reliable carcinogenicity studies with regard to the dose, the method of application, the observation period, or the animal numbers. The currently extant published carcinogenicity studies on carbon nanotubes are also not harmonised, as a result of the standards and criteria of accepted fibre-carcinogenicity testing, and it is highly questionable whether the studies reflect the exposure of humans at all. Data from these studies may only help to raise concerns and to highlight the need for long-term studies which reflect the natural exposure route by which these xenobiotics enter the human body. Mechanistic *in vitro* and *in vivo* studies demonstrating that the suggested testing strategies for fibrous particles are applicable for carbon nanotubes are necessary. The currently available data may at best be considered as raising concern and indicating the need for further studies with well-characterised, well-dispersed fibre samples. Carcinogenic effects must be reproducibly observed in independent experiments and the reasons for controversial results must be explained plausibly.

References

Acheson ED, Gardner MJ, Pippard EC, and Grime LP. 1982. Mortality of two groups of women who manufactured gas masks from chrysotile and crocidolite asbestos: A 40-year follow-up. *Brit J Ind Med* **39**: 344–348.
Bernstein D, Castranova V, Donaldson K, *et al.* 2005. Testing of fibrous particles: Short-term assays and strategies. *Inhal Toxicol* **17**: 497–537.

Bernstein DM, Riego Sintes JM, Ersboell BK, and Kunert J. 2001a. Biopersistence of synthetic mineral fibers as a predictor of chronic intraperitoneal injection tumor response in rats. *Inhal Toxicol* **13**: 851–875.

Bernstein DM, Riego Sintes JM, Ersboell BK, and Kunert J. 2001b. Biopersistence of synthetic mineral fibers as a predictor of chronic inhalation in rats. *Inhal Toxicol* **13**: 823–849.

Bernstein DM. 2007. Special-purpose fiber type 475: Toxicological assessment. *Inhal Toxicol* **19**: 149–159.

Churg A, Warnock ML, and Bensch KG. 1978. Malignant mesothelioma arising after direct application of asbestos and fiber glass to the pericardium. *Am Rev Respir Dis* **118**: 419–424.

Cistrulli CA, Sorger T, Marsella JM, Vaslet CA, and Kane AB. 1996. Spontaneous p53 mutation in murine mesothelial cells: Increased sensitivity to DNA damage induced by asbestos and ionizing radiation. *Toxicol Appl Pharmacol* **141**: 264–271.

Davis JM, Addison J, Bolton RE, Donaldson K, and Jones AD. 1986. Inhalation and injection studies in rats using dust samples from chrysotile asbestos prepared by a wet dispersion process. *Br J Exp Pathol* **67**: 113–129.

Donaldson K, Stone V, Seaton A, *et al*. 2008. Re: Induction of mesothelioma in p53$^{+/-}$ mouse by intraperitoneal application of multi-wall carbon nanotube. Letter to the editor. *J Toxicol Sci* **33**: 385–388.

Driscoll KE, Deyo LC, Carter JM, *et al*. 1997. Effects of particle exposure and particle-elicited inflammatory cells on mutation in rat ahhhlveolar epithelial cells. *Carcinogenesis* **18**: 423–430.

Dycaico JM, Provost GS, Kretz PL, *et al*. 1994. The use of shuttle vectors for mutation analysis in transgenic mice and rats. *Mutat Res* **307**: 461–478.

Gelzleichter TR, Bermudez E, Mangum JB, *et al*. 1999. Comparison of pulmonary and pleural responses of rats and hamsters to inhaled refractory ceramic fibers. *Toxicol Sci* **49**: 93–101.

Grimm HG, Bernstein DM, Attia M, Richard J, and de Reydellet A. 2002. Experience from a long-term carcinogenicity study with intraperitoneal injection of biosoluble synthetic mineral fibers. *Inhal Toxicol* **14**: 855–882.

Ichihara G, Castranova V, Tanioka A, and Miyazawa K. 2008. Letter to the editor. *J Toxicol Sci* **33**: 381–384.

Kitamura F, Araki S, Suzuki Y, *et al*. 2002. Assessment of the mutations of p53 suppressor gene and Ha- and Ki-ras oncogenes in malignant mesothelioma in relation to asbestos exposure: A study of 12 American patients. *Ind Health* **40**: 175–181.

Marsella JM, Liu BL, Vaslet CA, and Kane AB. 1997. Susceptibility of p53-deficient mice to induction of mesothelioma by crocidolite asbestos fibers. *Environ Health Perspect* **105**: 1069–1072.

Mercer RR, Hubs AF, Scabilloni JF, *et al*. 2010. Distribution and persistence of pleural penetrations by multi-walled carbon nanotubes. *Part Fibre Toxicol* **7**: 28.

Muller J, Delos M, Panin N, *et al*. 2009. Absence of carcinogenic response to multi-wall carbon nanotubes in a 2-year bioassay in the peritoneal cavity of the rat. *Toxicol Sci* **110**: 442–448.

Muller J, Huaux F, Fonseca A, *et al*. 2008. Structural defects play a major role in the acute lung toxicity of multiwall carbon nanotubes: Toxicological aspects. *Chem Res Toxicol* **21**: 1698–1705.

Muller J, Huaux F, Moreaux N, *et al*. 2005. Respiratory toxicity of multi-wall carbon nanotubes. *Toxicol Appl Pharmacol* **207**: 221–231.

Nelson HH and Kelsey KT. 2002. The molecular epidemiology of asbestos and tobacco in lung cancer. *Oncogene* **21**: 7284–7288.

Ni Z, Liu Y, Keshava N, Zhou G, Whong W, and Ong T. 2000. Analysis of K-ras and p53 mutations in mesotheliomas from humans and rats exposed to asbestos. *Mutat Res* **468**: 87–92.

Oberdörster G. 1995. Lung particle overload: Implications for occupational exposures of particles. *Regul Toxicol Pharmacol* **21**: 123–135.

Pauluhn J. 2010. Multi-walled carbon nanotubes (Baytubes®): Approach for derivation of occupational exposure limit. *Reg Toxicol Pharmacol* **57**: 78–89.

Poland CA, Duffin R, Kinloch I, *et al.* 2008. Carbon nanotubes introduced into the abdominal cavity of mice show asbestos-like pathogenicity in a pilot study. *Nat Nanotechnol* **3**, 423–428.

Pott F, Huth F, and Friedrichs KH. 1974. Tumorigenic effect of fibrous dusts in experimental animals. *Environ Health Perspect* **9**: 313–315.

Pott F, Roller M, Kamino K, and Bellmann B. 1994. Significance of durability of mineral fibers for their toxicity and carcinogenic potency in the abdominal cavity of rats in comparison with the low sensitivity of inhalation studies. *Environ Health Perspect* **102**: 145–150.

Rhin B, Coulais C, Kauffer E, *et al.* 2000. Inhaled crocidolite metagenicity in lung DNA. *Environ Health Perspect* **108**: 341–346.

Roller M, Pott F, Kamino K, Althoff GH, and Bellmann B. 1996. Results of current intraperitoneal carcinogenicity studies with mineral and vitreous fibres. *Exp Toxicol Pathol* **48**: 3–12.

Ryman-Rasmussen JP, Cesta MF, Brody AR, *et al.* 2009. Inhaled carbon nanotubes reach the sub-pleural tissue in mice. *Nat Nanotechnol* **4**: 747–751.

Sakamoto Y, Nakae D, Fukumori N, *et al.* 2009. Induction of mesothelioma by a single intrascrotal administration of multi-wall carbon nanotube in intact male Fischer 344 rats. *J Toxicol Sci* **34**: 65–76.

Saracci R. 1977. Asbestos and lung cancer: An analysis of the epidemiological evidence on the asbestos smoking interaction. *Int J Cancer* **20**: 323–331.

Schürkes C, Brock W, Abel J, and Unfried K. 2004. Induction of 8-hydroxydeoxyguanosine by man-made vitreous fibres and crocidolite asbestos administered intraperitoneally in rats. *Mutat Res* **553**: 59–65.

Shvedova AA, Kisin ER, Mercer R, *et al.* 2005. Unusual inflammatory and fibrogenic pulmonary responses to single-walled carbon nanotubes in mice. *Am J Physiol Lung Cell Mol Physiol* **289**: L698–L708.

Shvedova AA, Kisin E, Murray AR, *et al.* 2008. Inhalation vs. aspiration of single-walled nanotubes in C57BL/6 mice: Inflammation, fibrosis, oxidative stress, and mutagenesis. *Am J Physiol Lung Cell Mol Physiol* **295**: L552–L565.

Stanton MF, Layard M, Tegeris A, *et al.* 1981. Relation of particle dimension to carcinogenicity in amphibole asbestoses and other fibrous material. *J Natl Cancer Inst* **67**: 965–975.

Tabet L, Bussy C, Setyan A, *et al.* 2011. Coating carbon nanotubes with a polystyrene-based polymer protects against pulmonary toxicity. *Part Fibre Toxicol* **8**: 3.

Takagi A, Hirose A, Nishimura T, *et al.* 2008. Induction of mesothelioma in p53[+/−] mouse by intraperitoneal application of multi-wall carbon nanotube. *J Toxicol Sci* **33**: 105–116.

Topinka J, Loli P, Georgadis P, *et al.* 2004. Mutagenesis by asbestos in the lung of λ-lacI transgenic rats. *Mutat Res* **553**: 67–78.

Topping DC and Nettesheim P. 1980. Two-stage carcinogenesis studies with asbestos in Fischer 344 rats. *J Natl Canc Inst* **65**: 627–630.

Turci F, Tomatis M, Lesci IG, Roveri N, and Fubini B. 2011. The iron-related molecular toxicity mechanism of synthetic asbestos nanofibres: A model study for high-aspect-ratio nanoparticles. *Chem Eur J* **17**: 350–358.

Unfried K, Roller M, Pott F, Friemann J, and Dehnen W. 1997. Fiber-specific molecular features of tumors induced in rat peritoneum. *Environ Health Perspect* **105**: 1103–1108.

Unfried K, Schürkes C, and Abel J. 2002. Distinct spectrum of mutations induced by crocidolite: Clue for 8-hydroxy-deoxyguanosine-dependent mutagenesis in vivo. *Cancer Res* **62**: 99–104.

Van der Meeren a, Fleury J, Nebut M, *et al.* 1992. Mesothelioma in rats following intrapleural injection of chrysotile and phosphorylated chrysotile. *Int J Cancer* **50**: 937–942.

Varga C and Szendi K. 2010. Carbon nanotubes induce granulomas but not mesotheliomas. *In Vivo* **24**: 153–156.

Vaslet CA, Messier NJ, and Kane AB. 2002. Accelerated progression of asbestos-induced mesotheliomas in heterozygous p53 mice. *Toxicol Sci* **68**: 331–338.

WHO. 1985. *Reference Method for Measuring Airborne Man-Made Mineral Fibers (MMMF)*. Environmental Health Report no. 4 (Copenhagen: World Health Organization, Regional Office for Europe).

7

Fate and effects of carbon nanotubes following inhalation

JESSICA P. RYMAN-RASMUSSEN, MELVIN E. ANDERSEN, JAMES
C. BONNER

Disclaimer: The views expressed in this document are those of the authors and do not necessarily reflect the views and policies of the US Environmental Protection Agency. Reference to specific commercial products or services does not constitute endorsement.

7.1 Introduction

Carbon nanotubes (CNTs) represent an important family of emerging nano-technologies because they have many potential uses in engineering, electronics, and medicine due to their small size, ease of functionalization, unusual strength, and electrical conductivity. However, these novel engineered nanostructures also represent a potential risk for the respiratory tract, due to the potential for inhalation exposure and evidence that the lung is a key target organ for hazardous effects (Bonner, 2010a; Card et al., 2008). In recent years, acute and subchronic inhalation studies in rats and mice have become available. These studies both support and extend our understanding of results obtained via other methods of pulmonary exposure, such as intratracheal instillation and oropharyngeal aspiration, and results obtained from in vitro cell culture systems. Several inhalation studies show inhaled CNTs deposit within the distal regions of the lungs in both rats and mice, where they can persist for weeks or longer after cessation of exposure, and result in inflammatory and/or fibrotic effects. CNTs rapidly migrate to the subpleural regions, where they could pose a carcinogenic threat to the mesothelial lining (Ryman-Rasmussen et al., 2009a). There is also evidence that individuals with pre-existing respiratory disease may be at greater risk from inhaling CNTs, since inhaled CNTs have been shown to exacerbate pre-existing allergic airway inflammation in mice (Ryman-Rasmussen et al., 2009b). Finally, inhaled CNTs localize in lymphoid tissues (Ma-Hock et al., 2009; Pauluhn, 2010) and have immune effects that reach beyond the lung to influence systemic targets such as the spleen (Mitchell *et al.*, 2007, 2009). Toxic effects of inhaled CNTs should be considered along with their potential for therapeutic and other societal benefits. The aim of this chapter is to review the fate and effects of inhaled CNTs in order to facilitate careful evaluation of their known and potential effects, as they continue to be pursued as a vital part of the emerging nanotechnology revolution.

7.2 Pulmonary deposition and migration of inhaled carbon nanotubes

The size and aerodynamic properties of inhaled CNTs are key factors that determine the sites of CNT deposition in the respiratory tract, which, in turn play an integral role in the resulting physiological effects. However, the majority of studies investigating the effects of CNTs on the lung have not been done via the inhalation route, but by intratracheal instillation or oropharyngeal aspiration in mice. The decisions on dosing route are based on a combination of the practical limitations associated with inhalation exposures in general and with CNTs in particular. Inhalation studies are costly and there is limited availability of technical expertise and specialized facilities. Larger amounts of CNTs are required for inhalation than for instillation or aspiration, which can be problematic for scarce or expensive CNTs. Finally, aerosolization of CNTs can be difficult because they have a tendency to form non-respirable agglomerates. Nevertheless, inhalation studies are critical in evaluating the true nature of deposition patterns for particles and fibers, which cannot be adequately mimicked by delivery of these substances as an aqueous bolus dose. In recent years, the technical difficulties of performing CNT inhalation exposures in rodents have largely been overcome utilizing various methods (Baron *et al.*, 2008; Mitchell *et al.*, 2007; Ryman-Rasmussen *et al.*, 2009a, 2009b), and aerosols of CNTs from dry powders or from nebulized solution have been achieved in which the majority of nanoparticles or nanoparticle aggregates reach the alveolar region of the lung after inhalation. Figure 7.1 shows multi-walled CNTs that have been aerosolized in nebulized saline containing a Pluronic® surfactant and the resulting aggregates of nanotubes collected on filters at the distal port of the aerosol-generating system. These CNT aggregates represent the size of material that passes into a nose-only inhalation apparatus to expose mice in a single 6-hour exposure scenario. The use of surfactant-containing media has improved dispersal of CNT agglomerates for inhalation, instillation, and aspiration delivery to the lungs of rats and mice. Alternatively, improved dry milling of CNTs prior to aerosol generation has been improved and is probably the best way to recreate a real-world occupational exposure to CNT dust.

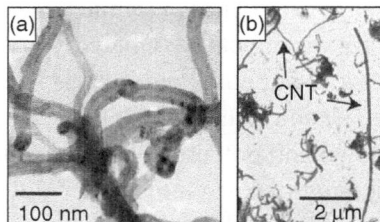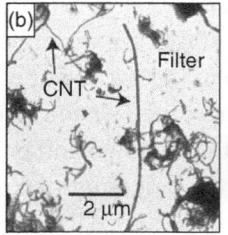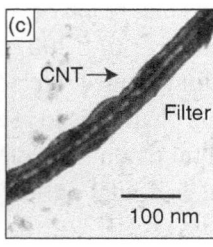

Figure 7.1 Electron microscope images of bulk and aerosolized multi-walled carbon nanotubes (CNTs). (a) Scanning electron microscope (SEM) image of bulk CNTs used for inhalation studies. (b) and (c) Low- and high-magnification transmission electron microscope (TEM) images showing the appearance of the MWCNTs shown in (a) after aerosolization and collection on a polycarbonate filter. (Images reprinted from Ryman-Rasmussen *et al.*, 2009b with the permission of the American Thoracic Society. Copyright © 2011 American Thoracic Society.)

Inhalation studies with CNTs have been performed by several groups, with durations of exposure ranging from acute to subchronic. Inhaled CNTs localize in the alveolar region in both rats (Ma-Hock *et al.*, 2009; Pauluhn, 2010) and mice (Mitchell *et al.*, 2007; Ryman-Rasmussen *et al.*, 2009a, 2009b; Shvedova *et al.*, 2008). As with other inhaled fibers, CNTs reach the distal regions of the lung and are intercepted at alveolar duct bifurcations. However, small aggregates and individual CNTs also deposit on the surface of the type I epithelium within alveolar spaces. The majority of these CNTs are rapidly engulfed and found within alveolar macrophages. Some inhaled CNTs reach the pleural region of the lungs and embed within the subpleural tissue (Ryman-Rasmussen *et al.*, 2009b). CNTs also reach the pleura via macrophages, which exit the lung via the pleural lymphatic system. Inhaled CNTs have also been found in lung-associated lymph nodes (Ma-Hock *et al.*, 2009; Pauluhn, 2010). As shown in Figure 7.2, transmission electron microscropy (TEM) shows that multi-walled CNTs deposit within the distal lung and are found within 1 day post-exposure on the alveolar epithelium, within alveolar macrophages throughout the lung, and within subpleural tissue. Most of these multi-walled CNTs are engulfed by macrophages as discussed below, but many individual CNTs or small aggregates evade macrophage uptake and presumably migrate throughout the lung via macrophage-independent processes. The migration of inhaled CNTs from the airways to the alveolar region to the subpleural tissue after inhalation exposure is illustrated in Figure 7.3. This illustration also shows possible outcomes of disease pathogenesis at each of these regions, including airway fibrosis, interstitial fibrosis and granuloma formation at the alveolar region, and pleural immune and fibrotic reactions at the pleura.

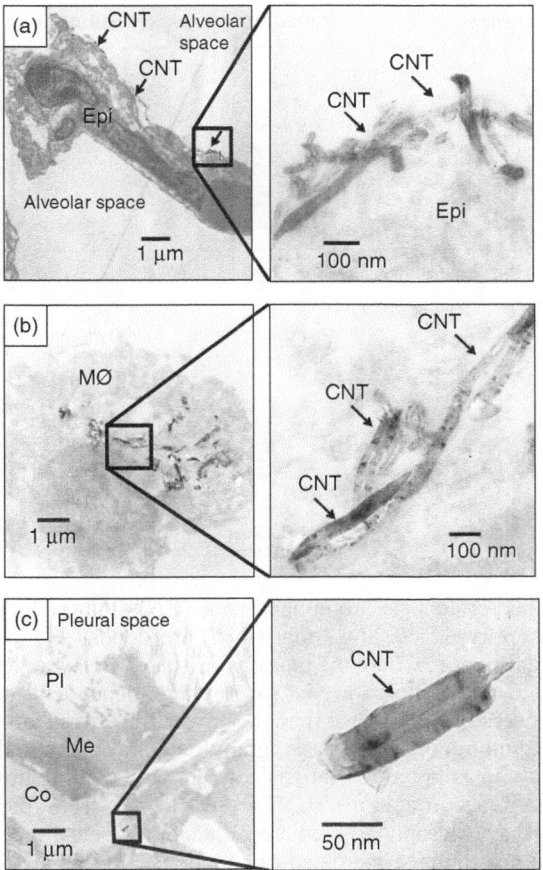

Figure 7.2 Transmission electron microscope (TEM) images showing multi-walled carbon nanotube (CNT) deposition and uptake in the lungs of mice after a single, brief inhalation exposure. (a) Low- and high-magnification images showing deposition of CNTs on type I alveolar epithelium 1 day post-inhalation. (Image reprinted from Bonner, 2010a with the permission of the American Thoracic Society. Copyright © 2011 American Thoracic Society.) (b) Low- and high-magnification images of CNTs engulfed by an alveolar macrophage. (c) Low- and high-magnification images of a CNT embedded within extracellular collagen (Co) matrix beneath the surface of the pleura (Pl) lined by mesothelial cells (Me). (Images in (B) and (C) reprinted from Ryman-Rasmussen *et al.*, 2009a with permission. Copyright © 2011 Nature Publishing Group.)

7.3 Macrophage uptake and clearance of carbon nanotubes

Macrophages are the first line of immune defense to inhaled foreign particles. Some studies indicate that single-walled CNTs (SWCNT) delivered to the lungs of mice escape immune surveillance by macrophages in that they are not readily recognized

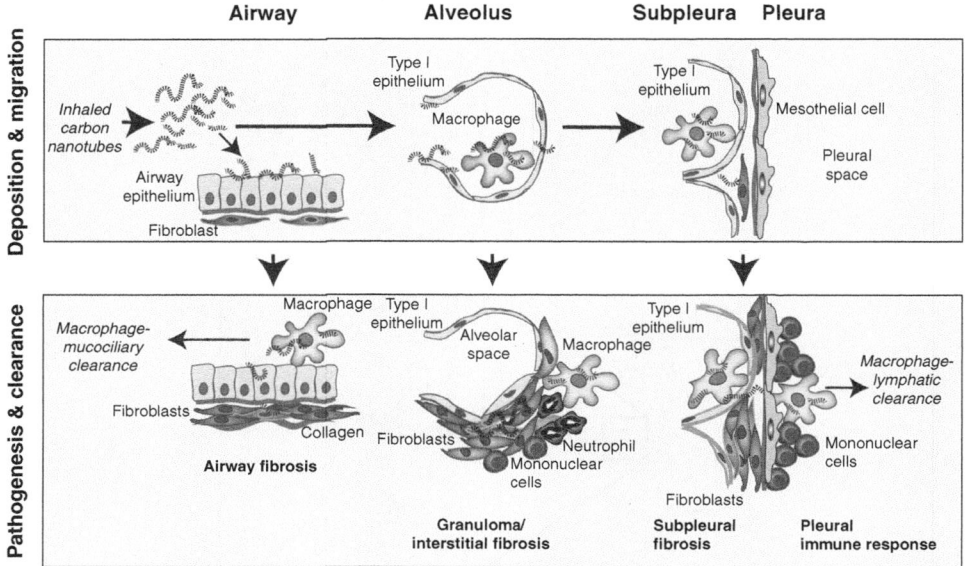

Figure 7.3 Fate and effects of carbon nanotubes (CNTs) in the lungs of mice after inhalation. The upper panel shows that CNTs are deposited along airways and within the alveolar region, where the majority are taken up by macrophages. Some CNTs migrate to the subpleura via macrophage-dependent or macrophage-independent processes. The lower panel shows the effects of CNTs on lung pathogenesis, including airway fibrosis, granuloma and interstitial fibrosis of the alveolar region, and subpleural fibrosis, along with pleural immune response (mononuclear cell accumulation). CNTs are cleared from the lung by two principal pathways: macrophage-mediated mucociliary clearance up and out of the airways, and macrophage-mediated lymphatic clearance across the pleura.

and taken up by phagocytosis unless the CNTs are treated with agents such as phosphatidylserine (Konduru *et al.*, 2009). However, other studies have shown that CNTs are readily recognized and avidly engulfed by macrophages. SWCNT uptake by rat alveolar macrophages occurs *in vivo* and there was unusual formation of SWCNT bridge structures that linked macrophages after 21 days of exposure (Mangum *et al.*, 2006). These bridge structures were not observed in rats exposed to 8 nm diameter carbon black nanoparticles, indicating that the carbon bridge structures between macrophages may be unique to SWCNTs. Carbon bridge structures might represent a specific molecular interaction between SWCNTs and cytoskeleton, although this remains to be determined. Whether or not SWCNTs are recognized by macrophages may depend on the aggregation state of the CNT preparation. It is likely that more dispersed SWCNTs would not be readily recognized, whereas aggregates of SWCNTs would be recognized as a larger particle and taken up by macrophages. Therefore, the aggregation status is likely an important determinant towards evaluating the toxicity of CNTs. CNTs that are engulfed by macrophages are removed from the lung through two principal avenues: the mucociliary escalator and the pulmonary lymphatic system. The mucociliary escalator transports macrophages and their engulfed payload up the airways of the lungs where they are ultimately swallowed or expelled through coughing (Bonner, 2008). Clearance of CNT-laden macrophages through the pulmonary lymphatic system could lead to adverse affects as many of these macrophages traffic across the pleural region where CNTs might have potential carcinogenicity similar to that of asbestos fibers (Bonner, 2010a).

The clearance of inhaled MWCNTs from the lungs and lung-associated lymph nodes (LALNs) of rats has been modeled and experimentally determined by measuring concentrations of residual cobalt (Co) catalyst in lung and LALT after 13 weeks of nose-only inhalation exposure for 6 hours/day x 5 days/week, and a 6-month post-exposure period (Pauluhn, 2010). A concentration-dependent increase in lung levels of Co was observed, although no time-dependent relationship was observed. The half-life of elimination was 151, 350, 318, and 375 days at 0.1, 0.4, 1.5, and 6 mg/m^3, respectively. In the LALNs, both a dose- and a time-dependent increase in Co content were observed after cessation of dosing. The author concluded that mild to moderate lung overload was present at 0.1 and 0.4 mg/m^3, with partial to complete overwhelming of clearance at 1.5 and 6 mg/m^3.

7.4 Fibrogenic reactions to inhaled carbon nanotubes

CNTs attracted the attention of respiratory toxicologists because of the fiber-like shape and aspect (length to width) ratio of CNTs that are similar to asbestos fibers, which are a known cause of fibrosis and mesothelioma in humans. The inhalation studies with CNTs that are available to date vary in a number of ways: different kinds of nanotubes were used (multi-walled or single-walled), there are different sources of nanotubes, different aerosolization methods have been utilized, different atmospheric conditions of nanotubes and durations of inhalation exposure have been used, the studies have been performed by different laboratories. In spite of these differences, there are remarkably common pulmonary effects, and these effects are similar to those obtained from studies in which pulmonary exposures occurred via other methods (e.g. intratracheal instillation or oropharyngeal aspiration).

The majority of inhalation studies with CNTs indicate that CNTs cause pulmonary inflammation (Li *et al.*, 2007; Ma-Hock *et al.*, 2009; Pauluhn, 2010; Ryman-Rasmussen *et al.*, 2009a, 2009b; Shvedova *et al.*, 2008). Pulmonary fibrosis has been reported by most of these same groups (Li *et al.*, 2007; Pauluhn, 2010; Ryman-Rasmussen *et al.*, 2009a; 2009b; Shvedova *et al.*, 2008), although in one study (Pauluhn, 2010) it was concluded that this effect was a result of pulmonary overload. In addition to effects in the deep lungs, effects in the upper respiratory tract of goblet cell hyperplasia and inflammation were also observed in the nasal cavity and upper airways (Ma-Hock *et al.*, 2009; Pauluhn, 2010). Pulmonary inflammation and fibrosis from inhaled CNTs are consistent with results that have been reported in studies using intratracheal instillation or oropharyngeal aspiration (Lam *et al.*, 2004; Mangum *et al.*, 2006; Muller *et al.*, 2005; Shvedova *et al.*, 2005; Warheit *et al.*, 2004). These studies show that CNTs, which aggregate into μm-sized bundles in aqueous media, stimulate the formation of inflammatory foci known as granulomas and fibrotic reactions within the lung parenchyma. Pulmonary fibrosis and/or granulomas have been reported in mice or rats within days after intratracheal instillation of SWCNTs (Shvedova *et al.*, 2005; Warheit *et al.*, 2004) or MWCNTs (Muller *et al.*, 2005). Granulomas are associated with large CNT aggregates in the lung that are easily visible by light microscopy, while interstitial pulmonary fibrosis appears to be associated with dispersed CNTs that are detected by electron microscopy (Mercer *et al.*, 2008).

Two studies have compared the difference in lung pathological effects between inhalation and other methods of pulmonary exposure. A study that compared instillation with inhalation of MWCNTs in mice showed a more diffuse pattern of deposition by inhalation than that observed after bolus intratracheal instillation (Li *et al.*, 2007). In this study, the lungs of mice that inhaled MWCNTs demonstrated thickening of the alveoli, whereas the instillation group demonstrated bronchiolar and alveolar inflammation and destruction of alveoli. Another study utilised single-walled CNTs and compared inhalation exposure with oropharyngeal aspiration of a highly dispersed single-walled CNT preparation (Shvedova *et al.*, 2008). The pulmonary effects were similar in both cases, but were more severe for the inhaled SWCNTs. Also, inhaled SWCNTs caused more pronounced alveolar wall thickening (i.e. interstitial fibrosis) as compared to instilled SWCNTs, which produced more granuloma formation. Therefore, it seems that granuloma formation caused by CNTs could be due to the delivery of a bolus dose of CNTs in aqueous media. Together, these studies indicate that the degree of CNT dispersion, together with the influence of particle mass and aerodynamic properties, contribute to the differences in potency and physiological effects observed when comparing inhalation to other routes of pulmonary exposure such as intratracheal instillation or oropharyngeal aspiration.

Aside from the method of pulmonary exposure and the degree of particle dispersion, there are two other major factors that determine the toxicity of CNTs. These are clearance and contamination by metal catalysts. First, longer CNTs impede clearance, and structures longer than 10 to 15 μm (the approximate width of an alveolar macrophage) are difficult to clear from lung tissues via macrophage-mediated mechanisms. This is especially relevant to MWCNTs, which are more rigid than SWCNTs. Long SWCNTs are more likely to fold and be taken up by macrophages, whereas long MWCNTs are more likely to cause frustruated phagocytosis and impede macrophage clearance. Second, the composition of CNTs must be carefully considered. Metals such as nickel, cobalt, and iron are commonly used as catalysts in the manufacture of CNTs and these same metals are well known to cause pulmonary diseases in humans, including pulmonary fibrosis and asthma (Kelleher *et al.*, 2000). For example, nickel is known to cause occupational asthma and contact dermatitis, whereas iron and cobalt cause interstitial pulmonary fibrosis in occupations related to mining and metallurgy. Metal catalysts can be removed to some extent from CNTs by acid washing, although this process usually does not completely remove metals.

The cellular and molecular mechanisms that cause CNT-induced lung disease have not been clearly elucidated, but it is expected that many lessons can be learned from studies on larger particles, fibres, and metals. However, CNTs and other engineered nanomaterials may also operate through unique and unexpected mechanisms to cause disease, given that they can interact at the molecular level with cellular organelles, proteins, phospholipids, and nucleic acids. A variety of soluble mediators (growth factors, cytokines, and chemokines) that play important roles in fibrosis are induced in the lungs of rats or mice after exposure to CNTs (Bonner, 2010b). Some of these mediators and their relationship to lung region and cell type are shown in Figure 7.4. Several studies have shown that either SWCNTs or MWCNTs delivered to the lung by intratracheal instillation in rats or inhalation in mice increase mRNA and protein levels of platelet-derived growth factor (PDGF) (Cesta *et al.*, 2010; Mangum *et al.*, 2006; Ryman-Rasmussen *et al.*, 2009a). PDGF stimulates the replication, chemotaxis, and survival of lung fibroblasts to promote the pathogenesis of fibrosis (Bonner, 2004). SWCNTs delivered to the lungs of mice by oropharyngeal aspiration or inhalation cause induction of TGF-β1 (Shvedova *et al.*, 2005; 2008), a central mediator of collagen production by mesenchymal cells (fibroblasts and myofibroblasts). However, MWCNTs containing nickel catalyst produce very little or no detectable TGF-β1 in the lungs of mice exposed by inhalation or in the lungs of rats exposed by instillation (Cesta *et al.*, 2010; Ryman-Rasmussen *et al.*, 2009a). Other factors such as osteopontin (OPN) could serve to stimulate collagen deposition and fibrosis in the absence of TGF-β1. OPN mRNA levels are highly induced in the lungs of rats exposed to SWCNTs (Mangum *et al.*, 2006). Alveolar macrophages, as well as airway epithelial cells and fibroblasts, produce PDGF, TGF-β1, and OPN. Macrophages are also a rich source of IL-1β and IL-6 (Shvedova *et al.*, 2008), and these two cytokines promote tissue fibrogenesis and are increased in the lungs of mice exposed to inhaled SWCNTs. Several chemokines are also induced by CNT exposure and drive the inflammatory response in the lung. CXCL8 (IL-8), a potent neutrophil chemoattractant, is produced by a human bronchial epithelial cell line *in vitro* after exposure to MWCNTs (Hirano *et al.*, 2010). CCL2, also known as monocyte chemoattractant protein-1 (MCP-1), is

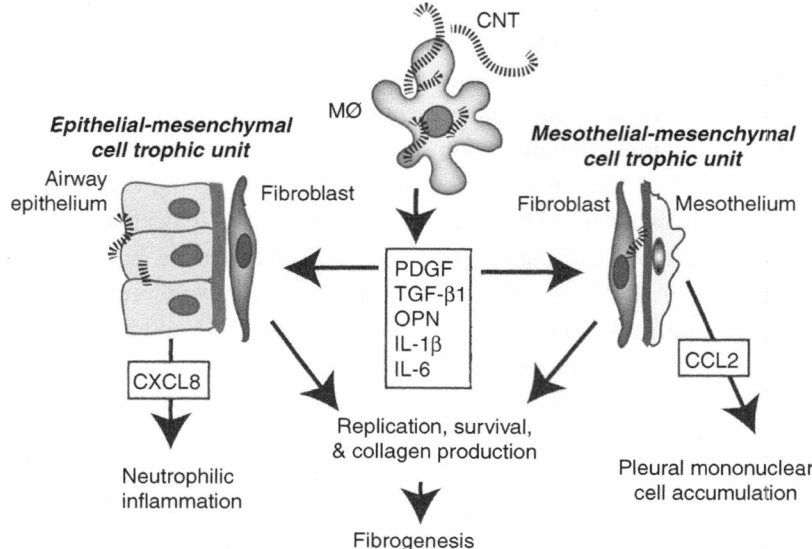

Figure 7.4 Cell–cell communication and soluble mediators implicated in the pathogenesis of CNT-induced lung disease. The majority of inhaled CNTs are taken up by alveolar macrophages (MO), which are stimulated to produce growth factors (PDGF, TGF-β1, OPN) and cytokines (IL-1β, IL-6) that promote fibroblast replication, survival, migration, and collagen deposition, resulting in fibrogenesis. CNTs also stimulate the airway epithelium to produce CXCL8 to cause neutrophil influx from the vasculature to promote lung inflammation. In addition, pleural mesothelial cells produce CCL2, which is proposed to mediate mononuclear cell accumulation at the pleural surface after CNT inhalation exposure. The interactions of epithelium and mesothelium with underlying fibroblasts (termed trophic units) via cell–cell contact or paracrine soluble factors are proposed as an important factor in the pathogenesis of airway or supleural fibrosis, respectively.

elevated in the lavage fluid of mice exposed to MWCNT inhalation (Ryman-Rasmussen *et al.*, 2009a). MCP-1 is also produced by pleural mesothelial cells (Visser *et al.*, 1998), and is a candidate that mediates mononuclear cell accumulation that occurs at the pleura of mice after inhalation of MWCNTs (Ryman-Rasmussen *et al.*, 2009b). In addition to soluble mediators, cell–cell interaction stimulated by CNTs could play a role in the pathogenesis of inflammatory or fibrogenic reactions. For example, understanding how CNTs affect the physical interaction between epithelial cells and underlying mesenchymal cells (termed the epithelial–mesenchymal cell trophic unit) could yield important insight into airway remodeling events. Likewise, since inhaled CNTs migrate to the subpleural tissues and remain there for months (Ryman-Rasmussen *et al.*, 2009b), it will be important to gain an understanding of how CNTs influence the interaction between mesothelial cells and subpleural mesenchymal cells.

7.5 Carbon nanotubes and pre-existing lung inflammation

Individuals with pre-existing respiratory diseases such as asthma, bronchitis, and chronic obstructive pulmonary disease (COPD) represent the populations most susceptible to nanoparticle exposure (Bonner, 2010a). Because CNTs are considered to be used for drug delivery (Prato *et al.*, 2008), careful consideration should be given to the possibility that they could potentially exacerbate pre-existing disease rather than have any beneficial effect. Recent evidence by Ryman-Rasmussen *et al.* (2009b) shows that MWCNTs cause little adverse pulmonary effect when delivered to the lungs of mice by inhalation except when the mice were first challenged with ovalbumin allergen. In this study, airway fibrosis was prominently increased by the combination of ovalbumin allergen and MWCNTs, whereas ovalbumin or MWCNTs alone caused only marginal increases in airway fibrosis. The implication of this study was that CNTs pose a hazard to individuals with allergic lung inflammatory diseases such as asthma. Inoue and colleagues (2009) reported that MWCNTs administered intra-tracheally significantly increased ovalbumin-induced T-lymphocyte proliferation and amplified lung Th2 cytokines and chemokines compared to ovalbumin exposure alone. More recent work by these investigators (Inoue *et al.*, 2010) also suggests that SWCNTs can exacerbate murine allergic airway inflammation via enhanced activation of Th immunity and increased oxidative stress, and that exacerbation may be partly through the inappropriate activation of antigen-presenting cells, including dendritic cells. MWCNTs delivered to the lungs of mice may induce allergic responses through B cell activation and production of IgE in the absence of any allergen pre-exposure (Park *et al.*, 2009). In addition to increased numbers of B cells, granuloma formation in the lung tissue and IgE production were also observed by MWCNTs in a dose-dependent manner. Nygaard *et al.* (2009) showed that SWCNTs as well as MWCNTs promote ovalbumin-induced allergic immune responses in the lungs of mice. Collectively, these mouse studies suggest that individuals with allergic asthma may be more susceptible to immune responses and airway remodeling caused by CNT exposure. Furthermore, CNTs alone (in the absence of a sensitizing agent) might serve as allergens upon repeated exposure. However, it is unknown whether carbon nanotubes will cause or exacerbate asthma in humans. Nevertheless, the data from studies with mice from several different laboratories all yielded similar results which indicate that CNTs promote allergen-induced airway inflammation. Therefore, CNTs are probably poor candidates for delivering drugs aimed at the treatment of asthma. However, other nanomaterials, particularly fullerenes, inhibit allergic airway inflammation in mice (Ryan *et al.*, 2007) and should be explored further in drug delivery for the treatment of asthma.

Bacterial lipopolysaccharide (LPS), a component of gram-negative bacteria cell walls, is a potent pro-inflammatory agent. LPS is ubiquitous in the environment and has been implicated in a number of occupational and environmental lung diseases in humans, including bronchitis, COPD, and asthma. SWCNTs or MWCNTs moderately exacerbated lung inflammation, pulmonary vascular permeability, and lung expression of pro-inflammatory cytokines induced by LPS (Inoue *et al.*, 2008). LPS pre-exposure also exacerbates the fibrogenic potential of CNTs when delivered to the lungs of rats (Cesta *et al.*, 2010). While LPS alone did not cause fibrosis when delivered by intranasal aspiration, it enhanced MWCNT-induced fibrosis and synergistically increased the production of macrophage and epithelial-derived platelet-derived PDGF. LPS also strongly induced PDGF alpha receptors (PDGF-Rα) on lung fibroblasts, and pre-exposure to LPS enhanced the proliferative and migratory responses of cells to PDGF that is up-regulated by CNTs. While not shown in these studies, it is also conceivable that LPS could adsorb to the surface of CNTs and be targeted to the distal airways and alveolar region of the lung upon inhalation exposure.

7.6 Carbon nanotubes and pleural disease

As mentioned earlier in this chapter, CNTs are cleared from the lungs via macrophage-mediated mechanisms; in one route macrophages remove particles from the lungs across the pleural lining via the lymphatic drainage. Therefore, CNTs have the opportunity to interact with the mesothelial lining that makes up the pleura. The durable nature of CNTs along with their fiber-like shape could pose a problem of long-term lung persistence, providing the possibility of asbestos-like behavior that could include carcinogencity (i.e. mesothelioma). Injection of long MWCNTs into the peritoneal cavity of mice (a surrogate for the pleura) has induced inflammation and granulmoma formation, suggesting that MWCNTs have asbestos-like pathogenicity (Poland *et al.*, 2008). This study was important because long MWCNTs, but not short MWCNTs, were found to cause granuloma formation at the pleura similar to long asbestos fibers. A more recent study wherein CNTs were directly instilled into the pleural cavity of mice also showed acute inflammation by long CNTs that was caused by length-dependent retention at the parietal pleura (Murphy *et al.*, 2011). A study using p53 heterzygous null mice, which are susceptible to the development of mesothelioma, showed that intraperitoneal injection of a high dose of MWCNTs increased the incidence of mesothelioma in the abdominal cavity (Takagi *et al.*, 2008). The delivery of MWCNTs to the lungs of mice by nose-only inhalation exposure (Ryman-Rasmussen *et al.*, 2009a) or by oropharyngeal aspiration (Porter *et al.*, 2010) demonstrated migration of CNTs to the pleura and throughout the lungs. The highest dose used in experiments by Ryman-Rasmussen *et al.* produced mononuclear cell accumulation at the pleural surface within 1 day after exposure and subpleural fibrotic lesions within 2 weeks, which resolved by 6 and 14 weeks, respectively. No evidence of mesothelioma was observed in the C57BL6 mice used in that study. The issue of whether CNTs are capable of causing mesothelioma remains a key topic of research that will have important implications for the responsible use of CNTs. So far it is unknown whether CNTs represent a new cancer risk factor for humans. However, a recent report by NIOSH set a recommended exposure limit for MWCNTs that was based on rodent studies (NIOSH, 2010). While carbon nanotubes cause lung-fibrotic reactions in the interstitium and pleura of mice, their carcinogenic potential has not been adequately addressed. Longer term, low-dose studies with CNTs will need to be undertaken to adequately address the potential carcinogenicity of CNTs. Elucidating the carcinogenic potential of CNTs at the pleural lining in rodents using relevant inhalation exposures with appropriate positive controls (e.g. asbestos fibers) will have important implications for the future use and development of CNTs for a variety of applications.

7.7 Genotoxicity caused by carbon nanotubes

CNTs cause genotoxic effects in lung cell types and in rodents *in vivo*. SWCNTs caused fragmented centrosomes, multiple mitotic spindle poles, anaphase bridges, and aneuploid chromosome number in cultured primary or immortalized human airway epithelial cell types (Sargent *et al.*, 2009). This work was the first to show disruption of the mitotic spindle by SWCNTs, and the authors noted that the similar size and geometry of SWCNT bundles might account for physical interaction of SWCNTs with the microtubules that form the mitotic spindle. MWCNTs cause DNA damage and apoptosis in mouse embryonic stem cells and activate the tumor suppressor protein p53, a marker of DNA damage (Zhu *et al.*, 2007). Finally, MWCNTs caused dose-dependent genotoxic effects of structural chromosome aberrations, DNA damage, and micronuclei formulation in mice when administered by intraperitoneal injection (Patlolla *et al.*, 2010). Collectively, these studies indicate that SWCNTs and MWCNTs have genotoxic effects in cultured cells and can reach sensitive sites *in vivo* to cause genotoxic effects.

7.8 Effects of carbon nanotubes on other organ systems

CNTs delivered to lungs by inhalation or aspiration can have adverse affects that extend beyond the boundries of the lung to impact distant organ systems, including the spleen and heart. This topic is mentioned only briefly in this chapter as it will be covered in more detail in Chapter 11. For instance, inhaled MWCNTs cause systemic immunosuppression and splenic oxidative stress (Mitchell *et al.*, 2007). The mechanism of splenic immunosuppression was further elucidated and involves the release of TGF-β1 from the lungs of MWCNT-exposed mice, which enters the bloodstream to signal COX-2-mediated increases in prostaglandin-E2 and IL-10 in the spleen, both of which play a role in suppressing T cell proliferation (Mitchell *et al.*, 2009). Either SWCNTs or MWCNTs deposited into the lung induced acute lung and systemic effects, suggesting that a prolonged systemic inflammatory response could trigger or exacerbate cardiovascular dysfunction and disease (Erdely *et al.*, 2009). Collectively, these studies suggest that inhaled CNTs could have significant effects on systemic targets.

7.9 Conclusions

Carbon nanotubes represent novel structural platforms for engineering, electronics, and drug delivery. However, the potential risks of inhaled CNTs will have to be more carefully considered during the manufacturing process and in the design of products that contain CNTs. Because of the infancy of the field of nanotechnology, there are few epidemiologic data to assess health hazards for the majority of nanomaterials. Some CNTs contain nanosized metals (e.g. nickel, cobalt, iron) that are present as residual catalysts and should clearly represent risk factors for lung diseases, as many of these metals in their native form are known to have fibrogenic or carcinogenic effects in humans. Moreover, the unique shape and dimensions of CNTs indicate that they will behave like similarly shaped toxic fibers and have the potential to cause fibrosis and possibly cancer. Also, the nanoscale properties of CNTs make it difficult to predict how these structures will interact with intracellular structures such as DNA, cell membranes, and cytoskeletal proteins.

References

Baron, P. A., Deye, G. J., Chen, B. T., *et al.* (2008). Aerosolization of single-walled carbon nanotubes for an inhalation study. *Inhal Toxicol* **20**(8), 751–760.

Bonner, J. C. (2010a). Nanoparticles as a potential cause of pleural and interstitial lung disease. *Proc Am Thorac Soc* **7**(2), 138–141.

Bonner, J. C. (2010b). Mesenchymal cell survival in airway and interstitial pulmonary fibrosis. *Fibrogenesis Tissue Repair* **3**, 15.

Bonner, J. C. (2004). Regulation of PDGF and its receptors in fibrotic diseases. *Cytokine Growth Factor Rev* **15**, 255–273.

Bonner, J. C. (2008). Respiratory toxicology. In Smart, R. C. and Hodgson, E., eds., *Molecular and Biochemical Toxicology*, 4th edn (New York: John Wiley & Sons), pp.639–670.

Card, J. W., Zeldin, D. C., Bonner, J. C., and Nestmann, E. R. (2008). Pulmonary applications and toxicity of engineered nanoparticles. *Am J Physiol Lung Cell Mol Physiol* **295**(3), L400–L411.

Cesta, M. F., Ryman-Rasmussen, J. P., Wallace, D. G., *et al.* (2010). Bacterial lipopolysaccharide enhances PDGF signaling and pulmonary fibrosis in rats exposed to carbon nanotubes. *Am J Respir Cell Mol Biol* **43**(2), 142–151.

Erdely, A., Hulderman, T., Salmen, R., *et al.* (2009). Cross-talk between lung and systemic circulation during carbon nanotube respiratory exposure: Potential biomarkers. *Nano Lett* **9**(1), 36–43.

Hirano, S., Fujitani, Y., Furuyama, A., and Kanno, S. (2010). Uptake and cytotoxic effects of multi-walled carbon nanotubes in human bronchial epithelial cells. *Toxicol Appl Pharmacol* **249**(1), 8–15.

Inoue, K., Koike, E., Yanagisawa, R., *et al.* (2009). Effects of multi-walled carbon nanotubes on a murine allergic airway inflammation model. *Toxicol Appl Pharmacol* **237**(3), 306–316.

Inoue, K., Takano, H., Koike, E., *et al.* (2008). Effects of pulmonary exposure to carbon nanotubes on lung and systemic inflammation with coagulatory disturbance induced by lipopolysaccharide in mice. *Exp Biol Med* **233**(12), 1583–1590.

Inoue, K., Yanagisawa, R., Koike, E., Nishikawa, M., and Takano, H. (2010). Repeated pulmonary exposure to single-walled carbon nanotubes exacerbates allergic inflammation of the airway: Possible role of oxidative stress. *Free Radic Biol Med* **48**(7), 924–934.

Kelleher, P. Pacheco, K., and Newman, L. S. (2000). Inorganic dust pneumonias: The metal-related parenchymal disorders. *Environ Health Perspect* **108**(suppl. 4), 685–696.

Konduru, N. V., Tyurina, Y. Y., Feng, W., *et al.* (2009). Phosphatidylserine targets single-walled carbon nanotubes to professional phagocytes in vitro and in vivo. *PLoS One*, **4**(2), e4398.

Lam, C. W., James, J. T., McCluskey, R., and Hunter, R. L. (2004). Pulmonary toxicity of single-wall carbon nanotubes in mice 7 and 90 days after intratracheal instillation. *Toxicol Sci* **77**, 126–134.

Li, J. G., Li, W. X., Xu, J. Y., *et al.* (2007). Comparative study of pathological lesions induced by multiwalled carbon nanotubes in lungs of mice by intratracheal instillation and inhalation. *Environ Toxicol* **22**, 415–421.

Ma-Hock, L., Trenmann, S., Strauss, V., *et al.* (2009) Inhalation toxicity of multiwall carbon nanotubes in rats exposed for 3 months. *Toxicol Sci* **112**, 468–481.

Mangum, J. B., Turpin, E. A., Antao-Menezes, A., *et al.* (2006). Single-walled carbon nanotube (SWCNT)-induced interstitial fibrosis in the lungs of rats is associated with increased levels of PDGF mRNA and the formation of unique intercellular carbon structures that bridge alveolar macrophages in situ. *Part Fibre Toxicol* **3**, 15.

Mercer, R. R., Scabilloni, J., Wang, L., *et al.* (2008). Alteration of deposition pattern and pulmonary response as a result of improved dispersion of aspirated single-walled carbon nanotubes in a mouse model. *Am J Physiol Lung Cell Mol Physiol* **294**, L87–L97.

Mitchell, L. A., Gao, J., Wal, R. V., *et al.* (2007). Pulmonary and systemic immune response to inhaled multiwalled carbon nanotubes. *Toxicol Sci* **100**, 203–214.

Mitchell, L. A., Lauer, F. T., Burchiel, S. W., and McDonald, J. D. (2009). Mechanisms for how inhaled multiwalled carbon nanotubes suppress systemic immune function in mice. *Nature Nanotech* **4**(7), 451–456.

Muller, J., Huaux, F., Moreau, N., *et al.* (2005). Respiratory toxicity of multi-wall carbon nanotubes. *Toxicol Appl Pharmacol* **207**, 221–231.

Murphy, F. A, Poland, C. A., Duffin, R., *et al.* (2011). Length-dependent retention of carbon nanotubes in the pleural space of mice initiates sustained inflammation and progressive fibrosis on the parietal pleura.*Am J Pathol* **178**(6), 2587–2600.

NIOSH (2010). *Occupational Exposure to Carbon Nanotubes and Nanofibers. Current Intelligence Bulletin* (Washington DC: National Institute for Occupational Safety and Health).

Nygaard, U. C., Hansen, J. S., Samuelsen, M., *et al.* (2009). Single-walled and multi-walled carbon nanotubes promote allergic immune responses in mice. *Toxicol Sci* **109**(1), 113–123.

Park, E. J., Cho, W. S., Jeong, J., *et al.* (2009). Pro-inflammatory and potential allergic responses resulting from B cell activation in mice treated with multi-walled carbon nanotubes by intratracheal instillation. *Toxicology* **259**(3), 113–121.

Patlolla, A. K., Hussain, S. M., Schlager, J. J., Patlolla, S., and Tchounwou, P. B. (2010). Comparative study of the clastogenicity of functionalized and nonfunctionalized multi-walled carbon nanotubes in bone marrow cells of Swiss-Webster mice. *Environ Toxicol* **25**(6), 608–621.

Pauluhn, J. (2010). Subchronic 13-week inhalation exposure of rats to multiwalled carbon nanotubes: Toxic effects are determined by density of agglomerate structures, not fibrillar structures. *Toxicol Sci* **113**(1): 226–242.

Poland, C. A., Duffin, R., Kinloch, I., *et al.* (2008). Carbon nanotubes introduced into the abdominal cavity of mice show asbestos-like pathogenicity in a pilot study. *Nat Nanotechnol* **3**(7), 423–428.

Porter, D. W., Hubbs, A. F., Mercer, R. R., *et al.* (2010). Mouse pulmonary dose- and time course-responses induced by exposure to multi-walled carbon nanotubes. *Toxicology* **269**, 136–147.

Prato, M., Kostarelos, K., and Bianco, A. (2008). Functionalized carbon nanotubes in drug design and discovery. *Acc Chem Res* **41**(1), 60–68.

Ryan, J. J., Bateman, H. R., Stover, A., *et al.* (2007). Fullerene nanomaterials inhibit the allergic response. *J Immunol* **179**, 665–672.

Ryman-Rasmussen, J. P., Cesta, M. F., Brody, A. R., *et al.* (2009a). Inhaled multi-walled carbon nanotubes reach the subpleural tissue in mice. *Nat Nanotechnol* **4**(11), 747–751.

Ryman-Rasmussen, J. P., Tewksbury, E. W., Moss, O. R., *et al.* (2009b). Inhaled multiwalled carbon nanotubes potentiate airway fibrosis in a murine model of allergic asthma. *Am J Resp Cell Mol Biol* **40**(3), 349–358.

Sargent, L. M., Shvedova, A. A., Hubbs, A. F., *et al.* (2009). Induction of aneuploidy by single-walled carbon nanotubes. *Environ Mol Mutagen* **50**(8), 708–717.

Shvedova, A. A., Kisin, E R., Mercer, R., *et al.* (2005). Unusual inflammatory and fibrogenic pulmonary responses to single-walled carbon nanotubes in mice. *Am J Physiol Lung Cell Mol Physiol* **289**, L698–L708.

Shvedova, A. A., Kisin, E., Murray, A. R., *et al.* (2008). Inhalation vs. aspiration of single-walled carbon nanotubes in C57BL/6 mice: Inflammation, fibrosis, oxidative stress, and mutagenesis. *Am J Physiol Lung Cell Mol Physiol* **295**(4), L552–L565.

Takagi, A., Hirose, A., Nishimura, T., *et al.* (2008). Induction of mesothelioma in p53[+/−] mouse by intraperitoneal application of multi-wall carbon nanotube. *J Toxicol Sci* **33**, 105–116.

Visser, C. E., Tekstra, J., Brouwer-Steenbergen, J. J., *et al.* (1998). Chemokines produced by mesothelial cells: huGRO-alpha, IP-10, MCP-1 and RANTES. *Clin Exp Immunol* **112**(2), 270–275.

Warheit, D. B., Laurence, B. R., Reed, K. L., *et al.* (2004). Comparative pulmonary toxicity assessment of single-wall carbon nanotubes in rats. *Toxicol Sci* **77**, 117–125.

Zhu, L., Chang, D. W., Dai, L., and Hong, Y. (2007). DNA damage induced by multi-walled carbon nanotubes in mouse embryonic stem cells. *Nano Lett* **7**(12), 3592–3597.

8

Responses to pulmonary exposure to carbon nanotubes

VINCENT CASTRANOVA, ROBERT R. MERCER

Disclaimer: The findings and conclusions in this report are those of the authors and do not necessarily represent the views of the National Institute for Occupational Safety and Health.

8.1 Introduction

Nanotechnology is the manipulation of matter on a near-atomic scale to produce nanoparticles with unique physicochemical properties which can be incorporated into new structures, materials, and devices with a wide range of commercial applications. Carbon nanotubes (CNT) discovered in 1991 by Iijima (1), are carbon atoms arranged in a crystalline graphene lattice with a tubular morphology. CNT can be produced as a single tubular structure to form a single-walled carbon nanotube (SWCNT), as a tube within a tube forming a double-walled carbon nanotube (DWCNT), or as multiple tubes within a tube forming a multi-walled carbon nanotube (MWCNT). SWCNT have a diameter of 1–2 nm, while MWCNT can be synthesized with diameters ranging from 10 to 100 nm depending upon the number of encapsulated tubes forming the CNT structure. CNT can range in length from 0.5 to over 20 μm. Therefore, CNT exhibit high aspect ratios and can be classified as man-made fibrous materials.

CNT are resistant to acid or heat treatment, exhibit high tensile strength, possess unique electrical properties, and can be easily functionalized. Therefore, applications as structural materials, in electronics, as heating elements, in the production of conductive fabric, for bone grafting and dental implants, in drug delivery systems, and as non-corrosive coatings on metals are being developed. With the projected increase in the synthesis and commercialization of CNT, human and environmental exposure during production, distribution, use and disposal is anticipated. Maynard et al. (2) have reported that vortexing SWCNT results in aerosolization of peak respirable dust levels as high as 53 μg/m^3. Han et al. (3) reported peak aerosolization levels of 210–430 μg/m^3 of total dust in a MWCNT synthesis laboratory during

The Toxicology of Carbon Nanotubes, ed. Ken Donaldson, Craig A. Poland, Rodger Duffin and James Bonner.
Published by Cambridge University Press. © Cambridge University Press 2012.

weighing, blending, mixing, and spraying procedures. Peak total particulate levels as high as 320 $\mu g/m^3$ have been reported in MWCNT facilities associated with oven opening, spraying, and sonication (4). Lastly, laboratory procedures, such as weighing or sonication of MWCNT, have been reported to aerosolize 123×10^3 or 42×10^3 particles/l, respectively (5). These studies indicate that worker inhalation of CNT is possible. Therefore, it is critical to determine the bioactivity of CNT and characterize the dose and time dependence of possible adverse health effects of exposure to inform risk assessment and development of prevention strategies.

8.2 Pulmonary response to CNT

8.2.1 SWCNT

The pulmonary effects of exposure of rodents to SWCNT were first reported in 2004. Warheit *et al.* (6) exposed rats by intratracheal instillation to a raw form of SWCNT (0.25–1.25 mg/rat) with the CNT sample containing 30–40% amorphous carbon, 5% nickel, and 5% cobalt. Although suspended in phosphate-buffered saline (PBS) containing 1% Tween 80®, the SWCNT were highly agglomerated. Pulmonary exposure to SWCNT resulted in a rapid but transient inflammatory and injury response as evidenced by increased bronchoalveolar lavage (BAL) levels of neutrophils, lactate dehydrogenase, and protein. Granulomas, mainly in the terminal bronchioles, were reported 1 week post-exposure and persisted through 3 months post-exposure. A 15% mortality within 1 day post-exposure was reported and was due to physical blockage of conducting airways by large SWCNT agglomerates. Lam *et al.* (7) also reported rapid and persistent granulomas following intratracheal instillation of mice to 0.1–0.5 of SWCNT/mouse. They reported no mortality due to SWCNT exposure. Mangum *et al.* (8) exposed rats by pharyngeal aspiration to purified SWCNT (0.5 mg/rat; 2.6% Co and 1.7% Mo) suspended in 1% Pluronic®. They reported no inflammatory responses. However, cell proliferation and platelet-derived growth factor (PDGF) were significantly increased 1 day post-exposure and significant interstitial fibrosis was noted at 21 days post-exposure. They also noted the formation of CNT structures bridging alveolar macrophages. Shvedova *et al.* (9) exposed mice by pharyngeal aspiration to purified SWCNT (10–40 μg/mouse). The suspended SWCNT preparation contained micrometer-sized agglomerates as well as smaller, nanorope structures. They reported rapid and transient inflammation and damage. They also reported granulomatous lesions and interstitial fibrosis within 7 days post-exposure which lasted through the 59-day course of the study. Initiation of the fibrotic response was associated with a peak in transforming growth factor beta (TGF-β) levels in BAL fluid at 7 days post-exposure. Granulomas were associated with the deposition of agglomerates in the terminal bronchioles and proximal alveoli, while interstitial fibrosis was associated with deposition of more dispersed SWCNT structures in the distal alveoli. At equivalent mass lung burdens, nano carbon black failed to cause any significant pulmonary responses. Therefore, persistent granulomas and interstitial fibrosis were viewed as CNT-specific responses. Shvedova *et al.* (10) reported the pulmonary response of mice to inhalation of SWCNT (5 mg/m^3, 5hr/day, 4 days). Qualitatively, short-term inhalation of mice produced pulmonary responses similar to bolus exposure by pharyngeal aspiration, i.e. transient inflammation and damage but persistent granulomas and interstitial fibrosis. The development and progression of fibrosis in response to pulmonary exposure to SWCNT appears to involve production of reactive oxygen species, since SWCNT-induced fibrosis is enhanced in mice on a vitamin E–deficient diet and is decreased in NADPH oxidase knockout mice (11, 12).

8.2.2 MWCNT

Muller *et al.* (13) exposed rats by intratracheal instillation to MWCNT (0.5–5 mg/rat) suspended in 1% Tween 80®. They reported inflammation, granulomas, and fibrosis with the unground MWCNT (6 μm length) being more potent than ground MWCNT (0.7 μm length). At 60 days post-exposure, 81% of the unground MWCNT were uncleared compared to 36% for the short MWCNT. Exposure of rats by intratracheal instillation of purified MWCNT (0.25–1.75 mg/rat) suspended in 1% Tween 80® resulted in a rapid but transient inflammatory response and persistent alveolar wall thickening (14). Ma-Hock *et al.* (15) reported pulmonary responses of rats to inhalation of MWCNT (0.1–2.5 mg/m^3, 6 hours/day, 5 days/week, 13 weeks; resultant burden 47–1170 μg/rat). The aerosolized MWCNT were well dispersed. Pulmonary responses included in lung weight, neutrophilic inflammation, and granulomatous inflammation. No fibrosis was reported. However, the authors did not use a specific collagen strain for the histopathologic analysis, so fibrosis may have been underscored. Kobayashi *et al.* (16) exposed rats by intratracheal instillation to a well-dispersed suspension of MWCNT (10–250 μg/rat). They reported transient inflammation and damage and a granulomatous response. They found no fibrosis but failed to use a collagen stain for histopathology. In contrast, Porter *et al.* (17) reported a rapid and persistent fibrotic response in mice after aspiration of a well-dispersed suspension of purified MWCNT (10–80 μg/mouse), using Sirius red staining for collagen. They also reported transient inflammation and damage with persistent granulomas at sites of agglomerate deposition. Likewise, Aiso *et al.* (18) reported transient inflammation and damage and persistent granulomas and alveolar wall fibrosis in rats after intratracheal instillation of MWCNT (40–160 μg/rat).

8.2.3 Comparison of pulmonary responses to SWCNT and MWCNT

Although studies used different modes of exposure (aspiration, intratracheal instillation, or inhalation) and different species (mice or rats), there is striking coherence among these studies. In general, studies with both SWCNT and MWCNT report qualitatively similar pulmonary responses, described as follows:

- a rapid but transient inflammation and lung injury response
- a granulomatous response of rapid onset which is persistent at deposition sites of CNT agglomerates
- a rapid and persistent/progressive interstitial fibrotic response.

Such a similar set of pulmonary responses to CNT exposure has allowed risk assessment across available animal studies using granulomatous inflammation or interstitial fibrosis as the pulmonary endpoint of health significance (19). A first step in this process was to determine the benchmark dose which resulted in a 10% risk of obtaining an adverse pulmonary response in a given rodent study. Benchmark lung burdens were then normalized to alveolar epithelial surface area (0.05 m^2/lung for mice, 0.4 m^2/lung for rats, and 102 m^2/lung for humans) to allow cross-species comparisons (20). This allowed estimation of the lung burden in workers which would result in a 10% risk of developing granulomas or interstitial fibrosis. Lastly, the National Institute for Occupational Safety and Health (NIOSH) in the United States is using such analyses to calculate workplace exposure levels which would result in the benchmark lung burden in a CNT worker over a working lifetime (8 hous/day, 5 days/week, 50 weeks/year, 45 years). Table 8.1 gives benchmark exposure limits calculated from three SWCNT and three MWCNT studies. NIOSH

Table 8.1 *Calculation of benchmark workplace levels for human exposure.*[a,b,c]

Study	Exposure	CNT	Benchmark exposure level (mg/m^3)
Lam *et al.*, 2004 (7)	Intratracheal (mice)	SWCNT	10.0
Shvedova *et al.*, 2005 (9)	Aspiration (mice)	SWCNT	1.80
Shvedova *et al.*, 2008a (10)	Inhalation (4 days) (mice)	SWCNT	0.11
Muller *et al.*, 2005 (13)	Intratracheal (rat)	MWCNT	18.0
Porter *et al.*, 2010 (17)	Aspiration (mice)	MWCNT	0.61
Ma-Hock *et al.*, 2009 (15)	Inhalation (13 weeks) (rat)	MWCNT	0.50

[a] Rodent benchmark lung burdens were calculated as the CNT burden which would result in a 10% risk of a significant granulomatous inflammatory or interstitial fibrotic response.
[b] Benchmark lung burdens were normalized across species as deposited CNT/alveolar epithelial surface area.
[c] Benchmark exposure level was calculated as lung burden = air level × ventilation × duration × deposition fraction, assuming worker ventilation = 9.6 m3/day; duration = 8 hours/day, 5 days/week, 50 weeks/year, 45 years; deposition fraction = 10%.

has used this information to propose a recommended exposure limit for CNT of 7 µg/m³, which is the current limit of detection for airborne CNT using elemental carbon as the detection method (21).

Although pulmonary responses to SWCNT and to MWCNT are qualitatively similar, resulting in transient inflammation but persistent granulomatous lesions and fibrosis, quantitative differences in pulmonary responses have been reported. In mice exposed to CNT by pharyngeal aspiration (10 µg/mouse), SWCNT caused a greater inflammatory response than MWCNT at 1 day post-exposure, as shown in Table 8.2 (9, 17, 22). Morphometric analyses indicate that well-dispersed SWCNT are not well recognized by alveolar macrophages (only 10% of the alveolar burden being within alveolar macrophages), while 90% of dispersed SWCNT structures rapidly cross alveolar epithelial cells and enter the interstitium (22). In contrast, 70% of MWCNT in the respiratory zone enter alveolar macrophages and 8% migrate into the alveolar septa (23, 24). This difference in pulmonary fate of SWCNT versus MWCNT is shown in Figure 8.1. As a result, well-dispersed SWCNT are more

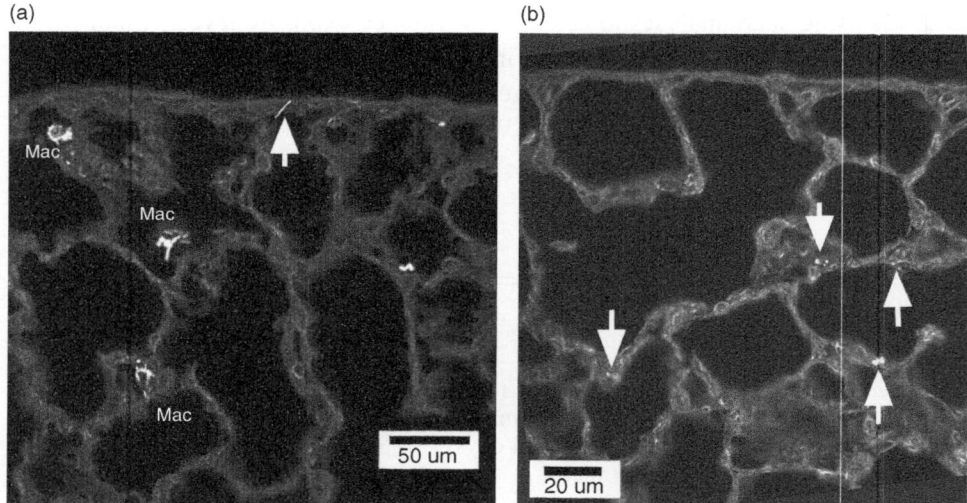

Figure 8.1 Enhanced darkfield image of CNT-exposed lungs. Panel (a) shows the general distribution of MWCNT in the lungs 7 days after aspiration (40 µg dose) with the section oriented with the pleural space running along the top. CNTs scatter light with high efficiency and thus produce the bright white structures in these enhanced darkfield images, while nuclei and other tissues produce a significantly duller image. Arrow points to an individual MWCNT in subpleural tissue. Alveolar macrophages (Mac) are foci for MWCNT. However, a few submicron MWCNT structures can be found in the alveolar interstitium throughout the section. Panel (b) gives a comparison image from a mouse lung exposed to a highly dispersed preparation of SWCNT (aspiration 10 µg dose, 7 days). In the case of dispersed SWCNT, the majority of CNT structures are rapidly incorporated into the alveolar interstitium (arrows).

potent in causing interstitial fibrosis on an equal mass lung burden basis than MWCNT (22, 24). This results in a lower calculated benchmark exposure level for SWCNT than for MWCNT (Table 8.1), using data from the Shvedova *et al.* (10) and Porter *et al.* (17) studies, respectively.

Another difference between pulmonary responses to SWCNT and to MWCNT is the ability to enter the intrapleural space. Both SWCNT and MWCNT have been reported in the subpleural tissue of the lung (22, 25). However, strong evidence for penetration of the visceral pleura and translocation to the intrapleural space (Figure 8.2) has been reported only for MWCNT (23).

Table 8.2 *Inflammatory potency of SWCNT versus MWCNT.*[a,b]

Exposure	PMN	References
SWCNT	$2.72 \pm 0.14 \times 10^5$	Shvedova *et al.*, 2005 (9)
SWCNT	$2.30 \pm 0.55 \times 10^5$	Mercer *et al.*, 2008 (22)
MWCNT	$1.33 \pm 0.46 \times 10^5$	Porter *et al.*, 2010 (17)

[a] Mice were exposed by pharyngeal aspiration to 10 μg of CNT.
[b] Bronchoalveolar lavage was performed 1 day post-exposure and the number of polymorphonuclear neutrophilic leukocytes (PMN) in the lavage fluid was determined as an indicator of pulmonary inflammation.

(a)   (b)

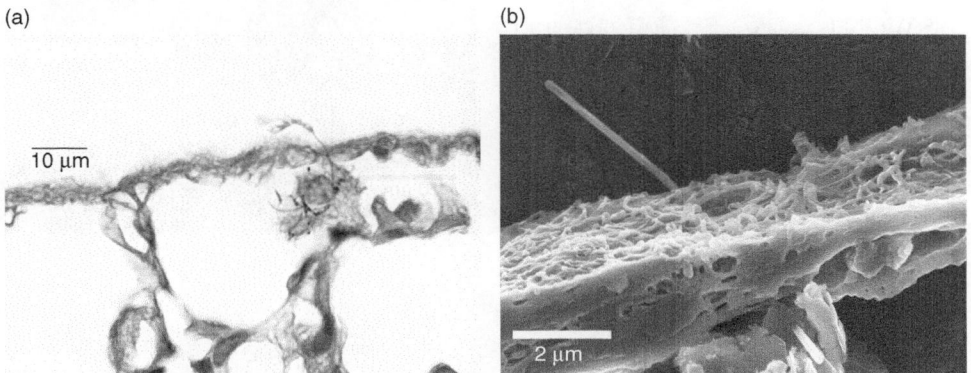

Figure 8.2 Penetration of the visceral pleura by MWCNT. Mice were exposed to well-dispersed MWCNT by pharyngeal aspiration. (a) A light micrograph of Sirius red-stained lung tissue 28 days after exposure to 40 μg MWCNT. Note a MWCNT leaving a macrophage and entering the intrapleural space. (b) Field-emission electron micrograph showing a MWCNT (50 nm × 5 μm) entering the intrapleural space (Figure 8.2(b) used with permission from Mercer *et al.* (23)).

8.2.4 Mechanism for CNT-induced pulmonary fibrosis

Both SWCNT and MWCNT have been reported to enter the alveolar septa, as in shown in Figure 8.3, and induce interstitial fibrosis (22, 17). Classic fibrogenic particles, such as crystalline silica and asbestos, cause persistent inflammation and lung injury, which results in parenchymal damage and scarring, i.e. fibrosis. Bronchoalveolar markers of inflammation and damage increase within 1–3 days after exposure to SWCNT or MWCNT, but decline toward control levels after 1 week post-exposure (9, 10, 17). Therefore, CNT-induced interstitial fibrosis persists and progresses over several months post-exposure in the absence of persistent pulmonary inflammation (9, 22, 17). Therefore, CNT are acting by a different mechanism than crystalline silica or asbestos to cause fibrosis. *In vitro* studies with lung fibroblasts indicate that SWCNT and MWCNT can directly enhance fibroblast proliferation and collagen production (26, 27, 28). Dispersed MWCNT also induce release of the fibrogenic factor TGF-β from lung epithelial cells in culture (28). Therefore, CNT appear to rapidly enter the alveolar interstitium (Figure 8.3) and lay down a substrate upon which fibroblast proliferate, leading to interstitial fibrosis. A similar scaffolding effect on osteoblast proliferation has led to the use of CNT in bone grafting and dental implants (29, 30).

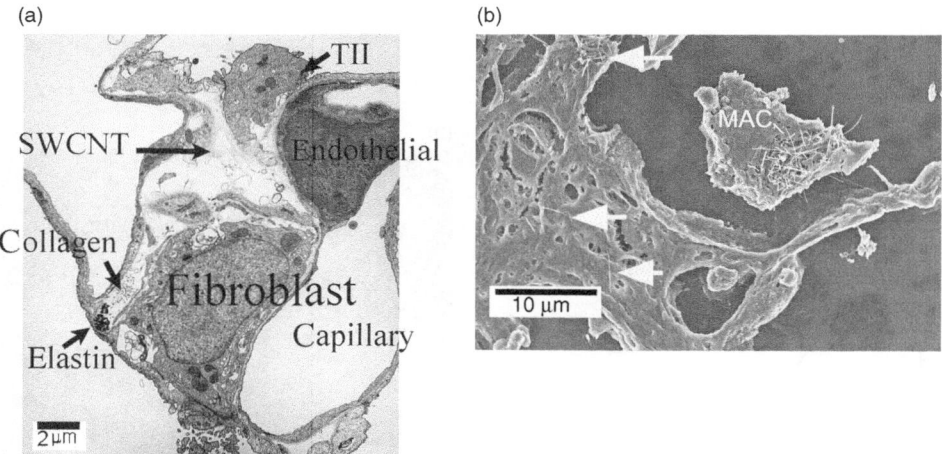

Figure 8.3 CNT within the alveolar septa. Mice were exposed to SWCNT or MWCNT by pharyngeal aspiration. (a) An electron micrograph of alveolar wall 3 days after aspiration of SWCNT, showing nanotubes in the alveolar interstitium. (b) A field-emission scanning electron micrograph of an alveolar wall 28 days after aspiration of MWCNT. Arrows indicate nanotubes in the alveolar interstitium. A MWCNT-loaded alveolar macrophage (MAC) is present in the neighboring airspace.

Table 8.3 *Inflammatory potential of raw versus purified SWCNT.*[a,b]

	PMN (fold increase)
Raw SWCNT (30% Fe)	51 ± 10
Purified SWCNT (0.3% Fe)	45 ± 1

[a] Mice were exposed to 10 µg of raw or purified SWCNT.
[b] At 1 day post-exposure, inflammatory response was determined by measuring polymorphonuclear neutrophilic leukocytes (PMN) in bronchoalveolar lavage fluid.

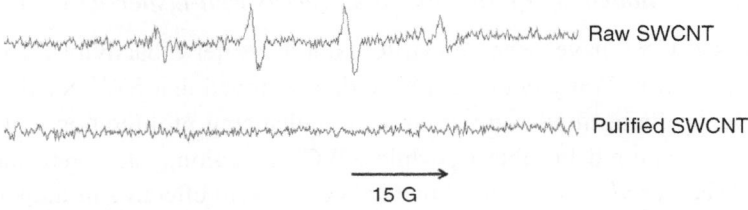

Figure 8.4 Acellular generation of hydroxyl radicals by CNT. In the presence of hydrogen peroxide, raw SWCNT generate hydroxyl radicals while purified SWCNT do not.

8.2.5 Comparison of raw versus purified CNT

As synthesized, raw CNT contain as much as 30% catalytic metals. Catalytic metals, such as iron, can generate hydroxyl radicals in the presence of hydrogen peroxide or cells (31). These catalytic metals can be removed by acid treatment or by high temperature to yield purified CNT with low metal content. As shown in Figure 8.4, removal of catalytic metals abolishes the ability of SWCNT (MWCNT not shown) to generate hydroxyl radicals *in vitro*. The *in vitro* cytotoxicity of CNT has been linked to oxidant injury (32). However, *in vivo* the pulmonary bioactivity of SWCNT does not appear to be affected by the presence or absence of catalytic metals. Lam *et al.* (7) compared the pulmonary response of mice to intratracheal instillation of raw (containing 25% metal catalyst) versus purified ($\approx$ 2% iron) SWCNT and found the granulomatous reaction was not dependent on metal contamination. Likewise, the acute inflammatory reaction of mice after aspiration of raw (30% iron) versus purified (1% iron) SWCNT was not affected by metal content, as shown in Table 8.3 (9, 10).

8.2.6 Effect of functionalization of CNT

CNT are highly hydrophobic. Functionalization of MWCNT with COOH or OH makes these CNT highly water-soluble. Han *et al.* (33) exposed mice by aspiration to carboxylic- and hydroxyl-functionalized MWCNT (20–40 μg/mouse). These functionalized MWCNT were well dispersed in PBS. Like other reports, transient inflammation, cytokine production, and lung injury were found. However, these authors did not compare the bioactivity of functionalized versus non-functionalized CNTs. Wang *et al.* (28) have reported that carboxylation of MWCNT partially decreased the ability of CNT to stimulate fibroblast proliferation *in vitro*. In addition, COOH-functionalized MWCNT have been reported to cause a significantly lower inflammatory response in mouse lungs at 1 and 7 days post-aspiration than unmodified MWCNT (34).

8.2.7 Bioactivity of agglomerated versus well-dispersed CNT

Well-dispersed CNT have been shown to exhibit greater bioactivity than agglomerated CNT *in vitro*. Wang *et al.* (26) have demonstrated that SWCNT dispersed in diluted alveolar lining fluid were effective in enhancing proliferation and collagen production by cultured fibroblasts, while SWCNT agglomerates were ineffective. Similarly, Wang *et al.* (28) reported that MWCNT were effective in inducing TGF-β production from bronchial epithelial cells and proliferation of fibroblasts *in vitro* only when well-dispersed using an artificial diluted alveolar lining fluid. Mercer *et al.* (22) gold-labeled poorly-dispersed and well-dispersed preparations of SWCNT prior to aspiration by mice (Figure 8.5). They demonstrated that poorly-dispersed SWCNT agglomerates deposited in the terminal bronchioles and proximal alveoli where they induce granulomatous lesions. In contrast, well-dispersed SWCNT structures deposited in the distal alveoli and rapidly migrated into the alveolar septa where they induce progressive interstitial fibrosis. In addition, transient inflammation and persistent interstitial fibrosis were four-fold greater on an equal mass burden basis for well-dispersed SWCNT than for poorly-dispersed SWCNT. Shvedova *et al.* (10) also reported an increased pulmonary response to a more dispersed preparation of SWCNT. Aerosolization of dry SWCNT resulted in smaller structures (count mode aerodynamic diameter (CMAD) ≈ 220 nm) than suspension of SWCNT for aspiration. Quantitatively, the dispersed SWCNT aerosol was four-fold more potent in causing inflammation and fibrosis after inhalation than aspiration of less-dispersed SWCNT (10). This increased pulmonary bioactivity of well-dispersed versus agglomerated SWCNT is summarized in Table 8.4. Pauluhn (35) exposed rats by inhalation (0.1–6 mg/m^3, 6 hours/day, 5 days/week, 13 weeks) to Baytube® MWCNT. Aerosolized structures were large, compact agglomerates with a mass median aerodynamic diameter (MMAD) ≈ 3 μm. Sub-chronic inhalation resulted in persistent inflammation, lung damage, granulomas, alveolar wall thickening, and a small increase in interstitial collagen staining. In general, the interstitial fibrotic response in the Pauluhn study was less than that reported after

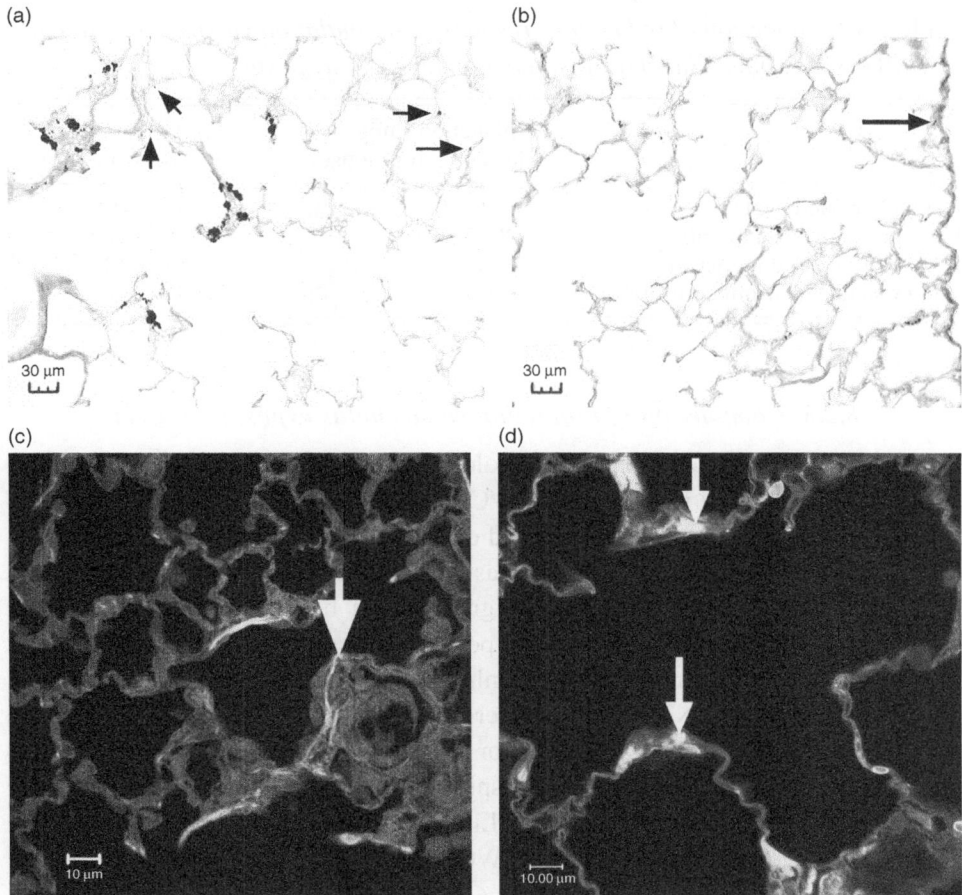

Figure 8.5 Pulmonary distribution and response after aspiration of gold-labeled SWCNT in mice. (a) Deposition of poorly-dispersed SWCNT agglomerates in the terminal bronchioles and proximal alveoli results in granulomatous lesions (Figure 8.5(c)). Arrows indicate deposition of small SWCNT structures in the distal alveolar walls. (b) Well-dispersed SWCNT within alveolar septa of the distal lung and resulting interstitial fibrosis (Figure 8.5(d)). (c), (d) Arrows in confocal micrographs indicate collagen stained by Lucifer yellow. Figures from Mercer *et al.* (22) with permission from the American Physiological Society.

aspiration of well-dispersed MWCNT preparation (17). It should be noted that individual MWCNT were observed within the alveolar septa in the Porter study (17), while no individual MWCNT were reported in the Pauluhn study (35). Because of the non-fibrous structure of these MWCNT agglomerates in the Pauluhn study, responses were attributed to volumetric overload (36).

Table 8.4 *Comparison of pulmonary responses to inhalation versus aspiration of raw SWCNT by mice. Superscripts indicate days post-exposure.*

	Inhalation (lung burden 5 μg/mouse)	Aspiration (10 μg/mouse)
PMN (fold increase)[1]	136 ± 20	51 ± 10
BAL protein (↑ from control)[1]	$68 \pm 3\%$	$35 \pm 3\%$
BAL TGF–β (fold increase)[7]	$7.9 \pm 7\%$	2.0 ± 0.1
Lung collagen (↑ from control)[7]	$127 \pm 7\%$	$53 \pm 1\%$

8.2.8 Comparison of inhalation versus bolus exposure to CNT

Shvedova *et al.* (10) compared the biological response resulting from a bolus aspiration and a 4-day inhalation of SWCNT. Both exposures resulted in qualitatively similar transient inflammation and damage and rapid but persistent fibrosis. However, on an equal mass burden basis, inhalation causes inflammatory and fibrotic responses that were four-fold greater than bolus exposure of mice by aspiration (Table 8.4). This difference in potency was most likely due to differences in the structure size distribution of the inhaled versus the aspirated SWCNT, with inhalation of dry SWCNT having smaller structures (CMAD ≈ 220 mm) than the fluid-suspended SWCNT sample used for aspiration. When mice aspirated a well-dispersed SWCNT preparation, bolus aspiration produced a quantitatively similar degree of fibrosis as inhalation (22, 10). Li *et al.* (37) also compared the pulmonary response of mice exposed to purified MWCNT by intratracheal instillation (50 μg/ mouse bolus dose) with inhalation (32.6 mg/m^3, 6 hours/day, 5–15 days). The MWCNT aerosolized from dry material were much less agglomerated than when suspended in 1% Tween 80®. Intratracheal instillation resulted in granulomas with some alveolar wall thickening, while inhalation resulted predominately in alveolar wall thickening and cell proliferation. Lastly, Wolfarth *et al.* (38) compared the degree of pulmonary inflammation and damage 1 day after aspiration of a well-dispersed MWCNT preparation to responses 24 hours after a 4-day inhalation exposure, which resulted in the same lung burden (Figure 8.6). Results indicate that the degree of pulmonary response to a bolus versus a short-term inhalation exposure was not significantly different.

8.3 Systemic responses to pulmonary exposure to CNT

Li *et al.* (39) reported that multiple aspirations of SWCNT (20 μg/mouse, every 2 weeks, for 2 months) in Apo E −/− mice caused a 71% increase in aortic plaques.

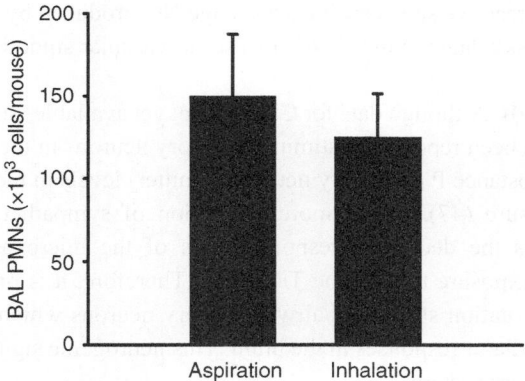

Figure 8.6 Acute inflammatory response to aspiration versus inhalation of MWCNT. Mice were exposed to a 10 μg lung burden of well-dispersed MWCNT by aspiration or by inhalation (10 mg/m³, 5 hours/day, 4 days). At 1 day post-exposure, inflammation was determined by counting the polymorphonuclear neutrophilic leukocytes (PMN) in bronchoalveolar lavage fluid. There was no significant difference in the response to aspiration versus short-term inhalation.

Inhalation of MWCNT (26 mg/m³ for 5 hours; lung burden of 22 μg) results in a 92% depression of the responsiveness of coronary arterioles to dilators 24 hours post-exposure (40). Furthermore, pharyngeal aspiration of MWCNT (80 μg/mouse) results in induction of mRNA for certain inflammatory mediators and markers of blood/brain barrier damage in the olfactory bulb, frontal cortex, midbrain, and hippocampus 24 hours post-exposure (41). Several possible mechanisms have been put forward to explain systemic responses to pulmonary exposure to CNT:

• **Translocation of CNT to systemic sites**: Translocation of intraperitoneally instilled MWCNT from the abdominal cavity to the lung has been reported (42). However, thus far there is no evidence that the systemic effects reported above are associated with translocation of CNT from the lung to the affected tissue. Indeed, aspirated gold-labeled SWCNT were not found in any systemic organ 2 weeks post-exposure (43).

• **Systemic inflammation**: Pulmonary exposure to particles causes localized inflammation at the sites of particle deposition in the alveoli. Erdely *et al*. (44) reported that aspiration of SWCNT or MWCNT (40 μg/mouse) induced a small but significant increase in blood neutrophils and mRNA expression and protein levels for certain inflammatory markers in the blood at 4 hours post-exposure, but not at later times. Such pulmonary CNT exposure also significantly elevated gene expression for mediators, such as Hif–3α and S100a, in the heart and aorta at 4 hours post-exposure. Evidence also exists that pulmonary exposure to particles alters systemic microvascular function by potentiating polymorpho-nuclear neutrophilic leukocytes (PMN) as they flow through pulmonary capillaries in close proximity to affected alveoli. These potentiated blood PMN adhere to microvessel walls and release reactive species which scavenge NO produced by endothelial cells (45, 46). Therefore, less dilator-induced NO diffuses to vascular smooth muscle, resulting in less dilation.

- **Neurogenic signals**: Although data for CNT are not yet available, pulmonary exposure to ultrafine TiO_2 has been reported to stimulate sensory neurons in the lung as indicated by an increase in Substance P (a sensory neurotransmitter) levels in the nodose ganglion 24 hours after exposure (47). Furthermore, inhibition of sympathetic input to systemic arterioles reverses the decreased responsiveness of the microvasculature to dilators after pulmonary exposure to ultrafine TiO_2 (48). Therefore, it is proposed that particle-induced airway irritation stimulates airway sensory neurons which send a signal to the brain, causing mediator responses in the brain. This neurogenic signal is then transmitted to the cardiovascular system.

8.4 Summary

In general, pulmonary exposure of rats or mice by intratracheal instillation, pharyngeal aspiration, or inhalation of SWCNT or MWCNT results in transient inflammation and lung damage. Granulomatous lesions and interstitial fibrosis, which are of rapid onset and persistent in nature, have also been a common report. The presence (raw) or absence (purified) of catalytic metals does not appear to greatly affect pulmonary response. The degree of agglomeration does affect deposition site and response. Large agglomerates tend to deposit at the terminal bronchioles and proximal alveoli and induce a granulomatous response, while more dispersed structures can deposit in the distal alveoli and cause interstitial fibrosis. The lung may be exposed to more dispersed CNT structures by inhalation than by exposure to a suspension of CNT. When this is the case, the response to short-term inhalation is often greater than that for the same lung burden given as a bolus dose. Pulmonary exposure to CNT has also been shown to cause changes in cardiovascular and central nervous system function. Mechanisms, such as inflammatory signals or neurogenic pathways, causing these systemic responses to pulmonary CNT exposure are under investigation. In light of the pulmonary and systemic responses reported in rodent studies, it appears prudent to minimize worker exposure to CNT. Normal industrial hygiene practices, i.e. containment, ventilation, and personal protective equipment, appear effective in controlling exposure (3, 49).

References

1. S. Iijima. Helical microtubules of graphite carbon. *Nature*, **354** (1991), 56–58.
2. A. D. Maynard, P. A. Baron, M. Foley, *et al.* Exposure to carbon nanotube material: Aerosol release during the handling of unrefined single walled carbon nanotube material. *J. Toxicol. Environ. Health A*, **67** (2004), 87–107.

3. J. H. Han, E. J. Lee, J. H. Lee, *et al.* Monitoring multiwalled carbon nanotube exposure in carbon nanotube research facility. *Inhal. Toxicol.*, **20** (2008), 741–749.

4. J. H. Lee, S.-B. Lee, G. N. Bae, *et al.* Exposure assessment of carbon nanotube manufacturing workplaces. *Inhal. Toxicol.*, **22** (2010), 369–381.

5. D. R. Johnson, M. M. Methner, A. J. Kennedy, and J. A. Steevens. Potential for occupational exposure to engineered carbon-based nanomaterials in environmental laboratory studies. *Environ. Health Perspect.*, **118** (2010), 49–54.

6. D. B. Warheit, B. R. Laurence, K. L. Reed, *et al.* Comparative pulmonary toxicity assessment of single-wall carbon nanotubes in rats. *Toxicol. Sci.*, **77** (2004), 117–125.

7. C. W. Lam, J. T. James, R. McCluskey, and R. L. Hunter. Pulmonary toxicity of single-wall carbon nanotubes in mice 7 and 90 days after intratracheal instillation. *Toxicol. Sci.*, **77** (2004), 125–134.

8. J. B. Mangum, E. A. Turpin, A. Antao-Menezes, *et al.* Single-walled carbon nanotube (SWCNT)-induced interstitial fibrosis in the lungs of rats is associated with increased levels of PDGF mRNA and the formation of unique intercellular carbon structures that bridge alveolar macrophages in situ. *Part. Fibre Toxicol.*, **3** (2006), 15.

9. A. A. Shvedova, E. R. Kisin, R. Mercer, *et al.* Unusual inflammatory and fibrogenic pulmonary responses to single walled carbon nanotubes in mice. *Am. J. Physiol. Lung Cell Mol. Physiol.*, **289** (2005), L698–L708.

10. A. A. Shvedova, E. Kisin, A. R. Murray, *et al.* Inhalation versus aspiration of single walled carbon nanotubes in C57BL/6 mice: Inflammation, fibrosis, oxidative stress and mutagenesis. *Am. J. Physiol. Lung Cell Mol. Physiol.*, **295** (2008a), L552–L565.

11. A. A. Shvedova, E. R. Kisin, A. R. Murray, *et al.* Vitamin E deficiency enhances pulmonary inflammatory response and oxidative stress induced by single-walled carbon nanotubes in C57BL/6 mice. *Toxicol. Appl. Pharmacol.*, **221** (2007), 339–348.

12. A. A. Shvedova, E. R. Kisin, A. R. Murray, *et al.* Increased accumulation of neutrophils and decreased fibrosis in the lungs of NADPH oxidase-deficient C57BL/6 mice exposed to carbon nanotubes. *Toxicol. Appl. Pharmacol.*, **231** (2008b), 235–240.

13. J. Muller, F. Huaus, N. Moreau, *et al.* Respiratory toxicity of multi-wall carbon nanotubes. *Toxicol. Appl. Pharmacol.*, **207** (2005), 221–231.

14. A. Liu, K. Sun, J. Yang, and D. Zhao. Toxicological effects of multi-wall carbon nanotubes in rats. *Nanopart. Res.*, **10** (2008), 1303–1307.

15. L. Ma-Hock, S. Trenmann, V. Strauss, *et al.* Inhalation toxicity of multiwall carbon nanotubes in rats exposed for 3 months. *Toxicol. Sci.*, **112** (2009), 468–481.

16. N. Kobayaski, M. Naya, M. Ema, *et al.* Biological response an morphological assessment of individually dispersed multi-walled carbon nanotubes in the lung after intratracheal instillation in rats. *Toxicol.*, **276** (2010), 143–153.

17. D. W. Porter, A. Hubbs, R. R. Mercer, *et al.* Mouse pulmonary dose- and time course-response induced by exposure to multi- walled carbon nanotubes. *Toxicol.*, **269** (2010), 136–147.

18. S. Aiso, K. Yamazaki, Y. Umeda, *et al.* Pulmonary toxicity of intratracheally instilled multiwall carbon nanotubes in male Fischer 344 rats. *Ind. Health*, **48** (2010), 783–795.

19. E. Kuempel and V. Castranova. Hazard and risk assessment of workplace exposure to engineered nanoparticles: Methods, issues, and carbon nanotube case study. In G. Ramachandran, ed., *Assessing Nanoparticle Risks to Human Health* (New York: Elsevier, in press).

20. K. Stone, R. R. Mercer, P. Gehr, B. Stockstill, and J. D. Crapo. Allometric relationships of cell numbers and size in the mammalian lung. *Am. J. Respir. Cell Mol. Biol.*, **6** (1992), 235–243.

21. NIOSH. *Occupational Exposure to Carbon Nanotubes and Nanofibers.* Current Intelligence Bulletin (Washington: National Institute for Occupational Safety and Health), 2010. http://www.cdc.gov/niosh/docket/review/docket161A/pdfs/carbon NanotubeCIB_PublicReviewOfDraft.pdf

22. R. R. Mercer, J. F. Scabilloni, L. Wang, *et al.* Alteration of deposition pattern and pulmonary response as a result of improved dispersion of aspirated single-walled carbon nanotubes in a mouse model. *Am. J. Physiol. Lung Cell Mol. Physiol.,* **294** (2008), L87–L97.

23. R. R. Mercer, A. F. Hubbs, J. F. Scabilloni, *et al.* Distribution and persistence of pleural penetrations by multi- walled carbon nanotubes. *Part. Fibre Toxicol.,* **7** (2010), 28.

24. R. R. Mercer, A. F. Hubbs, J. F. Scabilloni, *et al.* Pulmonary fibrotic response to sub-chronic multi-walled carbon nanotube exposure. *The Toxicologist,* **120** (2011), A56.

25. J. P. Ryman-Rasmussen, M. F. Cesta, A. R. Brody, *et al.* Inhaled carbon nanotubes reach the subpleural tissue in mice. *Nat. Nanotechnol.,* **4** (2009), 747–751.

26. L. Wang, V. Castranova, A. Mishra, *et al.* Dispersion of single-walled carbon nanotubes by a natural lung surfactant for pulmonary *in vitro* and *in vivo* toxicity studies. *Part. Fibre Toxicol.,* **7** (2010), 31.

27. A. Mishra, Y. Rojanasakul, V. Castranova, R. Mercer, and L. Wang. Assessment of fibrogenic biomarkers induced by multi wall carbon nanotubes. *The Toxicologist,* **120** (2011), A1183.

28. X. Wang, T. Xia, S. A. Ntim, *et al.* Quantitative techniques for assessing and controlling the dispersion and biological effects of multiwalled carbon nanotubes in mammalian tissue culture cells. *ACS Nano* (in press).

29. X. Li, H. Gao, M. Uo, *et al.* Maturation of osteoblast-like SaoS2 induced by carbon nanotubes. Biomed. Mater., **4** (2009), 015005; doi: 10.1088/1748–6041/4/1/015005.

30. E. M. Christenson, K. S. Anseth, J. J. J. P. von den Beucken, *et al.* Nanobiomaterial applications in orthopedics. *J. Orthop. Res.,* **25** (2007), 11–22.

31. V. E. Kagan, Y. Y. Tyurina, V. A. Tyurina, *et al.* Direct and indirect effects of single walled carbon nanotubes on RAW 264.7 macrophages: Role of iron. *Toxicol. Lett.,* **165** (2006), 88–100.

32. A. A. Shvedova, E. R. Kisin, A. R. Murray, *et al.* Exposure to carbon nanotube material: Assessment of the biological effects of nanotube materials using human keratinocytes. *J. Toxicol. Environ. Health A,* **66** (2003), 1901–1926.

33. S. G. Han, R. Andrews, and C. G. Gairola. Acute pulmonary response of mice to multi-walled carbon nanotubes. *Inhal. Toxicol.,* **22** (2010), 340–347.

34. T. Sager, M. Wolfarth, D. Porter, *et al.* Effects of surface modification on the bioavailability and inflammatory potential of multi-walled carbon nanotubes. *The Toxicologist,* **120** (2010), A1178.

35. J. Pauluhn. Subchronic 13-week inhalation exposure to multiwalled carbon nanotubes: Toxic effects are determined by density of agglomerate structures, not fibrillar structures. *Toxicol. Sci.,* **113** (2010), 226–242.

36. J. Pauluhn. Poorly soluble particulates searching for a unifying denominator of nanoparticles and fine particles for DNEL estimation. *Toxicol.,* **270** (2011), 176–188.

37. J.-G. Li, W.-Y. Li, J.-Y. Xu, *et al.* Comparative study of pathological lesions induced by multiwalled carbon nanotubes in lungs of mice by intratracheal instillation and inhalation. *Environ. Toxicol.,* **22** (2007), 415–421.

38. M. G. Wolfarth, W. McKinney, B. T. Chen, V. Castranova, and D. W. Porter. Acute pulmonary responses to MWCNT inhalation. *The Toxicologist,* **120** (2011), A53.

39. Z. Li, T. Hulderman, R. Salmen, *et al.* Cardiovascular effects of pulmonary exposure to single-wall carbon nanotubes. *Environ. Health Perspect.,* **115** (2007), 77–82.

40. P. A. Stapleton, V. Minarchick, A. Cumpston, *et al.* Time-course of improved coronary arteriolar endothelium-dependent dilation after multi-walled carbon nanotube inhalation. *The Toxicologist,* **120** (2011), A194.
41. K. Sriram, D. W. Porter, A. M. Jefferson, *et al.* Neuro inflammation and blood-brain barrier changes following exposure to engineered nanomaterials. *The Toxicologist,* **108** (2009), A2197.
42. G. Liang, L. Yin, J. Zhang, *et al.* Effects of subchronic exposure to multi-walled carbon nanotubes in mice. *J. Toxicol. Environ. Health A,* **73** (2010), 463–470.
43. R. R. Mercer, J. F. Scabilloni, L. Wang, L. A. Battelli, and V. Castranova. Use of labeled single walled carbon nanotubes to study translocation from the lungs. *The Toxicologist,* **108** (2009), A2192.
44. A. Erdely, T. Hulderman, R. Salmen, *et al.* Cross-talk between lung and systemic circulation during carbon nanotube respiratory exposure: Potential biomarkers. *Nano. Lett.,* **9** (2009), 36–43.
45. T. R. Nurkiewicz, D. W. Porter, M. Barger, *et al.* Systemic microvascular dysfunction and inflammation after pulmonary particulate matter exposure. *Environ. Health Perspect.,* **114** (2006), 412–419.
46. T. R. Nurkiewicz, D. W. Porter, A. F. Hubbs, *et al.* Pulmonary nanoparticle exposure disrupts systemic microvascular nitric oxide signaling. *Toxicol. Sci.,* **110** (2009), 191–203.
47. H. Kan, Z. X. Wu, S.-H. Young, *et al.* Nanoparticle inhalation enhances cardiac protein phosphorylation and neurotransmitter synthesis in the nodose ganglia of rats. *The Toxicologist,* **120** (2011), A1459.
48. T. L. Knuckles, D. G. Frazer, J. L. Cumpston, *et al.* Nanoparticle inhalation modulates arteriolar sympathetic constriction: Role of nitric oxide, prostanoids, and α-adrenergic receptors. *The Toxicologist,* **118** (2010), A1728.
49. S. Regasamy, W. King, B. Eimer, and R. Shaffer. Filtration performance of NIOSH-approved N95 and P100 filtering face mask respirators against 4–30 nanometer-size nanoparticles. *J. Occup. Environ. Hyg.,* **5** (2008), 556–564.

9

Genotoxicity of carbon nanotubes

ROEL P. F. SCHINS, CATRIN ALBRECHT, KIRSTEN GERLOFF,
DAMIEN VAN BERLO

9.1 Introduction

It is well accepted that damage to the genomic DNA can have major pathophysiological implications for cells, tissues and organisms. DNA damage plays an important role in carcinogenicity and genetic diseases, and genotoxicity testing forms an indispensable aspect of regulatory risk assessment. Currently, a large number of genotoxicity assays are available for the identification of potential mutagenic and carcinogenic hazards of chemicals as well as to evaluate the underlying molecular genotoxic mechanisms. This includes assays that screen for gene mutations, chromosomal mutations, aneugenic effects (aneuploidy), DNA strand breaks, oxidised DNA bases and bulky DNA adducts, as well as assays that measure DNA repair (McGregor et al., 1999).

With the rapidly growing demand for nanoparticles in various industries and consumer products, genotoxicity tests have also become increasingly used and debated in nanotoxicology research (Knaapen et al., 2004; Schins and Knaapen, 2007; Gonzalez et al., 2008; Landsiedel et al., 2009; Donaldson et al., 2010; Warheit and Donner, 2010). In this context, carbon nanotubes (CNT) represent a rather specific group of nanomaterials. Because of their unique physicochemical and electrical properties, CNT are being introduced – and already used – within various industrial, medical and consumer products. However, several of the properties of CNT show similarities with specific physicochemical characteristics of asbestos fibres, and to a lesser extent with crystalline silica particles, which represent two classified human carcinogens. First, like CNT, asbestos and quartz are highly insoluble materials. The biopersistence of both latter materials is considered a critical determinant of chronic health effects such as lung fibrosis and cancer (Oberdörster, 2002; Donaldson and Tran, 2004; Knaapen et al., 2004). Moreover, specific toxic effects of these materials have been shown to depend on the amount and speciation of transition metal contaminants (Kamp et al., 1995; Fenoglio et al.,

The Toxicology of Carbon Nanotubes, ed. Ken Donaldson, Craig A. Poland, Rodger Duffin and James Bonner.
Published by Cambridge University Press. © Cambridge University Press 2012.

2001). The role of metal impurities of CNT has been a major topic of recent investigation, as will be outlined in this chapter. Last but not least, the obvious morphological similarity with asbestos fibres has sparked research interest into the potential harmfulness of CNT in relation to their diameter, length, aspect ratio and rigidity (Kane and Hurt, 2008; Jaurand *et al.*, 2009). A possible carcinogenic potential of CNT has emerged from a number of recent toxicological investigations in experimental animal models (discussed in Chapters 3 and 6). These studies have also raised the specific concern that, in concordance with asbestos, CNT may pose a risk for mesothelioma (Poland *et al.*, 2008; Takagi *et al.*, 2008; Jaurand *et al.*, 2009).

This chapter provides an overview of major genotoxicity studies which have been performed with CNT and a discussion about the underlying mechanisms of action that are demonstrated in such investigations and/or postulated on the basis of earlier genotoxicity studies with other types of (nano)particles. Therefore, we first briefly discuss the principal mechanisms of DNA damage and how specific genotoxicity assays can detect the various types of damage, either directly or via the measurement of features that result from, and/or lead to, DNA damage and mutations. We then provide an overview of the most important genotoxicity studies with CNT and put these into perspective with observations from various genotoxicity studies performed with asbestos, silica and specific other particulate toxicants. On the basis of two major features of CNT, i.e. their small size and elongated shape, it is addressed whether CNT represent a cluster of nanomaterials of specific concern regarding the assessment of genotoxic hazard and risk to humans. This chapter concludes with a discussion on issues of dosimetry and exposure routes, potential flaws and limitations of specific genotoxicity assays for CNT testing, and future research needs.

9.2 DNA damage and genotoxicity

The spectrum of genotoxicity tests that is currently available allows for the detection of (1) agents that cause direct damage to the genomic DNA, and/or (2) agents that compromise genome integrity by affecting cellular constituents or pathways that control this integrity, such as mitosis and DNA repair (Gonzalez *et al.*, 2008). When DNA damage takes place, a number of scenarios are possible on the level of the cell, which depend on the type of damage, its extent and persistence. In most cases, the damage is rapidly removed by the extensive DNA repair machinery present within eukaryotic cells (Wood *et al.*, 2001). When these protective mechanisms are incapable of restoring the DNA in its original form, cell death may occur. DNA damage can lead to necrotic cell death as well as to the specific activation of caspase-dependent or caspase-independent pathways of apoptosis, which are mediated by p53 and AIF signalling, respectively (Borges *et al.*, 2008). When this cellular last stand fails, a DNA mutation can become fixed and is replicated with each subsequent cellular mitosis. DNA mutations affecting key loci (i.e. tumour promoter and tumour suppressor genes, involved in processes like DNA repair and proliferation) can give rise to the development of a malignant tumour. Mutations can occur within a single gene, a cluster of genes or a complete chromosome.

To date, several genotoxicity assays have been used in studies with CNT. These include the alkaline comet assay, also known as the single-cell gel electrophoresis (SCGE) assay; the micronucleus (MN) test; and the chromosomal aberration (CA) assay. Each of these assays can be used *in vitro* after treatment of primary cells or cell lines as well as *in vivo* in exposed animals. However, it is important to know that in many cases studies have employed such assays without meeting the currently recommended guideline criteria for these methods (Tice *et al.*, 2000; Kirsch-Volders *et al.*, 2003; Stone *et al.*, 2009).

The assay principle of the comet assay is based on the detection of damaged DNA fragments within individual cells that, when evaluated using fluorescence micro-scopy upon cell lysis and subsequent denaturation and electrophoresis of its DNA content, appear as 'comets'. The comet assay allows for DNA damage detection independent of the cell cycle stage. Such damage may have been the result of a direct DNA breaking action and/or incomplete or erroneous DNA repair induced by a testing compound. Modified versions of this assay have been developed to enable specific evaluation of, for instance, oxidative DNA adducts such as 8-hydroxy-deoxyguanosine (by the detection of formamidopyrimidine DNA glycosylase sensitive sites), DNA damage-repair effects or DNA double-strand breaks (i.e. neutral comet assay).

Irreparable DNA double strand breaks are typically detected by the chromosomal aberration test and the MN assay. However, these assays also allow for the specific identification of numerical chromosome changes (i.e. aneuploidy). The CA test is based on the microscopical evaluation of structural and numerical alterations of stained chromosomes in metaphase cells which can be enriched by successive treatment of cell cultures by a mitogen (e.g. PHA) and the metaphase arresting compound colcemid. The MN assay is based on the microscopical detection of a chromosome or chromosome fragment from a cell, which has failed to integrate into the nucleus of its daughter cell after division. In the so-called cytokinesis block micronucleus assay, the actin-inhibitor cytochalasin B is applied to the cell culture after treatment with the testing compound. This approach allows for the quantifica-tion of background MN levels as well as cell proliferation, by distinguishing mononuclear from binucleated cells, i.e. cells which, respectively, did not or did undergo division during the cell culture. Using fluorescent *in situ* hybridisation (FISH) with probes targeted to the centromere region, one can determine whether a specific micronucleus represents an acentric chromosome fragment (i.e. resulting from a clastogenic event) or holds an entire chromosome (i.e. aneugenic effect).

Although it does not represent a classic genotoxicity test, measurement of the phosphorylation of the genomic caretaker and tumour suppressor protein H2AX has also been applied to evaluate potential DNA double strand breakage effects by CNT. Phosphorylation of H2AX, resulting in the formation of gamma-H2AX foci around DNA double strand breaks, represents an early response to this type of DNA damage and is typically detected by immunofluorescence microscopy. Recent investigations with a panel of reference genotoxicants have confirmed that phosphorylation of H2AX is a valid biomarker for genotoxicity (Watters *et al.*, 2009).

Genotoxicity of CNT has also been addressed by the evaluation of gene mutations, either in bacterial systems (i.e. Salmonella Reverse Mutation Assay) or in mammalian cells and tissues (e.g. Hpgrt-gene, k-ras gene sequencing). The Salmonella Reverse Mutation Assay (also known as the Ames test) represents the most widely applied *in vitro* genotoxicity assay subject to well-defined international recommendations (i.e. OECD Test Guideline 471) and has been used for the testing of various nanoparticles, including CNT. However, the relevance of bacterial assays for genotoxicity testing of (nano)particles has been under increasing debate. In particular, the relevance of the bacterial cell wall is under discussion in relation to known and discussed pathways of nanoparticle uptake by mammalian cells (Jaurand *et al.*, 2009; Landsiedel *et al.*, 2009; Stone *et al.*, 2009). This has in fact also been discussed for many years for asbestos, which mostly tested negative in bacterial assays whereas in sensitive mammalian cell systems genotoxicity was evident (Park and Aust, 1998; Unfried *et al.*, 2002; Xu *et al.*, 2007; Jaurand *et al.*, 2009).

9.3 Genotoxicity investigations with carbon nanotubes

In this section, emphasis will be put on studies in which specific genotoxic effects of CNT have been identified. However, this by no means implies that the results of specific studies, obtained with a specific sample, can be quantitatively, or qualitatively, extrapolated to all types of CNT. As discussed in Chapter 1, CNT can be produced by various methods with numerous specific modifications, and therefore represent a heterogeneous group of nanomaterials in terms of their primary structure (i.e. single-walled, multi-walled), specific dimensions (i.e. length, diameter, aspect ratio, formation of bundles and tangles), surface properties (e.g. functionalised CNT), the presence of metals (used as catalyst) and of carbonaceous impurities (e.g. graphitic materials, fullerenes, amorphous carbon) (Maynard *et al.*, 2004; Aitken *et al.*, 2006; Han *et al.*, 2008; Yeganeh *et al.*, 2008; Sanchez *et al.*, 2009; Wang *et al.*, 2009; Sargent *et al.*, 2010). However, in this paragraph we have merely made a subdivision of available studies between the two major structural clases of CNT, i.e. single-walled carbon nanotubes (SWCNT) and multi-walled carbon nanotubes (MWCNT). Moreover, one should keep in mind that observed effects in specific studies may also depend on the study design and the selected experimental conditions. The potential impact of sample storage, handling and application methods (e.g. choice of suspending vehicle) on biological readouts is well known in the classical particle toxicology field and has meanwhile also become a subject of discussion in nanotoxicology (Stone *et al.*, 2009; Donaldson *et al.*, 2010). In particular, the role of exposure route-dependent 'corona' effects has sparked major interest in recent years (Lynch *et al.*, 2006; Donaldson *et al.*, 2010). The phenomenon of variation introduced through CNT manufacturing and testing is well appreciated but not dealt with in the subsequent sections, in view of the limited number of currently available CNT genotoxicity studies.

9.3.1 In vitro *genotoxicity findings with carbon nanotubes*

Table 9.1 lists a number of studies that identified genotoxic effects *in vitro* with single-walled carbon nanotubes (SWCNT), whereas *in vitro* studies with MWCNT are listed in Table 9.2. Although these investigations have been performed with specific samples and under specific assay conditions, as discussed above, it is justified to assume that CNT may elicit a variety of genotoxic effects, as categorised below.

9.3.1.1 *Carbon nanotubes can cause DNA single strand breakage and oxidative DNA damage*

Depending on treatment time and concentrations, MWCNT as well as SWCNT have been shown to cause DNA single strand breaks. Such effects have been observed in various cell types including lung epithelial cells, macrophages and (embryonal) fibroblasts (Karlsson *et al.*, 2008; Lindberg *et al.*, 2009; Yang *et al.*, 2009; Patlolla *et al.*, 2010a; Yamashita *et al.*, 2010; Di Giorgio *et al.*, 2011). Importantly, SWCNT have also been evaluated for their DNA-damaging potential in mesothelial cells. These investigations have revealed that this type of CNT can cause DNA damage in normal as well as malignant human mesothelial cells (Pacurari *et al.*, 2008).

Carbon nanotubes have also been shown to induce oxidative DNA damage, measured using modified comet assay. In RAW264.7 mouse macrophages, enhanced oxidation of purines and pyrimidines could be detected after treatment with SWCNT as well as MWCNT (Migliore *et al.*, 2010). In a mouse lung epithelial cell line used in parallel for mutagenicity testing (FE1 Muta™Mouse cells), SWCNT caused a significant induction of oxidative DNA damage. Remarkably, however, no increased DNA strand breakage could be detected under the same treatment conditions (Jacobsen *et al.*, 2008).

Table 9.1 *In vitro* genotoxicity studies with single-walled CNT in mammalian cells. Studies in which no significant effect was observed are italicised.

Endpoint	Cells	Reference
DNA strand breakage (comet assay)	Normal and malignant human mesothelial cells	Pacurari *et al.*, 2008
	Primary mouse embryo fibroblast cells	Yang *et al.*, 2009
	BEAS-2B human bronchial epithelial cells	Lindberg *et al.*, 2009[a]
	RAW264.7 mouse macrophages	Di Giorgio *et al.*, 2011
	FE1 Muta™Mouse lung epithelial cells	*Jacobsen et al.*, 2008
	A549 human lung epithelial cells	*Yamashita et al.*, 2010
DNA base oxidation	FE1 Muta™Mouse lung epithelial cells	Jacobsen *et al.*, 2008
	RAW264.7 mouse macrophages	Migliore *et al.*, 2010
Micronuclei	RAW264.7 mouse macrophages	Di Giorgio *et al.*, 2011
	HDMEC normal human dermal fibroblasts	Cveticanin *et al.*, 2010
	BEAS-2B human bronchial epithelial cells	Lindberg *et al.*, 2009
	RAW264.7 mouse macrophages	Migliore *et al.*, 2010
Chromosomal aberrations	RAW264.7 mouse macrophages	Di Giorgio *et al.*, 2011
Formation of γH2AX foci	Normal and malignant human mesothelial cells	Pacurari *et al.*, 2008
	HDMEC normal human dermal fibroblasts	Cveticanin *et al.*, 2010
Mutant frequencies	*FE1 Muta™Mouse lung epithelial cells*	*Jacobsen et al.*, 2008

[a] The SWCNT sample in this study had a purity of about 50%; the remaining fraction contained other nanotubes.

9.3.1.2 Carbon nanotubes can cause DNA double strand breaks

The potential of CNT to induce these more severe lesions is indicated by various studies. The formation of γ-H2AX foci, a sensitive indicator of DNA double strand breaks, has been observed in human umbilical vein endothelial cells (HUVEC) as well as in mouse embryonic fibroblasts following treatment with MWCNT (Zhu *et al.*, 2007; Guo *et al.*, 2011). SWCNT were shown to cause H2AX phosphorylation in normal as well as malignant human mesothelial cells (Pacurari *et al.*, 2008). In normal human dermal fibroblasts, H2AX phosphorylation has been shown both for MWCNT and SWCNT (Cveticanin *et al.*, 2010). Importantly, it must be taken into account that these severe DNA-damaging effects occurred at concentrations which

Table 9.2 *In vitro* genotoxicity studies with multi-walled CNT in mammalian cells. Studies in which no significant effect was observed are italicised.

Endpoint	Cells	Reference
DNA strand breakage (comet assay)	A549 human lung epithelial cells	Karlsson *et al.*, 2008
	A549 human lung epithelial cells	Yamashita *et al.*, 2010
	RAW264.7 mouse macrophages	Di Giorgio *et al.*, 2011
	NHDF normal human dermal fibroblasts	Patlolla *et al.*, 2010a
DNA base oxidation	RAW264.7 mouse macrophages	Migliore *et al.*, 2010
	A549 human lung epithelial cells	*Karlsson et al., 2008*
Micronuclei	RAW264.7 mouse macrophages	Di Giorgio *et al.*, 2011
	MCF-7 human lung epithelial cells	Muller *et al.*, 2008a
	RLE rat lung epithelial cells	Muller *et al.*, 2008b
	HDMEC normal human dermal fibroblasts	Cveticanin *et al.*, 2010
	RAW264.7 mouse macrophages	Migliore *et al.*, 2010
	V79 cells (in presence or absence of S9 mix)	*Wirnitzer et al., 2009*
	CHL/IU Chinese hamster lung cells	*Asakura et al., 2010*
Chromosomal aberrations	RAW264.7 mouse macrophages	Di Giorgio *et al.*, 2011
	CHL/IU Chinese hamster lung cells[a]	*Asakura et al., 2010*
	V79 cells in presence or absence of S9 mix	*Wirnitzer et al., 2009*
Formation of γH2AX foci	HDMEC normal human dermal fibroblasts	Cveticanin *et al.*, 2010
	HUVEC human umbilical vein endothelial cells	Guo *et al.*, 2011
	Mouse embryonic stem cells	Zhu *et al.*, 2007
Mutant frequencies	Mouse embryonic stem cells (Aprt)	Zhu *et al.*, 2007
	CHL/IU Chinese hamster lung cells (Hpgrt)	*Asakura et al., 2010*

[a] Polyploidy was observed in the absence of structural chromosome aberrations.

also showed cytotoxicity and apoptosis (Cveticanin *et al.*, 2010; Guo *et al.*, 2011), and as such may not lead to mutations. However, results from MN and CA assays provide further evidence for chromosome-breaking effects, as described below.

9.3.1.3 Carbon nanotubes can cause micronucleus formation and chromosomal aberrations

Increased MN frequencies have been observed in BEAS-2B human lung epithelial cells and in RLE rat lung epithelial cells following treatment with SWCNT and

MWCNT, respectively (Muller *et al.*, 2008b; Lindberg *et al.*, 2009). Both types of nanotubes have also been found to generate MN in human lymphocytes cultures (Cveticanin *et al.*, 2010) and in RAW264.7 macrophages (Di Giorgio *et al.*, 2011). SWCNT and MWCNT also caused CA in RAW264.7 cells (Di Giorgio *et al.*, 2011). However, MWCNT failed to induce MN and CA in V79 Chinese hamster lung fibroblasts (Wirnitzer *et al.*, 2009) and in CHL/IU Chinese hamster lung cells (Asakura *et al.*, 2010). In another study with V79 cells, a significantly increased MN frequency was found upon treatment with another SWCNT sample (Kisin *et al.*, 2011).

9.3.1.4 Carbon nanotubes can cause aneuploidy

Findings in specific studies indicate that, besides clastogenic effects, CNT can also elicit aneugenic effects. Aneugenicity has been observed in studies by MN assay as well as CA evaluations. Using FISH with a pancentrometric probe in MCF-7 human epithelial cells, inductions of centromere negative as well as centromere positive MN could be detected, demonstrating that the specific MWCNT sample used caused both clastogenic and aneugenic effects (Muller *et al.*, 2008a). From these findings it has been hypothesised that CNT, and likely other specific types of nanoparticles, may target constituents of the mitotic spindle apparatus (Gonzalez *et al.*, 2008). Aneugenic effects of carbon nanotubes have also been described in macrophages. Treatment of RAW264.7 mouse macrophages with MWCNT or SWNCT resulted in chromosomal breakage and de-condensation as well as numerical chromosome alterations (Di Giorgio *et al.*, 2011). In SWCNT-treated human airway epithelial cells, numerical chromosome changes could be identified, as well as fragmented centrosomes, multiple mitotic spindle poles and anaphase bridges. Notably, confocal microscropy revealed SWCNT within the nucleus in association with cellular and mitotic tubulin as well as the chromatin (Sargent *et al.*, 2009). Accordingly, it has been hypothesised that SWCNT may interact with the mitotic spindle as a result of their toughness and similarity in dimensions to cellular microtubuli (Sargent *et al.*, 2010). The ensuing disruption of centrosomes and the mitotic spindle can lead to centrosome division of the chromosomes and thus aneuploidy (Sargent *et al.*, 2010).

9.3.1.5 Carbon nanotubes can cause gene mutations

Measuring the marker gene adenine phosphoribosyltransferase (Aprt), treatment of MWCNT has been found to result in a two-fold increase in mutation frequency in mouse embryonic stem cells (Zhu *et al.*, 2007). In contrast, no increased Hpgrt-mutation frequencies were observed in CHL/IU Chinese Hamster lung cells after treatment with MWCNT (Asakura *et al.*, 2010). In FE1 Muta™Mouse lung epithelial cells, SWNCT failed to increase mutation frequency in cII gene

(Jacobsen *et al.*, 2008). Also, when subjected to bacterial reverse mutation assays, several types of MWCNT have consistently tested negative (Di Sotto *et al.*, 2009; Wirnitzer *et al.*, 2009; Kim *et al.*, 2010).

9.3.2 In vivo *genotoxicity findings with carbon nanotubes*

The *in vivo* genotoxic potential of CNT has been investigated in several studies. Important findings are listed in Table 9.3 with inclusion of the characteristics of the investigated material, the animal model, the route of application and the applied dose. As inhalation is considered to be the most relevant route of exposure with regard to potential human health risks of carbon nanoparticles, the earliest *in vivo* genotoxicity investigations have focused on genotoxicity in lung cells and tissues. Muller *et al.* (2008a) determined micronucleus frequencies in lung epithelial cells isolated from rat lungs, 3 days after a single intratracheal instillation of MWCNT. The sample used in their study, which had an average length of about 0.7 μm, caused a significant dose-dependent increase in micronuclei. Importantly, the treatment also resulted in a clear dose-dependent pulmonary inflammation (Muller *et al.*, 2008a). The approach used within their study can be considered highly relevant for the testing of genotoxicity of inhalable (nano)particles, as isolated lung epithelial cells are known targets in the rat lung for the tumourigenic effects of particles such as quartz (Knaapen *et al.*, 2002). Using various markers of pulmonary toxicity, Shvedova *et al.* (2008) have investigated the toxicity of a SWCNT sample (length 0.1–1 μm) after a single pharyngeal aspiration and, by comparison, after 5 consecutive days of inhalation in mice. The genotoxic potential of the sample was indicated by observed increases in k-ras mutations in the lung tissue DNA. Notably, mutations were found at various time points post-instillation as well as post-inhalation. In another study, SWCNT (length <1 μm) were found to cause DNA single strand breakage in cells obtained from bronchoalveolar lavage of mice, 3 hours after their intratracheal instillation (Jacobsen *et al.*, 2009).

In more recent years, genotoxicity of CNT has also been investigated after oral administration. In SWCNT-treated rats no enhanced oxidative DNA damage could be detected in colon mucosa (Folkmann *et al.*, 2009). In contrast, oxidative DNA damage was found to be increased in a dose-dependent manner in the lung and liver tissues of the same animals. The authors explained the lack of increased oxidative DNA damage in the gut by the high cell turnover in the colon mucosa, leading to a rapid loss of damaged cells (Folkmann *et al.*, 2009). In another study, urinary genotoxicity was measured in rats after oral treatment to SWCNT and MWCNT. However, no mutagenicity was detected by the Ames test, MN test or the sister chromatid exchange test (Szendi and Varga, 2008). Patlolla and coworkers (Patlolla *et al.*, 2010b) evaluated the genotoxic potential of MWCNT after repeated

Table 9.3 *In vivo* studies with CNT showing genotoxicity.

Material	Effect	Model	Reference
MWCNT, 0.7 µm length; 11.3 nm diameter	Significant dose-dependent increase in MN formation in lung epithelial cells collected from CNT- or saline-treated rats	Female Wistar rats (200–250g); single intratracheal instillation (0.5 or 2 mg CNT/rat)	Muller *et al.*, 2008a
SWCNT, 0.1–1 µm length; 0.8–1.2 nm diameter; elemental carbon (82%), Fe (17.7%), Cu (0.16%), Cr (0.049%) and Ni (0.046%). BET surface 508 m²/g	K-ras mutations in lung tissue DNA from mice that were exposed by inhalation, but not in mice exposed by pharyngeal aspiration; effects were observed at various time points after exposure	Female C57BL/6 mice (8–10 weeks): (1) whole-body inhalation (5 mg/m³), 4 consecutive days, 5 hours/day exposure; (2) single pharyngeal aspiration (5, 10 or 20 µg/mouse); lungs of the animals were analysed 1, 7 and 28 days post-inhalation.	Shvedova *et al.*, 2008
SWCNT, <1 µm length; 0.9–1.7 nm diameter	Increased DNA strand breakage (comet assay) in bronchoalveolar lavage cells, 3 hours after instillation of SWCNT	Apolipoprotein E (ApoE-/-)-deficient mice (8 weeks); single intratracheal instillation	Jacobsen *et al.*, 2009
SWCNT, same material as used in Jacobsen *et al.*, 2009	Dose-dependent increase in oxidative DNA damage (8-OHdG adducts) in lung and liver tissue; no significant increase in oxidative DNA damage found in colon mucosa	Female Fischer 344 rats (9 weeks); oral gavage (0.064 and 0.64 mg /kg b.w.)	Folkmann *et al.*, 2009
MWCNT (pristine), 15–20 µm length; functionalised MWCNT (2–7% COOH)	Increased CA and MN in bone marrow cells; increased DNA damage in leukocytes (measured by comet assay); purified functionalised MWCNT demonstrated a stronger genotoxic potential than the non-functionalised MWCNT	Male Swiss-Webster mice (5–7 weeks); repeated intraperitoneal injections (5 days, single injection/day, 0.25, 0.5 or 0.75 mg/kg b.w.)	Patlolla *et al.*, 2010b

intraperitoneal injections in mice. The nanotubes used in this study had an estimated length of about 15–20 μm. Significant genotoxic effects could be indentified in the treated mice by several assays. Increased DNA strand breaks were identified by comet assay analysis of leukocyte DNA and enhanced MN frequencies and chromosomal aberrations were detected in bone-marrow cells. Taken together, investigations in mice and rats have revealed that specific samples of SWCNT as well as MWCNT have some genotoxic potential *in vivo*.

9.4 The variable hazard of carbon nanotubes

From decades of research in the field of classical particle toxicology it has been established that modifications of the physicochemical properties of particles and fibres can have a major impact on their toxicity and hence have major implications for their hazard and risk assessment (e.g. Donaldson and Borm, 1998). Considerable variations in potency have also been shown specifically for genotoxicity. For instance, different samples of quartz powders have been shown to display marked differences in DNA damaging potency *in vitro* (Cakmak *et al.*, 2004). Surface modification of a relatively toxic quartz sample (DQ12) with a synthetic polymer poly-2-vinylpyridine-N-oxide (PVNO), or with aluminium ions, has been shown to lead to an almost total abrogation of genotoxicity in lung epithelial cells, both *in vitro* and *in vivo* (Knaapen *et al.*, 2002; Schins *et al.*, 2002; Albrecht *et al.*, 2009). Marked contrasts in genotoxicity have also been well established for different types of asbestos, as well as for specifically modified individual asbestos samples (Lu *et al.*, 1994; Jaurand, 1997; Liu *et al.*, 2000; Fan *et al.*, 2001; Takata *et al.*, 2009; Srivastava *et al.*, 2010).

Also for engineered nanoparticles, it is accepted that toxicity is driven by specific physicochemical properties (Oberdörster *et al.*, 2005; Fubini *et al.*, 2010). The available literature suggests that this also holds true for the potential genotoxic properties of CNT. Interesting studies in this regard are those in which pristine nanotubes have been compared with surface-functionalised nanotubes. Functionalisation of CNT is recognised as an important tool for the further improvement of specific beneficial properties of CNT, e.g. in drug-delivery strategies. However, various effects of functionalisation of CNT in biological environments (e.g. de-aggregation, stabilisation, translocation across tissue barriers, uptake and subcellular re-localisation in target cells) may likely also impact on their toxicity.

The effect of functionalisation and purification of MWCNT on genotoxicity has been nicely demonstrated by the *in vivo* investigations of Patlolla *et al.* (2010b), in which the pristine material overall showed a lower genotoxicity (i.e. comet assay in leukocytes, and MN and CA in bone marrow cells) after intraperitoneal injection than the modified nanotubes. Also, *in vitro* studies indicate that functionalisation can have a considerable impact on the genotoxic potency of CNT. Cveticanin *et al.* (2010) observed a higher induction of MN frequencies in human dermal fibroblasts by an amide-modified sample of SWCNT in a cytokinesis-block micronucleus assay, as compared to the effect of the pristine material. A remarkable observation in their study was that the proliferation rate of the CNT-treated cells, measured by cytokinesis-block proliferation index (CBPI), was not affected by the functionalisation. This suggests that proliferation and genotoxicity effects of SWCNT occur via different mechanisms (Cveticanin *et al.*, 2010). In contrast, the observed effects on MN frequencies and γH2AX foci formation did not markedly differ in response to pristine versus amide-functionalised SWCNT (Cveticanin *et al.*, 2010). Important findings on the variability of genotoxic responses were also obtained from a study in which effects of grinding and/or heat treatment of a MWCNT sample were investigated. Exposure of the sample to high temperatures (600°C or 2400°C), to modify its structure (i.e. annealing of defects) and metal content, was shown to reduce its ability to elicit MN formation in rat lung epithelial cells. However, grinding of the heated samples restored its genotoxic potential, which indicated that this effect was driven by structural defects (Muller *et al.*, 2008b).

9.5 The role of oxidant species

The generation of oxidants has been considered an important feature to explain the genotoxic potential of many types of particles as well as fibres (Jaurand, 1997; Schins and Hei, 2007; Schins and Knaapen, 2007). These oxidants include reactive oxygen species (ROS) such as hydrogen peroxide, hydroxyl radicals, superoxide and singlet oxygen, as well as reactive nitrogen species (RNS) including nitric oxide and peroxynitrite (Knaapen *et al.*, 2004). The relevance of oxidants in particle genotoxicity is supported by their ability to cause oxidative DNA damage, which may result in base-pair mutations, deletions as well as insertions. Moreover, ROS and RNS are also known to attack lipids and proteins, and this has been implicated in the DNA-damaging effects of, for example, silica and asbestos (Shi *et al.*, 1994; Howden and Faux, 1996). Exocyclic etheno-adducts can be formed during lipid peroxidation with associated formation of mutagenic electrophiles such as malondialdehyde and 4-hydroxynonenal (Knaapen *et al.*, 2004). There are three principle ways whereby particles may generate oxidants (Schins and Knaapen, 2007; Unfried *et al.*, 2007):

- Intrinsic generation of oxidants from the particles (i.e. in the absence of cells)
- Oxidant generation resulting from the interaction of particles with their target cell
- Generation of oxidants during inflammation, triggered by a particle-driven recruitment and/or activation of inflammatory cells (macrophages, neutrophils).

Several findings from the literature indicate that oxidant generation may also be relevant for carbon nanotubes in relation to genotoxicity. Importantly, one should discriminate between studies that have merely shown oxidant generation and genotoxic effects in parallel, and studies in which antioxidant treatment has actually been shown to reduce or even abrogate genotoxicity.

9.5.1 Intrinsic oxidant generation by carbon nanotubes

Several types of particles have been demonstrated to generate oxidants in a cell-free environment, and these properties have been considered useful for the prediction of their acute *in vivo* inflammatory potential (Oberdörster *et al.*, 2005; Duffin *et al.*, 2007; Unfried *et al.*, 2007; Rushton *et al.*, 2010). For asbestos fibres as well as crystalline silica particles, intrinsic ROS-generating properties have been found to be associated with genotoxicity (Jaurand, 1997; Schins and Hei, 2007; Schins and Knaapen, 2007). However, whether cell-free oxidant generation is also relevant for the genotoxic potential of carbon nanotubes is not known. Investigations in which the cell-free oxidant generation of CNT has been investigated have often revealed, in line with studies with C60 fullerenes, that these materials do not generate oxidants, but rather possess a marked ROS-quenching activity (e.g. Muller *et al.*, 2009).

9.5.2 Oxidant generation resulting from interactions of CNT with their target cells

Increased ROS formation could be detected in mesothelial cell suspensions co-incubated with SWCNT, and was identified as hydroxyl radicals by electron spin resonance (ESR) (Pacurari *et al.*, 2008). The effects could be reduced in the presence of the antioxidant catalase or the iron chelator deferoxamine. The latter finding is in line with the hypothesis that residual metal impurities in the nanotube sample contribute to the SWCNT-triggered ROS generation, similar to the role of iron shown to be involved in ROS generation by asbestos (Lund and Aust, 1992; Kagan *et al.*, 2006). Further evidence for the generation of ROS in the nanotube-treated cells was provided using the superoxide-specific and H_2O_2-specific dyes dihydroethidium (DHE) and dichlorofluorescein diacetate (DCFH-DA). The observed effects with these ROS markers could be abolished by pre-treatment of the cells with catalase or superoxide dismutase (SOD) (Pacurari *et al.*, 2008). Taken together, these findings suggest a role for ROS in the observed DNA single and double strand breakage effects (see Table 9.1) of the SWCNT sample used in their study. Increased ROS generation in relation to genotoxicity observations was also shown by DCFH-DA measurements in RAW264.7 macrophages after treatment with MWCNT or SWCNT (Di Giorgio *et al.*, 2011).

More detailed investigations on the actual contribution of oxidants to DNA damage were carried out by Guo *et al.* (2011) in HUVEC cells. In the MWCNT-treated cells, increased ROS levels were detected by DCFH-DA assay. Also, increased levels of the lipid peroxidation product malondialdehyde (MDA) as well as changes in the activities of the SOD and glutathione peroxidase (GSH-Px) were found. In cells that were pre-treated with the antioxidant N-acetyl-cysteine (NAC), a marked inhibition of MWCNT-induced ROS formation as well as γH2AX foci formation could be observed (Guo *et al.*, 2011). As such, this study demonstrated that DNA double strand breakage induction by this MWCNT sample occurred in an oxidant-dependent manner.

9.5.3 Oxidant generation during CNT-driven inflammation

The inflammogenic potential of inhaled biopersistent particles is considered to be of major importance for the development of subsequent pathology. In quartz-exposed lungs, the excessive and persistent formation of ROS generated by activated phagocytes (i.e. macrophages, neutrophils) during inflammation is considered to cause oxidative stress in silicotic lungs, and has been implicated in oxidative DNA damage induction and mutagenesis (Knaapen *et al.*, 2004). Chronic inflammation has also been considered to explain the formation of tumours in the lungs of rats after high and long-term exposures to poorly soluble low-toxicity dusts such as TiO_2 and carbon black. Beyond this, for asbestos, the 'frustrated phagocytosis' and impaired clearance of long, biopersistent fibres by macrophages is considered responsible for the development of a chronic inflammatory status in which inflammatory mediators including ROS and RNS are thought to drive fibrosis and cancer (Donaldson and Tran, 2004). *In vitro* studies have confirmed that ROS generated from phagocytes can induce DNA damage and mutagenesis in various cell types including lung epithelial cells (reviewed in Knaapen *et al.*, 2006).

It can therefore be hypothesised that the *in vivo* genotoxic effects observed in carbon nanotube–treated mice or rats (see Table 9.3) involve phagocyte-mediated oxidant formation. Indeed, the lungs of carbon nanotube–exposed animals in which genotoxicity was detected also displayed an influx of inflammatory phagocytes (Muller *et al.*, 2008b; Shvedova *et al.*, 2008; Jacobsen *et al.*, 2009), an increased MDA level, and decreases in GSH level and total antioxidant status (Shvedova *et al.*, 2008). Since similar (or the same) CNT samples were also found to induce genotoxic effects *in vitro* (Jacobsen *et al.*, 2008; Muller *et al.*, 2008b; Pacurari *et al.*, 2008; Kisin *et al.*, 2011), the true contribution of phagocyte-derived ROS to the observed *in vivo* genotoxicity cannot be established. However, a comparison of the typically applied *in vitro* treatment concentrations with the applied *in vivo* treatment doses, or the estimated deposition rate after inhalation, indicates that a contribution from inflammation is likely. Importantly, besides ROS and RNS, other inflammatory mediators (e.g. cytokines, growth factors) are believed to contribute to particle-induced mutagenesis by affecting proliferation, apoptosis or DNA repair processes (Schins and Hei, 2007). The role of inflammation in carbon nanotube-induced genotoxicity requires further investigation. This will be especially relevant for mesothelial cells as well in relation to the established relation between nanotube length and inflammation at the mesothelial surface (Poland *et al.*, 2008; Murphy *et al.*, 2011).

9.6 Uptake and subcellular translocation of carbon nanotubes

With regard to the importance of CNT uptake in relation to genotoxicity, one should discriminate between uptake by professional phagocytes of the innate immune system and uptake by potential target cells for carcinogenic effects, e.g. lung epithelial cells and mesothelial cells.

9.6.1 Uptake by professional phagocytes

Particle uptake by professional phagocytes serves as a major defence mechanism to remove foreign material from the body and hence may offer protection against mutagenic and carcinogenic effects of toxic particles. However, as mentioned above, particle uptake can lead to their activation, leading to formation of ROS and RNS as well as further mediators of inflammation and proliferation. Activated alveolar macrophages and neutrophils are known to generate excessive amounts of oxidants, and are considered to drive genotoxic, mutagenic and proliferative effects in tumourigenesis associated with chronic inflammatory diseases (Knaapen *et al.*, 2006). Thus, uptake of CNT may lead to DNA damage within the phagocyte itself, but in addition may also elicit genotoxic and mutagenic effects in neighbouring target cells (e.g. lung epithelial cell, mesothelial cell) (see Section 9.5.2).

   In a recent study, uptake and ROS formation in RAW264.7 macrophages were comparatively investigated for asbestos, SWCNT and carbon nanofibres. The latter material is similar in composition to MWCNT with regard to its hollow core, but in contrast to nanotubes the graphene plane is not aligned with the fibre axis (Kisin *et al.*, 2011). Using transmission electron microscopy (TEM) and ESR, respectively, it was found that, unlike asbestos and carbon nanofibres, the SWCNT were not taken up and also did not cause significant ROS production. In contrast, Di Giorgio *et al.* (2011) observed uptake and ROS generation for both SWCNT and MWCNT using RAW264.7 cells (see also Section 9.5.2). These and other contrasting observations are likely to be explained by the different materials and/or specific treatment conditions used. Further investigations are needed to determine the relationship between uptake of CNT by macrophages and genotoxicity/mutagenicity. As already mentioned, an important aspect will be to address this in relation to the dimensions of the materials and the role of 'frustrated phagocytosis' (Donaldson and Tran, 2004).

9.6.2 Uptake by target cells for carcinogenesis

In vitro studies with asbestos, quartz and other types of particles have revealed that uptake into target cells may be of major importance for their potential genotoxic effects (Jaurand, 1997; Schins and Hei, 2007). Uptake may lead to the formation of ROS by activation of membrane-associated NADPH oxidase enzymes. Although not as excessive as observed during the 'classical' oxidative burst of professional phagocytes, particle-induced ROS formation inside target cells is considered to contribute to oxidative stress and DNA damage (Knaapen *et al.*, 2004, 2006). Mitochondria are considered as a major source of intracellular ROS in target cells; studies with asbestos and quartz have shown that the DNA-damaging effect of these materials depends on mitochondrial function, which suggests that ROS leaking from these organelles upon their interaction with the particles are involved (Kamp *et al.*, 2002; Li *et al.*, 2007).

9.6.3 Nuclear translocation of carbon nanotubes

Although the nuclear envelope is considered to represent a major barrier to maintain genome integrity, specific types of nanoparticles have been shown to be able to cross this membrane, depending on their size (Chen and von Mikecz, 2005; Geiser *et al.*, 2005). This aspect has also been discussed and, to some extent, already investigated with CNT (Cheng *et al.*, 2009; Cveticanin *et al.*, 2010; Sargent *et al.*, 2010). Furthermore, the specific functionalisation of nanoparticles has been applied as a tool for their effective delivery into the nucleus, e.g. by specific modification with cell-penetrating peptides or nuclear localisation sequences (Tkachenko *et al.*, 2003; Nativo *et al.*, 2008). Translocation into the nucleus of cells has also been achieved with functionalised CNT, and these have therefore been proposed as a nanovector in various pharmaceutical and biomedical applications (Pantarotto *et al.*, 2004; Cheng *et al.*, 2008).

The translocation of CNT into the nucleus, or their penetration of the nuclear membrane, may have serious consequences for the integrity of the genome. DNA damage may be caused by a direct interaction with this macromolecule or by the generation of increased levels of ROS in its close proximity, as has been discussed for asbestos as well as crystalline silica (Daniel *et al.*, 1995; Jaurand, 1997). Notably, during cell division stages when the nuclear envelope is absent, CNT may also interact directly with the chromosomes. Such interaction has been shown recently for SWCNT (Sargent *et al.*, 2009). Cveticanin *et al.* (2010) have suggested that the generation of DNA double strand breaks which they could observe with SWCNT was related to the negative charge and corresponding electrochemical characteristics of this material.

9.6.4 *Interaction with the mitotic spindle during cell division*

Last but not least, apart from direct interactions with the DNA, CNT may also interfere with cellular constituents that guide chromosome segregation during cell division, e.g. the mitotic spindle. As already discussed, this mechanism of action has been proposed from the observed aneugenic effects as seen in the MN assay and CA assay (see Section 9.3.1). In addition, this mode of action has been recognised for many years as potentially relevant for the genotoxicity of asbestos fibres (Jaurand, 1997), and more recently for engineered nanoparticles (Gonzalez *et al.*, 2008). Recent confocal microscopy studies have revealed SWCNT localised within the nucleus of lung epithelial cells in association with cellular and mitotic tubulin (Sargent *et al.*, 2009, 2010).

9.7 Summary and conclusions

The available studies suggest that carbon nanotubes are potentially capable of eliciting genotoxic effects *in vivo* as well as *in vitro*. Effects that were identified *in vitro* include induction of oxidative DNA damage, generation of DNA single and double strand breaks as well as chromosome breaks and aneugenic effects (See Figure 9.1). In line with the oxidative stress paradigm in particle toxicology, uptake of CNT into cells and the likely associated generation of ROS appear to play dominant roles in CNT-associated genotoxicity. Based on observations on the lack of ability of CNT to generate ROS in cell-free assays, one may suggest that oxidative DNA-damaging properties of CNT are the result of oxidants originating from cellular sources, e.g. NADPH oxidase enzymes or mitochondria (Unfried *et al.*, 2007). The high aspect ratio of CNT is a further likely determinant of genotoxicity, in line with literature findings on asbestos fibres (e.g. Moolgavkar *et al.*, 2001). Long and thin (and biopersistent) tubes are associated with the process of frustrated phagocytosis and the associated induction of inflammation, and ROS formation is considered to drive a persistent oxidative stress that leads to genotoxic and mutagenic outcomes (Donaldson and Tran, 2004; Knaapen *et al.*, 2006). Moreover, in specific studies it has been revealed that the genotoxic effects of CNT may involve physical interactions with chromosomes and/or mitotic spindle constituents (e.g. microtubules). Investigations are needed to further evaluate this interesting mode of action of nanotubes as well as of other nanoparticles

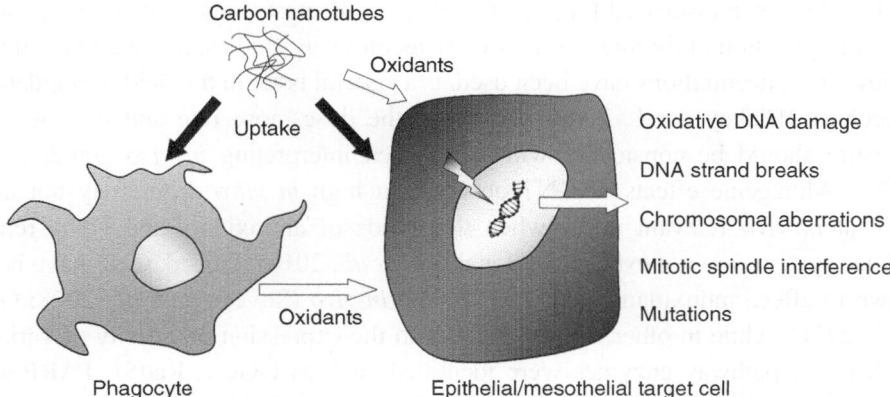

Figure 9.1 Proposed mechanisms of genotoxicity of carbon nanotubes. CNT may cause oxidative DNA damage induction, generation of DNA single and double strand breaks and mutations, as well as structural and numerical chromosome changes. The formation of reactive oxygen species (ROS) has been implicated in the DNA-damaging properties of CNT. These ROS are likely generated upon the direct interaction between CNT and the target cell for carcinogenicity (e.g. lung epithelial cells, mesothelial cells), or from activated phagocytes during CNT-induced inflammation and/or phagocyte activation. Uptake and subcellular translocation have also been implicated in CNT genotoxicity via ROS-dependent and -independent mechanisms (e.g. physical interactions with DNA, chromosomes and mitotic spindle apparatus).

(Gonzalez *et al.*, 2011) and the importance of specific physicochemical properties herein (e.g. size, length, rigidity, surface charge).

In support of the *in vitro* findings, genotoxic effects have been observed with CNT in a number of *in vivo* studies. However, it is important to emphasise that studies have also been published, both *in vitro* and *in vivo*, in which specific genotoxic effects were shown to be absent. Moreover, in specific studies where CNT samples of different length, purity or functionalisation (in association with aggregation) were comparatively investigated, a marked variation in genotoxic potency was noted. Altogether this indicates that the genotoxic hazard of CNT strongly depends on its physicochemical properties, and this provides a basis for the design of safer materials in the nanotechnology sector.

Last but not least, for an appropriate interpretation of findings on CNT genotoxicity for risk assessment purposes, it is pivotal to include the aspects of dose, target delivery and clearance. Interpretation of *in vitro* genotoxicity studies with lung epithelial cells should take into account inhalation and deposition kinetics of CNT at realistic exposure scenarios when addressing potential pulmonary carcinogenicity. Similarly, the interpretation of effects observed in mesothelial cells should be handled with caution in relation to deposition, translocation and clearance of

CNT, as has been discussed for asbestos fibres in the past. Increased concern about over-interpretation of findings from *in vitro* genotoxicity studies, in which maximum achievable concentrations have been used, is a general issue in the field of regulatory toxicology (Blakey *et al.*, 2008). Likewise, the dose, dose rate and the route of exposure should be considered with care when interpreting *in vivo* genotoxicity studies. Mutagenic effects of CNT observed at high *in vitro* doses may not take place at *in vivo*–relevant doses when thresholds of antioxidant and DNA repair defence systems are not exceeded (Donaldson *et al.*, 2010). Indeed, CNT have been shown to affect antioxidant status *in vitro* and *in vivo* (Shvedova *et al.*, 2008; Guo *et al.*, 2011), while in other studies changes in the expression or activity of various DNA repair pathway enzymes were identified such as OGG1, Rad51, PARP and Msh2 (Zhu *et al.*, 2007; Pacurari *et al.*, 2008; Haniu *et al.*, 2010). Investigations on the potential effects of CNT, as well as of engineered nanoparticles in general, on antioxidant and DNA repair pathways represent a challenging topic for future research.

In conclusion, the currently available literature indicates that exposure to carbon nanotubes could represent a genotoxic hazard. However, the inconsistent findings reported among different studies provide clear evidence that this hazard may be highly variable in its dependence on their heterogeneous physicochemical properties. However, as the toxic responses to CNT may also depend on the choice of dispersing agents and/or culture medium composition, no solid statement can be made at present regarding the actual hazard. The challenge for future research will be to address genotoxic effects of CNT in controlled experimental settings and the selection of an appropriate dose range on the basis of available human exposure data. Such testing should also take into account an extensive characterisation of the CNT under the applied assay conditions and, ideally, aim for the identification of the underlying mechanism of action.

References

Aitken, R. J., Chaudhry, M. Q., Boxall, A. B., and Hull M. (2006). Manufacture and use of nanomaterials: Current status in the UK and global trends. *Occup Med (Lond)* **56**(5): 300–6.

Albrecht, C., Knaapen, A. M., Demircigil, C., *et al.* (2009). Genomic instability in quartz dust exposed rat lungs: Is inflammation responsible? *J Phys Conf Ser* **151**: 012046.

Asakura, M., Sasaki, T., Sugiyama, T., *et al.* (2010). Genotoxicity and cytotoxicity of multi-wall carbon nanotubes in cultured Chinese hamster lung cells in comparison with chrysotile A fibers. *J Occup Health* **52**(3): 155–66.

Blakey, D., Galloway, S. M., Kirkland, D. J., and MacGregor, J. T. (2008). Regulatory aspects of genotoxicity testing: From hazard identification to risk assessment. *Mutat Res* **657**(1): 84–90.

Borges, H. L., Linden, R., and Wang, J. Y. (2008). DNA damage-induced cell death: Lessons from the central nervous system. *Cell Res* **18**(1): 17–26.

Cakmak, G., Schins, R. P. F., Shi, T., *et al.* (2004). In vitro genotoxicity assessment of commercial quartz flours in comparison to standard DQ12 quartz. *Int J Hyg Environ Health* **207**: 105–13.

Chen, M. and von Mikecz, A. (2005). Formation of nucleoplasmic protein aggregates impairs nuclear function in response to SiO_2 nanoparticles. *Exp Cell Res* **305**: 51–62.

Cheng, C., Müller, K. H., Koziol, K. K., *et al.* (2009). Toxicity and imaging of multi-walled carbon nanotubes in human macrophage cells. *Biomaterials* **30**(25): 4152–60.

Cheng, J., Fernando, K. A., Veca, L. M., *et al.* (2008). Reversible accumulation of PEGylated single-walled carbon nanotubes in the mammalian nucleus. *ACS Nano* **2**(10): 2085–94.

Cveticanin, J., Joksic, G., Leskovac, A., *et al.* (2010). Using carbon nanotubes to induce micronuclei and double strand breaks of the DNA in human cells. *Nanotechnology* **21**(1): 015102.

Daniel, L. N., Mao, Y., Williams, A. O., and Saffiotti, U. (1995). Direct interaction between crystalline silica and DNA: A proposed model for silica carcinogenesis, *Scand J Work Environ Health* **21**(suppl. 2): 22–6.

Di Giorgio, M. L., Bucchianico, S. D., Ragnelli, A. M., *et al.* (2011). Effects of single and multi walled carbon nanotubes on macrophages: Cyto and genotoxicity and electron microscopy. *Mutat Res* **722**(1): 20–31.

Di Sotto, A., Chiaretti, M., Carru, G. A., Bellucci, S., and Mazzanti, G. (2009). Multi-walled carbon nanotubes: Lack of mutagenic activity in the bacterial reverse mutation assay. *Toxicol Lett* **184**(3): 192–7.

Donaldson, K. and Borm, P. J. A. (1998). The quartz hazard: A variable entity. *Ann Occup Hyg* **42**(5): 287–94.

Donaldson, K., Poland, C. A., and Schins, R. P. F. (2010). Possible genotoxic mechanisms of nanoparticles: Criteria for improved test strategies. *Nanotoxicology* **4**: 414–20.

Donaldson, K. and Tran, C. (2004). An introduction to the short-term toxicology of respirable industrial fibres. *Mutat Res* **553**: 5–9.

Duffin, R., Tran, L., Brown, D., Stone, V., and Donaldson, K. (2007). Proinflammogenic effects of low-toxicity and metal nanoparticles in vivo and in vitro: Highlighting the role of particle surface area and surface reactivity. *Inhal Toxicol* **19**: 849–56.

Fan, J. G., Wang, Q. E., and Liu, S. J. (2001). Ameliorated chrysotile-induced DNA damage in human embryo lung cells by surface modification of chrysotile with rare earth compounds. *Biomed Environ Sci* **14**(3): 220–8.

Fenoglio, I., Prandi, L., Tomatis, M., and Fubini, B. (2001). Free radical generation in the toxicity of inhaled mineral particles: The role of iron speciation at the surface of asbestos and silica. *Redox Rep* **6**: 235–41.

Folkmann, J. K., Risom, L., Jacobsen, N. R., *et al.* (2009). Oxidatively damaged DNA in rats exposed by oral gavage to C60 fullerenes and single-walled carbon nanotubes. *Environ Health Perspect* **117**(5): 703–8.

Fubini, B., Ghiazza, M., and Fenoglio, I. (2010). Physico-chemical features of engineered nanoparticles relevant to their toxicity. *Nanotoxicology* **4**: 347–63.

Geiser, M., Rothen-Rutishauser, B., Kapp, N., *et al.* (2005). Ultrafine particles cross cellular membranes by non-phagocytic mechanisms in lungs and in cultured cells. *Environ Health Perspect* **113**(11): 1555–60.

Gonzalez, L., Corradi, S., Thomassen, L. C., *et al.* (2011). Methodological approaches influencing cellular uptake and cyto-(geno) toxic effects of nanoparticles. *J Biomed Nanotechnol* **7**(1): 3–5.

Gonzalez, L., Lison, D., and Kirsch-Volders, M. (2008). Genotoxicity of engineered nanomaterials: A critical review. *Nanotoxicology* **2**: 252–73.

Guo, Y. Y., Zhang, J., Zheng, Y. F., Yang, J., and Zhu, X. Q. (2011). Cytotoxic and genotoxic effects of multi-wall carbon nanotubes on human umbilical vein endothelial cells in vitro. *Mutat Res* **721**(2): 184–91.

Han, J. H., Lee, E. J., Lee, J. H., *et al.* (2008). Monitoring multiwalled carbon nanotube exposure in carbon nanotube research facility. *Inhal Toxicol* **20**(8): 741–9.

Haniu, H., Matsuda, Y., Takeuchi, K., *et al.* (2010). Proteomics-based safety evaluation of multi-walled carbon nanotubes. *Toxicol Appl Pharmacol* **242**(3): 256–62.

Howden, P. J. and Faux, S. P. (1996). Fiber-induced lipid peroxidation leads to DNA adduct formation in salmonella typhimurium TA104 and rat lung fibroblasts. *Carcinogenesis* **17**: 413–19.

Jacobsen, N. R., Møller, P., Jensen, K. A., *et al.* (2009). Lung inflammation and genotoxicity following pulmonary exposure to nanoparticles in ApoE-/- mice. *Part Fibre Toxicol* **6**: 2.

Jacobsen, N. R., Pojana, G., White, P., *et al.* (2008). Genotoxicity, cytotoxicity, and reactive oxygen species induced by single-walled carbon nanotubes and C(60) fullerenes in the FE1-Mutatrade markMouse lung epithelial cells. *Environ Mol Mutagen* **49**(6): 476–87.

Jaurand, M. C., Renier, A., and Daubriac, J. (2009). Mesothelioma: Do asbestos and carbon nanotubes pose the same health risk? *Part Fibre Toxicol* **6**: 16.

Jaurand, M. C. (1997). Mechanisms of fiber-induced genotoxicity. *Environ Health Perspect* **105**(suppl. 5): 1073–84.

Kagan, V. E., Tyurina, Y. Y., Yurin, V. A., *et al.* (2006). Direct and indirect effects of single walled carbon nanotubes on RAW 264.7 macrophages: Role of iron. *Toxicol Lett* **165**: 88–100.

Kamp, D. W., Israbian, V. A., Preusen, S. E., Zhang, C. X., and Weitzman, S. A. (1995). Asbestos causes DNA strand breaks in cultured pulmonary epithelial cells: Role of iron-catalyzed free radicals. *Am J Physiol* **268**(3 Pt 1): L471–80.

Kamp, D. W., Panduri, V., Weitzman, S. A., and Chandel, N. (2002). Asbestos-induced alveolar epithelial cell apoptosis: Role of mitochondrial dysfunction caused by iron-derived free radicals. *Mol Cell Biochem* **234**–235: 153–60.

Kane, A. B. and Hurt, R. H. (2008). Nanotoxicology: The asbestos analogy revisited. *Nat Nanotechnol* **3**: 378–9.

Karlsson, H. L., Cronholm, P., Gustafsson, J., and Möller, L. (2008). Copper oxide nanoparticles are highly toxic: A comparison between metal oxide nanoparticles and carbon nanotubes. *Chem Res Toxicol* **21**(9): 1726–32.

Kim, J. S., Lee, K., Lee, Y. H., *et al.* (2010). Aspect ratio has no effect on genotoxicity of multi-wall carbon nanotubes. *Arch Toxicol* **85**(7): 775–86.

Kirsch-Volders, M., Sofuni, T., Aardema, M., *et al.* (2003). Report from the In Vitro Micronucleus Assay Working Group. *Mutat Res* **540**(2): 153–63.

Kisin, E. R., Murray, A. R., Sargent, L., *et al.* (2011). Genotoxicity of carbon nanofibers: Are they potentially more or less dangerous than carbon nanotubes or asbestos? *Toxicol Appl Pharmacol* **252**(1): 1–10.

Knaapen, A. M., Albrecht, C., Becker, A., *et al.* (2002). DNA damage in lung epithelial cells isolated from rats exposed to quartz: Role of surface reactivity and neutrophilic inflammation. *Carcinogenesis* **23**: 1111–20.

Knaapen, A. M., Borm, P. J. A., Albrecht, C., and Schins, R. P. F. (2004). Inhaled particles and lung cancer, part A: Mechanisms. *Int J Cancer* **109**: 799–809.

Knaapen, A. M., Güngör, N., Schins, R. P. F., Borm, P. J., and Van Schooten, F. J. (2006). Neutrophils and respiratory tract DNA damage and mutagenesis: A review. *Mutagenesis* **21**: 225–36.

Landsiedel, R., Kapp, M. D., Schulz, M., Wiench, K., and Oesch, F. (2009). Genotoxicity investigations on nanomaterials: Methods, preparation and characterization of test material, potential artifacts and limitations – Many questions, some answers. *Mutat Res* **681**: 241–58.

Li, H., Haberzettl, P., Albrecht, C., *et al.* (2007). Inhibition of the mitochondrial respiratory chain function abrogates quartz induced DNA damage in lung epithelial cells. *Mutat Res* **617**: 46–57.

Lindberg, H. K., Falck, G. C., Suhonen, S., *et al.* (2009). Genotoxicity of nanomaterials: DNA damage and micronuclei induced by carbon nanotubes and graphite nanofibres in human bronchial epithelial cells in vitro. *Toxicol Lett* **186**(3): 166–73.

Liu, W., Ernst, J. D., and Broaddus, V. C. (2000). Phagocytosis of crocidolite asbestos induces oxidative stress, DNA damage, and apoptosis in mesothelial cells. *Am J Respir Cell Mol Biol* **23**(3): 371–8.

Lu, J., Keane, M. J., Ong, T., and Wallace, W. E. (1994). In vitro genotoxicity studies of chrysotile asbestos fibers dispersed in simulated pulmonary surfactant. *Mutat Res* **320**(4): 253–9.

Lund, L. G. and Aust, A. E. (1992). Iron mobilization from crocidolite asbestos greatly enhances crocidolite-dependent formation of DNA single-strand breaks in phi X174 RFI DNA. *Carcinogenesis* **13**: 637–42.

Lynch, I., Dawson, K. A., and Linse, S. (2006). Detecting cryptic epitopes created by nanoparticles. *Sci STKE* **2006**(327): pe14.

Maynard, A. D., Baron, P. A., Foley, M., *et al.* (2004). Exposure to carbon nanotube material: Aerosol release during the handling of unrefined single-walled carbon nanotube material. *J Toxicol Environ Health A* **67**(1): 87–107.

McGregor, D. B., Rice, J. M., and Venitt, S., eds. (1999). *The Use of Short and Medium-Term Tests for Carcinogens and Data on Genetic Effects 680 in Carcinogenic Hazard Evaluation.* IARC scientific publication no. 146 (Lyon: International Agency for Research on Cancer).

Migliore, L., Saracino, D., Bonelli, A., *et al.* (2010). Carbon nanotubes induce oxidative DNA damage in RAW 264.7 cells. *Environ Mol Mutagen* **51**(4): 294–303.

Moolgavkar, S. H., Brown, R. C., and Turim, J. (2001). Biopersistence, fiber length, and cancer risk assessment for inhaled fibers. *Inhal Toxicol* **13**(9): 755–72.

Muller, J., Decordier, I., Hoet, P. H., *et al.* (2008a). Clastogenic and aneugenic effects of multi-wall carbon nanotubes in epithelial cells. *Carcinogenesis* **29**(2): 427–33.

Muller, J., Huaux, F., Fonseca, A., *et al.* (2008b). Structural defects play a major role in the acute lung toxicity of multiwall carbon nanotubes: Toxicological aspects. *Chem Res Toxicol* **21**(9): 1698–705.

Muller, J., Delos, M., Panin, N., *et al.* (2009). Absence of carcinogenic response to multiwall carbon nanotubes in a 2-year bioassay in the peritoneal cavity of the rat. *Toxicol Sci* **110**(2): 442–8.

Murphy, F. A., Poland, C. A., Duffin, R., *et al.* (2011). Length-dependent retention of carbon nanotubes in the pleural space of mice initiates sustained inflammation and progressive fibrosis on the parietal pleura. *Am J Pathol* **178**(6): 2587–600.

Nativo, P., Prior, I. A., and Brust, M. (2008). Uptake and intracellular fate of surface-modified gold nanoparticles. *ACS Nano* **2**: 1639–44.

Oberdörster, G. (2002). Toxicokinetics and effects of fibrous and nonfibrous particles. *Inhal Toxicol* **14**: 29–56.

Oberdörster, G., Maynard, A., Donaldson, K., *et al.* (2005). Principles for characterizing the potential human health effects from exposure to nanomaterials: Elements of a screening strategy (ILSI Research Foundation/Risk Science Institute Nanomaterial Toxicity Screening Working Group). *Part Fibre Toxicol* **2**: 8.

Pacurari, M., Yin, X. J., Zhao, J., *et al.* (2008). Raw single-wall carbon nanotubes induce oxidative stress and activate MAPKs, AP-1, NF-kappaB, and Akt in normal and malignant human mesothelial cells. *Environ Health Perspect* **116**(9): 1211–7.

Pantarotto, D., Briand, J. P., Prato, M., and Bianco, A. (2004). Translocation of bioactive peptides across cell membranes by carbon nanotubes. *Chem Commun (Camb)* **1**: 16–7.

Park, S. H. and Aust, A. E. (1998). Participation of iron and nitric oxide in the mutagenicity of asbestos in hgprt-, gpt+ Chinese hamster V79 cells. *Cancer Res* **58**(6): 1144–8.

Patlolla, A. K., Knighten, B., and Tchounwou, P. B. (2010a). Multi-walled carbon nanotubes induce cytotoxicity, genotoxicity and apoptosis in normal human dermal fibroblast cells. *Ethn Dis* **20**(1 suppl. 1): 65–72.

Patlolla, A. K., Hussain, S. M., Schlager, J. J., Patlolla, S., and Tchounwou, P. B. (2010b). Comparative study of the clastogenicity of functionalized and nonfunctionalized multi-walled carbon nanotubes in bone marrow cells of Swiss-Webster mice. *Environ Toxicol* **25**(6): 608–21.

Poland, C. A., Duffin, R., Kinloch, I., *et al.* (2008). Carbon nanotubes introduced into the abdominal cavity of mice show asbestos-like pathogenicity in a pilot study. *Nat Nanotechnol* **3**: 423–8.

Rushton, E. K., Jiang, J., Leonard, S. S., *et al.* (2010). Concept of assessing nanoparticle hazards considering nanoparticle dosemetric and chemical/biological response metrics. *J Toxicol Environ Health A* **73**(5): 445–61.

Sanchez, V. C., Pietruska, J. R., Miselis, N. R., Hurt, R. H., and Kane, A. B. (2009). Biopersistence and potential adverse health impacts of fibrous nanomaterials: What have we learned from asbestos? *Wiley Interdiscip Rev Nanomed Nanobiotechnol* **1**(5): 511–29.

Sargent, L. M., Reynolds, S. H., and Castranova, V. (2010). Potential pulmonary effects of engineered carbon nanotubes: In vitro genotoxic effects. *Nanotoxicology* **4**: 396–408.

Sargent, L. M., Shvedova, A. A., Hubbs, A. F., *et al.* (2009). Induction of aneuploidy by single-walled carbon nanotubes. *Environ Mol Mutagen* **50**(8): 708–17.

Schins, R. P. F., Duffin, R., Höhr, D., *et al.* (2002). Surface modification of quartz inhibits toxicity, particle uptake, and oxidative DNA damage in human lung epithelial cells. *Chem Res Toxicol* **15**: 1166–73.

Schins, R. P. F. and Knaapen, A. M. (2007). Genotoxicity of poorly soluble particles. *Inhal Toxicol* **19**(suppl. 1): 189–98.

Schins, R. P. F. and Hei, T. K. (2007). Genotoxic effects of particles. In Donaldson, K. and Borm, P., eds., *Particle Toxicology* (Boca Raton FL: CRC Press), pp. 285–98.

Shi, X., Mao, Y., Daniel, L. N., *et al.* (1994). Silica radical-induced DNA damage and lipid peroxidation. *Environ Health Perspect* **102**(suppl. 10): 149–54.

Shvedova, A. A., Kisin, E., Murray, A. R., *et al.* (2008). Inhalation vs. aspiration of single-walled carbon nanotubes in C57BL/6 mice: Inflammation, fibrosis, oxidative stress, and mutagenesis. *Am J Physiol Lung Cell Mol Physiol* **295**(4): L552–65.

Srivastava, R. K., Lohani, M., Pant, A. B., and Rahman, Q. (2010). Cyto-genotoxicity of amphibole asbestos fibers in cultured human lung epithelial cell line: Role of surface iron. *Toxicol Ind Health* **26**(9): 575–82.

Stone, V., Johnston, H., and Schins, R. P. F. (2009). Development of in vitro systems for nanotoxicology: Methodological considerations. *Crit Rev Toxicol* **39**: 613–26.

Szendi, K. and Varga, C. (2008). Lack of genotoxicity of carbon nanotubes in a pilot study. *Anticancer Res* **28**(1A): 349–52.

Takagi, A., Hirose, A., Nishimura, T., *et al.* (2008). Induction of mesothelioma in p53$^{+/-}$ mouse by intraperitoneal application of multi-wall carbon nanotube. *J Toxicol Sci* **33**(1): 105–16.

Takata, A., Yamauchi, H., Toya, T., *et al.* (2009). Forsterite exposure causes less oxidative DNA damage and lung injury than chrysotile exposure in rats. *Inhal Toxicol* **21**(9): 739–46.

Tice, R. R., Agurell, E., Anderson, D., *et al.* (2000). Single cell gel/comet assay: Guidelines for in vitro and in vivo genetic toxicology testing. *Environ Mol Mutagen* **35**(3): 206–21.

Tkachenko, A. G., Xie, H., Coleman, D., *et al.* (2003). Multifunctional gold nanoparticle-peptide complexes for nuclear targeting. *J Am Chem Soc* **125**: 4700–1.

Unfried, K., Schürkes, C., and Abel, J. (2002). Distinct spectrum of mutations induced by crocidolite asbestos: Clue for 8-hydroxydeoxyguanosine-dependent mutagenesis in vivo. *Cancer Res* **62**(1): 99–104.

Unfried, K., Albrecht C, von Mikecz, A., Grether-Beck, S., and Schins, R. P. F. (2007). Cellular responses to nanoparticles: Target structures and mechanisms. *Nanotoxicology* **1**: 52–71.

Wang, X., Li, Q., Xie, J., *et al.* (2009). Fabrication of ultralong and electrically uniform single-walled carbon nanotubes on clean substrates. *Nano Lett* **9**(9): 3137–41.

Warheit, D. B. and Donner, E. M. (2010). Rationale of genotoxicity testing of nanomaterials: Regulatory requirements and appropriateness of available OECD test guidelines. *Nanotoxicology* **4**(4): 409–13.

Watters, G. P., Smart, D. J., Harvey, J. S., and Austin, C. A. (2009). H2AX phosphorylation as a genotoxicity endpoint. *Mutat Res* **679**(1–2): 50–8.

Wirnitzer, U., Herbold, B., Voetz, M., and Ragot, J. (2009). Studies on the in vitro genotoxicity of baytubes, agglomerates of engineered multi-walled carbon-nanotubes (MWCNT). *Toxicol Lett* **186**(3): 160–5.

Wood, R. D., Mitchell, C. E., Sgouros, J., and Lindahl, T. (2001). Human DNA repair genes. *Science* **191**: 1284–9.

Xu, A., Smilenov, L. B., He, P., *et al.* (2007). New insight into intrachromosomal deletions induced by chrysotile in the gpt delta transgenic mutation assay. *Environ Health Perspect* **115**(1): 87–92.

Yamashita, K., Yoshioka, Y., Higashisaka, K., *et al.* (2010). Carbon nanotubes elicit DNA damage and inflammatory response relative to their size and shape. *Inflammation* **33**(4): 276–80.

Yang, H., Liu, C., Yang, D., Zhang, H., and Xi, Z. (2009). Comparative study of cytotoxicity, oxidative stress and genotoxicity induced by four typical nanomaterials: The role of particle size, shape and composition. *J Appl Toxicol* **29**(1): 69–78.

Yeganeh, B., Kull, C. M., Hull, M. S., and Marr, L. C. (2008). Characterization of airborne particles during production of carbonaceous nanomaterials. *Environ Sci Technol* **42**(12): 4600–6.

Zhu, L., Chang, D. W., Dai, L., and Hong, Y. (2007). DNA damage induced by multiwalled carbon nanotubes in mouse embryonic stem cells. *Nano Lett* **7**(12): 3592–7.

10

Carbon nanotube–cellular interactions: macrophages, epithelial and mesothelial cells

VICKI STONE, ALI KERMANIZADEH, HELINOR JOHNSTON, MATTHEW BOYLES, JULIA VARET

10.1 Introduction

Previous chapters in this book have described carbon nanotubes and other high-aspect-ratio nanomaterials (HARN), and have dealt with their *in vivo* toxicology. These chapters provide evidence to suggest that HARN are likely to comply with the fibre pathogenicity paradigm (see Chapter 2), which means that their length (greater than 10–15 μm) and durability, as well as the degree to which the HARN are long straight fibres versus entangled structures, are key features in terms of their pathogenicity (Poland *et al.*, 2008) (Figure 10.1). Newly developed HARN are likely to vary with respect to composition, surface chemistry, surface reactivity and solubility. While dissolution has in the past suggested a relatively low potency for fibres, the impact of release of toxic components from fibres over a short or protracted time is not yet understood. Furthermore, the nanoscale-associated properties that make these materials useful for industrial exploitation could also impact upon bioreactivity.

In this chapter we will review the interactions between nanotubes and *in vitro* cellular models in order to further understand their potential mechanisms of toxicity following exposure via inhalation, as well as to discuss the potential for using such models as predictive alternatives to animal testing in the future.

Following inhalation there are a number of cell types that particles are likely to encounter. This review will focus firstly on the alveolar or bronchial epithelial cells at the site of deposition, secondly on immune cells such as macrophages as they attempt to clear the particles from the respiratory system, and then on mesothelial cells to represent the target most relevant following penetration into the pleural cavity.

The Toxicology of Carbon Nanotubes, ed. Ken Donaldson, Craig A. Poland, Rodger Duffin and James Bonner.
Published by Cambridge University Press. © Cambridge University Press 2012.

(a)   (b)

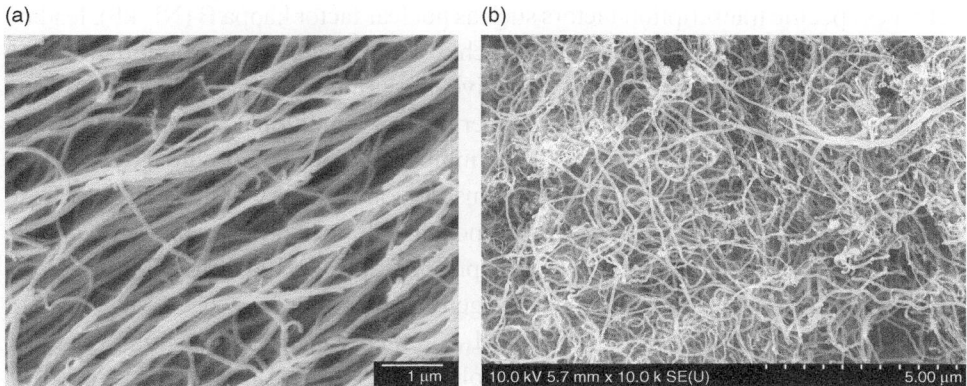

Figure 10.1 MWCNTs with (a) long straight or (b) entangled morphology.

10.2 Mechanisms of toxicity: oxidative stress and inflammation

There are a number of physical and chemical characteristics of particles that can influence the toxicity of particles in the human body. Study of inhaled particles has demonstrated that size, surface area, shape (aspect ratio), composition, dissolution/durability and crystal structure are factors that can influence the potential of particles to induce inflammation and toxicity (Stone and Kinloch, 2007). Pharmaceutical particle research has also demonstrated that surface charge can influence particle uptake and toxicity (Arvizo *et al.*, 2010). For fibres, clearly fibre length and durability determine pathogenicity because they lead to biopersistence. The pathogenicity arises from the chronic inflammation that develops, involving the release of reactive oxygen species (ROS) from phagocytic cells, as well as pro-inflammatory mediators (Donaldson *et al.*, 2011). Therefore, central to the pathogenicity of pathogenic fibres are their physicochemical characteristics.

Free radicals and other oxidants are produced as metabolic by-products in an oxygen-rich environment by a large number of physiological processes (Grimm *et al.*, 2010). The main sources of free radicals include electron leakage from the mitochondrial electron transport chain, the generation of hydroxyl radicals by Fenton-type reactions, and the production of superoxide anion radicals, hydrogen peroxide and hypochlorite as a consequence of several enzymatic reactions (Grimm *et al.*, 2010). An imbalance between the formation and detoxification of ROS can lead to oxidative stress (Grimm *et al.*, 2010), which can have damaging consequences for normal cell function.

The literature suggests that the consequences of oxidative stress are 'tiered' in relation to the severity of the oxidative insult, with relatively low levels inducing a protective response (e.g. antioxidant upregulation), while moderate levels initiate inflammation, and extreme levels induce an overwhelming of the antioxidant defences, leading to cell death (Nel *et al.*, 2006). At sub-lethal levels oxidative stress activates specific transcription factors such as nuclear factor kappa B (NF-kß), leading to their nuclear translocation and the switching on of genes that drive inflammation (Han *et al.*, 2010). NF-kß is a redox-sensitive transcription factor that is formed from two subunits (e.g. p65 and p50), and under normal conditions is located in the cell cytoplasm as a result of the action of the inhibitory subunit Ikß. On activation, the inhibitory subunit detaches from the transcription factor, which is then able to translocate to the nucleus where it is responsible for controlling gene expression, and in particular genes involved in cell proliferation and inflammation. Oxidant-mediated increases in pro-inflammatory gene expression have been observed for environmental particulate air pollution (PM_{10}) (Jimenez *et al.*, 2000), ultrafine carbon black (Brown *et al.*, 2004), nanoparticle polystyrene beads (Brown *et al.*, 2001) and also pathogenic fibres such as asbestos (Brown *et al.*, 1999).

10.3 Fibres and oxidative stress and inflammation

In vivo studies that have been conducted to assess the pulmonary toxicity of CNTs and other pathogenic fibres (such as asbestos) have provided the basis for *in vitro* investigations that assess the implications of CNT exposure on lung cells and investigate the mechanisms underlying any observed effects. The effects of CNTs within the lung are considered to be mediated, in part, by ROS. Both SWCNTs and MWCNTs have been shown to cause oxidative stress and inflammation *in vivo*. SWCNT exposure via the lungs has been shown to induce a number of markers of oxidative stress such as increased lipid peroxidation and depletion of the antioxidants glutathione and ascorbic acid (Shvedova *et al.*, 2007). This work demonstrated that intratracheal instillation of SWCNTs resulted in an inflammatory response within mouse lungs that was virtually eliminated by pre-treatment with the antioxidant ascorbic acid. The same authors also demonstrated that vitamin E deficiency within mice resulted in alterations in antioxidant status that increased susceptibility to oxidative stress, leading to a higher sensitivity to the SWCNT-induced inflammation and fibrosis (Shvedova *et al.*, 2007). Taken together these studies strongly suggest a role for an oxidant-mediated mechanism in the induction of pulmonary inflammation by the SWCNTs studied by Shvedova and colleagues.

MWCNTs have also been shown to produce ROS (Han *et al.*, 2010; Rothen-Rutishauser *et al.*, 2010; Thurnherr *et al.*, 2011) and to induce antioxidant defence mechanisms suggesting the stimulation of increased ROS production by macrophages in response to MWCNT exposure (Brown *et al.*, 2010). MWCNTs have also been shown to enhance pro-inflammatory mediator production, which was alleviated by the addition of antioxidants (Brown *et al.*, 2010) and ROS scavengers (Han *et al.*, 2010), suggesting that ROS were involved in the mechanism by which the MWCNT induced pro-inflammatory mediator production.

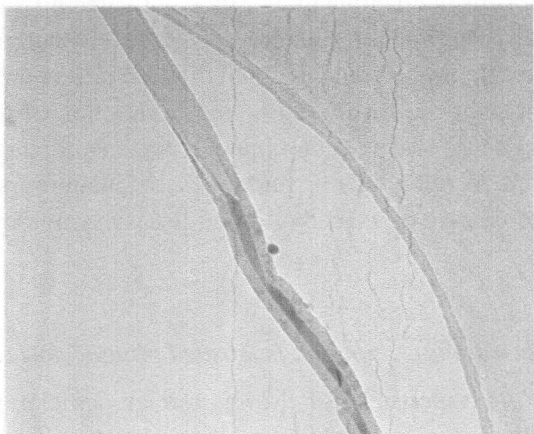

Figure 10.2 Iron encapsulated within a MWCNT. (SEM image provided courtesy of Dr. Ian Kinloch.)

Following the exposure of the lungs to CNTs it is therefore apparent that an inflammatory response is stimulated, which is typified by the recruitment of neutrophils and an increase in markers of inflammation such as pro-inflammtory cytokines (Muller *et al.*, 2005; Rothen-Rutishauser *et al.*, 2010). It is debatable whether there is a role for oxidative stress and ROS in driving the pulmonary inflammatory response stimulated by CNTs.

CNTs often contain metals because of their use as catalysts in the manufacturing process. It is possible in some CNT samples that the iron is not readily bioavailable as a result of its containment within the nanotube lumen. However a difference in toxicity has been observed for purified and non-purified CNT samples by a number of investigators (Kagan *et al.*, 2006; Pulskamp *et al.*, 2007). It has also been shown, using measurements of iron mobilisation and redox activity, that, although most iron remains enclosed and inaccessible within the CNT (Figure 10.2), a small quantity can be released into suspension fluids. This iron was sufficient to cause redox reactions leading to single strand breaks in DNA (Guo *et al.*, 2007). Guo *et al.* indicated that additional stresses, such as those encountered by CNTs post-production (oxidation, mechanical grinding and sonication) may also further increase iron availability. These studies suggest that the iron content of CNTs could also contribute to their toxicity.

10.4 Macrophages

Macrophages are long-lived leukocytes that are responsible for providing immune surveillance in tissues. They play a pivotal role in innate immunity as well as being crucial in contributing to the adaptive immune response if required (Gordon and Taylor, 2005). They can participate in an immune response as antigen-presenting cells and as effector cells in adaptive immunity if activated by T lymphocytes or by antibodies secreted by the B cells (Shalhoeb *et al.*, 2011). Macrophages derive from monocytes that circulate in the blood and that differentiate as they leave the bloodstream. They are responsible for ingestion and destruction of foreign antigens as well as for releasing a vast array of cytokines. Hence, macrophages are responsible for inducing and implementing inflammation, which if inappropriate in amplitude or duration can lead to disease or exacerbation of disease symptoms (Donaldson and Stone, 2003).

10.4.1 Macrophages: uptake, frustrated phagocytosis and ROS

Since macrophages are responsible for clearing foreign particles, they play a critical role in biological responses to inhaled fibres. Upon activation (e.g. during uptake of a particle) macrophages may instigate a phagocytic burst associated with formation of ROS (Chanock *et al.*, 1994; Valko *et al.*, 2007; Takahashi *et al.*, 2011). Since macrophages are only 10–15 μm in diameter, they are limited in terms of the size of the materials that they can ingest. Therefore, fibres of greater than 15 μm are not readily phagocytosed and cleared by macrophages (Tomatis *et al.*, 2010). If a macrophage fails to engulf its target fully (i.e. long fibres, >15 μm) in an action referred to as 'frustrated phagocytosis', the contents of its phagosome (mixture of anti-microbial peptides, lysosomal digestive enzymes (proteases, neutrophil elastase (ELA 2), cathepsins), prostoglandins and highly toxic oxygen and nitrogen compounds) are released to affect the function of surrounding cells (Dorger and Krombach, 2000; Poland *et al.*, 2008), as these agents have the potential to elicit damage to healthy cells and tissues (Wu and Sun, 2011). The importance of frustrated phagocytosis and the association between fibre length and enhancement of pro-inflammatory and oxidative conditions have been clearly demonstrated with exposures to asbestos fibres (chrysotile and crocidolite) (Mongan *et al.*, 2000), glass fibres (Ye *et al.*, 1999) and MWCNTs (Brown *et al.*, 2007) of different lengths (Figure 10.3). Because of their structural similarities to asbestos it is hypothesised that carbon nanotubes could potentially exert toxicity in a similar manner to that observed for asbestos, provided that the nanotubes are high in aspect ratio with relatively long length.

SWCNT uptake by macrophages has been studied utilising nanotubes filled with silver iodide to enable imaging via confocal microscopy (Porter *et al.*, 2007). Human macrophages were exposed to the SWCNTs for 2 or 4 days, at concentrations of up to 10 μg/ml. A cytotoxic effect was evident at concentrations above 0.3 μg/ml after 4 days of exposure, which was associated with a greater cellular uptake of SWCNTs. At 2 days SWCNTs were found within the lysosomes of the macrophages, while at day 4 SWCNT bundles were fused with the plasma membrane, which caused disruptions to membrane structure, to enable further uptake. SWCNTs

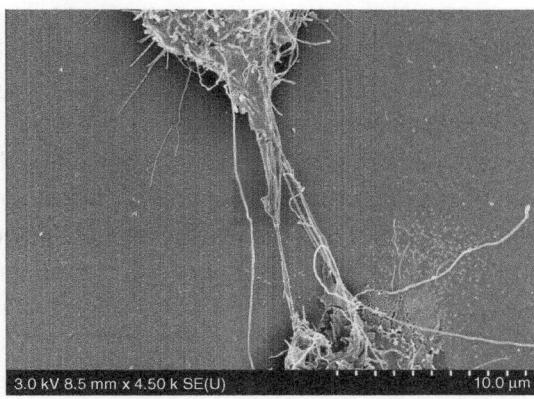

3.0 kV 8.5 mm x 4.50 k SE(U)   10.0 μm

Figure 10.3 Frustrated phagocytosis of MWCNT: two macrophages are attempting but failing to take up numerous MWCNTs.

in that particular study were relatively short (often less than 1 μm), which means that their uptake by macrophages was likely to be more manageable, in comparison to longer fibres.

Macrophages loaded with SWCNTs *in vivo* (in the mouse lung) and *in vitro* have been associated with a cytotoxic response. It has been demonstrated that SWCNTs can aggregate within alveolar macrophages subsequent to the intratracheal instillation of mice (Lam *et al.*, 2004). The SWCNT-laden macrophages were shown to cluster and form granulomas, which could later lead to impaired lung function and give rise to a fibrotic reaction and chronic inflammation. This highlights the importance of the uptake of SWCNTs and the eventual toxicological outcome (Lam *et al.*, 2004). These studies are not likely to be associated with frustrated phagocytosis because of the short length of the SWCNTs; however, the exposures were still associated with toxic responses, suggesting a non-fibre-mediated mechanism of toxicity.

The enhanced generation of ROS and subsequent development of oxidative stress has been demonstrated in response to SWCNTs, despite their short length, with increased lipid peroxidation, extracellular superoxide dismutase cleavage, decreased antioxidants glutathione and ascorbic acid (Shvedova *et al.*, 2007). However, in other studies in which macrophages were exposed to SWCNTs, there was no oxidative burst or production of nitric oxide (NO). In this study the authors reported that the macrophages did not attempt to phagocytose the fibres (Fiorito *et al.*, 2005; Kagan *et al.*, 2006).

Pulskamp *et al.* (2007) have also shown that intracellular nitric oxide is not produced in response to SWCNTs or MWCNTs. Additionally it was shown that lipopolysaccharide (LPS)-induced formation of NO is dampened with the inclusion of MWCNTs or SWCNTs. The authors suggested that this could be due to the binding of CNTs to the protein inducible nitric oxide synthase (iNOS). Therefore, although SWCNTs and MWCNTs can induce ROS, this does not necessarily translate to nitric oxide production.

MWCNTs can be generated in a variety of lengths, including those considered to be long (greater than 10–15 µm), suggesting that they could conform to the fibre pathogenicity paradigm. Poland *et al.*, (2008) observed frustrated phagocytosis by peritoneal macrophages isolated from mice exposed to long, straight MWCNTs. Furthermore, Brown *et al.* (2007) compared the uptake of long (50 µm), straight MWCNTs with that of shorter (10 µm) and entangled MWCNT samples *in vitro*, to assess their ability to induce frustrated phagocytosis. The uptake of the longer fibres was observed to be incomplete, while a visible proportion of the 10 µm sample protruded from the cell surface, and the entangled specimen was completely engulfed. This correlated with the superoxide anion production, with both the 50 µm and 10 µm samples stimulating frustrated phagocytosis. A recent study using a cell free system (Rothen-Rutishauser *et al.*, 2010) has demonstrated that high surface area of entangled MWCNTs and long straight MWCNTs can generate considerably more ROS than a short MWCNT with relatively small surface area. Upon incubation of MWCNTs with the phagocytic cell line J774.A1 it was confirmed that the entangled sample was rapidly phagocytosed by the macrophages, shown through the visualisation of dark aggregates of material within the cells, but also from the rapid ROS response, seen after only 30 minutes.

In a recent study (Hirano *et al.*, 2008), the uptake of MWCNTs by macrophages was associated with cell death, and was suggested to occur as a result of the mechanical rupture of the cell membrane. The direct interaction of MWCNTs with the plasma membrane was partly mediated by their association with the macrophage receptor with collagenous structure (MARCO) scavenger receptors, on the cell surface. The authors suggested that MWCNTs bind to macrophages via the MARCO receptor, which encouraged the plasma membrane to extend along the material, causing damage to the membrane and eventual cell death. However the cytotoxicity observed was not dose-dependent. At low concentrations (10–100 µg/ml) there was a decrease in cell viability, in a time-dependent manner (up to 32 hours). At higher MWCNT concentrations (100–1000 µg/ml), cytotoxicity to the cells was reduced compared to lower-concentration exposures (Hirano *et al.*, 2008). It was suggested that this unusual dose–response relationship is likely to derive from the agglomeration of the MWCNTs, which reduced their interaction with cells and therefore reduced their cytotoxicity. The authors also pre-treated the macrophages with an antioxidant (N-acetyl cysteine) and found that this treatment was unable to prevent the detrimental impact of MWCNTs on macrophage viability, suggesting that ROS were not involved in the cytotoxicity observed.

In another recent study (Jia *et al*., 2005), primary guinea pig alveolar macrophages were exposed to a panel of carbon-based nanomaterials consisting of SWCNTs (1.4 nm diameter, 1 μm length), MWCNTs (10–20 nm diameter, 0.5–40 μm length) and C_{60}. The SWCNTs were found to be most capable of inducing cytotoxicity, although MWCNTs were also demonstrated to elicit a similar response, albeit to a lesser extent. However, the authors suggest that the enhanced cytotoxicity of SWCNTs witnessed could be attributed to the higher impurity levels within the materials. Morphological assessment (using TEM) suggested that CNT-induced cytotoxicity occurred via apoptosis. The ability of macrophages to retain their phagocytic function was also assessed by determining the ability of macrophages to internalise latex beads, subsequent to particle exposure. SWCNTs were able to impair phagocytosis to the greatest extent, while MWCNTs and C_{60} required higher concentrations to induce such a response (Jia *et al*., 2005). In contrast Pulskamp *et al*. (2007) demonstrated that SWCNTs and MWCNTs elicited low acute cytotoxicity to alveolar macrophages (NR8383) at concentrations of 5–100 μg/ml for 24 hours. However, CNTs were internalised by macrophages and were observed to detrimentally affect mitochondrial function and enhance intracellular ROS production, but no inflammatory response was associated with exposure within either cell type (Pulskamp *et al*., 2007). The toxicity of CNTs was also related to their purity, so that a decrease in metal content of samples attenuated their (oxidative) toxic potency.

The authors of a recent study (Fiorito *et al*., 2009) aimed to verify whether MWCNT electrical properties could affect macrophage integrity. Human macrophages were challenged with a positively charged MWCNT sample. It was demonstrated that the charged MWCNTs were less cytotoxic to the macrophages but possess a higher inflammatory potential, as compared to control non-charged MWCNTs. Moreover, only charged MWCNTs significantly affected macrophage mitochondrial membrane polarity.

It has been suggested that cell signaling events instigated at the cell surface can lead to the activation of transcription factor NF-κβ. Accordingly, such a response may be responsible for mediating the inflammatory response associated with the exposure of macrophages to MWCNTs (Han *et al*., 2010). The inflammatory response associated with MWCNT exposure can potentially stimulate release of ROS (Kobayashi *et al*., 2010).

In an uptake study of SWCNTs by differentiated THP-1 monocytic cells (Chou *et al*., 2008), the particles were found to trigger the release of ROS which was responsible for the activation of the transcription factor NF-κβ, which in turn stimulation the production of pro-inflammatory cytokines and chemokines.

10.4.2 Macrophages and inflammatory signaling

The ability of macrophages to stimulate a pro-inflammatory and fibrotic response within macrophages has been an area of great interest because of evidence that CNTs can stimulate an inflammatory response *in vivo* (Shvedova *et al.*, 2005; Rothen-Rutishauser *et al.*, 2010). *In vitro* investigations have therefore been utilised in order to understand the mechanisms underlying the inflammatory response observed, and the type of response stimulated (through assessment of the cytokine production).

Macrophages are capable of producing a wide array of cytokines following exposure to, and internalisation of, an antigen. However there is contradictory evidence as to whether this is the case following exposure to CNTs. No discernable release of TNF-α IL-1β, IL-10 or TGF-β was observed upon *in vitro* exposure of RAW 264.7 macrophages to SWCNTs (Shvedova *et al.*, 2005); and no release of TNF-α was associated with RAW 264.7 macrophage cells in response to CNT exposure, although release was elicited through LPS stimulation. Of interest is that macrophages previously stimulated with LPS are not shown to increase their inflammatory response (TNF-α secretion) with secondary exposure of CNTs, nano-particle CB or quartz (Pulskamp *et al.*, 2007).

However, others have shown that MWCNTs and SWCNTs stimulate the release of IL-10, IL-12, IL-8 and TNF-α (Grecco *et al.*, 2011). Finally, a number of *in vivo* studies have demonstrated an inflammatory response following inhalation exposure of mice (Shvedova *et al.*, 2005) and rats (Muller *et al.*, 2005), as well as intraper-itoneal exposure of mice (Poland *et al.*, 2008) to CNTs. *In vitro*, MWCNT-exposed macrophages (Brown *et al.*, 2007) have also been shown to enhance production of pro-inflammatory mediators (TNF-α).

Shvedova *et al.* (2005) used the macrophage cell line (RAW 264.7) to evaluate the contribution of this cell type to the SWCNT-induced pulmonary fibrogenic responses previously witnessed in mice. They found that the production of the pro-fibrotic cytokine TGF-β was enhanced and was comparable to the levels induced by the classical stimulant zymosan, suggesting that macrophages are involved in the mediation of fibrotic responses on exposure to SWCNTs. However, production of TNF-α and IL-1β was markedly lower than that induced by zymosan; therefore the release of these pro-inflammatory cytokines was con-sidered to be a potentially less important component of macrophage response to SWCNTs (Shvedova *et al.*, 2005). One explanation for the relatively low uptake of SWCNTs by the RAW macrophage cell line could have been the fact that the SWCNTs were relatively well dispersed. Therefore, perhaps larger structures are better recognised and internalised by macrophages, and therefore more capable of stimulating a toxic response than their smaller counterparts (Shvedova *et al.*, 2005).

10.4.3 Additional factors influencing macrophage: CNT interactions

One very important factor which could influence the *in vitro* assessment of CNT-induced cellular effects relates to whether the nanomaterials come into direct contact with the cells. In an interesting study, Muller *et al.* (2005) exposed primary rat peritoneal macrophages to ground or unground MWCNTs for 6–24 hours, at concentrations of up to 100 µg/ml. While the ground MWCNTs were relatively well dispersed within the cell culture medium, the unground MWCNTs formed large aggregates that floated within the medium, and as a result were not in direct contact with the cells. It was also observed that the ground MWCNTs, chrysolite asbestos and ultrafine carbon black were all capable of inducing the production of TNF-α and stimulating cytotoxicity. These MWCNTs were found to be more pathogenic than ultrafine carbon black, and were either equivalent to, or less toxic than, asbestos. In contrast, the larger aggregated MWCNT samples had no effect on the macrophages. There is therefore evidence, both *in vivo* and *in vitro*, that the toxicity of MWCNTs is dictated by their characteristics and/or dispersion. The aggregation and agglomeration state of CNTs therefore has the ability to impact on their toxicity. As described for epithelial cells, processes are used in order to promote the dispersion of CNTs within the biological media utilised within *in vitro* tests. There is a need for compromise between obtaining a physiologically relevant sample (to reflect what cells are exposed to *in vivo*) and obtaining a well-dispersed sample that can be utilised to assess the properties of CNTs that drive any observed toxicity.

The purity of CNTs can play a crucial role in their ability to affect the response of macrophages. Of interest is that the presence of iron within asbestos is believed to partly contribute to its toxicity (Shukla *et al.*, 2003a). The potential of residual iron within CNTs derived from the production process therefore has the potential to contribute to the effects mediated by CNTs. The ability of metal contaminants to impact on the toxicity of CNTs has been demonstrated in a study in which purified CNT (reduced metal content) samples lacked the ability to initiate an oxidative burst or produce ROS (Pulskamp *et al.*, 2007), while non-purified samples with a high metal content were able to cause oxidative burst from the macrophages after a 10-minute exposure, and dose-dependently for up to 24 hours. Interestingly, exposure to CNT samples with reduced levels of amorphous carbon resulted in a delayed increase in ROS levels , irrespective of iron content (Pulskamp *et al.*, 2007). The role of iron is further illustrated in experiments in which macrophages exposed to CNTs with high iron content (26 wt%) generated more ROS, and depleted a greater proportion of antioxidant reserves compared to low iron CNTs (0.23 wt%) (Kagan *et al.*, 2006). Furthermore, Shvedova *et al.* (2008b) noted a depletion of intracellular antioxidants in rats exposed to 5 mg/m^3 short SWCNTs (<1 µm) with a high iron content (17 wt%) for 4 days. Measurements were taken from 1, 7 and 28 days after initial exposure; antioxidant depletion was significantly evident after 1 day post-exposure.

10.4.4 Macrophage summary

The structural resemblance of carbon nanotubes to pathogenic fibres such as asbestos has prompted concerns surrounding their safety. As such, the fibre patho-genicity paradigm has been fundamental in understanding the response of macro-phages to CNTs, and has allowed for investigation of the properties of CNTs that drive their toxicity, and directed investigations responsible for identifying the mechanisms by which CNTs elicit toxicity. Determining the uptake, production of ROS and stimulation of an inflammatory response has been of particular interest because of the importance of the induction of frustrated phagocyotsis within the response of macrophages to pathogenic fibres, which is a phenomenon which is known to be driven by fibre length. The uptake of CNTs by macrophages is dependent on their size, so that longer (>15 µm) CNTs cannot be effectively internalised by macrophages, which stimulates frustrated phagocytosis. CNTs are also able to stimulate inflammatory, oxidative and genotoxic consequences within macrophages that can culminate in cytotoxicity. However, it has been revealed that the response of macrophages to CNTs is driven by their chemical and physical properties. This derives from the fact that CNTs are a diverse population of materials that differ with respect to their morphology and surface characteristics (length, diameter, surface area, shape, contaminant metals and crystallinity), which can influence their toxicity. Furthermore, the experimental setup utilised is also able to influence the findings observed, including, for example, the model used, the con-centrations, exposure times and dispersion procedure. It is widely assumed that both direct and indirect effects of these fibres might be responsible for inducing their adverse affects *in vitro* and *in vivo*, and a greater understanding of the molecular mechanisms of CNTs would be extremely useful.

10.5 Epithelial cells

Epithelial cells provide a major barrier within the lung that impedes the uptake of inhaled particles and pathogens into the body. The phenotype of epithelial cells varies according to their location within the lung, and includes ciliated columnar cells such as those found in the bronchial regions and non-ciliated flattened cells such as those found in the alveolar regions. There are two main types of alveolar epithelial cells: type I cells are organised in order to ensure gas exchange, while type II cells are responsible for surfactant production and are progenitor cells for the alveolar

Table 10.1 *Examples of* in vitro *models of lung epithelial cells.*

Name	Origin
A549 cells	Human carcinoma alveolar epithelial cells type II
BEAS-2B cells	Transformed human bronchial epithelial cells
CaLu-1 cells	Human lung epidermoid carcinoma
CaLu-3 cells	Human lung adenocarcinoma
C10 cells	Murine alveolar epithelial cells
H1299 cells	Human lung carcinoma; non-small-cell lung cancer
H446 cells	Human carcinoma; small-cell lung cancer
H596 cells	Human lung adenosquamous carcinoma
NHBE cells (normal human bronchial epithelial cells)	Primary lung epithelial cells

epithelium (which are required, for example, to replace damaged cells). Regardless of their morphology, these cells provide the main barrier function of the lung, as well as participating in the inflammatory response. Damage to this layer and translocation of particles across the cell layer can lead to particle uptake into either the interstitium, lymphatics, pleura or blood. The translocation of particles from the lung to secondary target sites is likely to be dictated by the chemical and physical properties of the particles, and may be integral to their toxicity. This is exemplified by the finding that fibres have been demonstrated to migrate to the pleura, leading to mesothelial cell interaction and the potential to induce diseases such as mesothelioma, while inter-stitialisation may lead to fibrosis (e.g. asbestosis) (Miserocchi *et al.*, 2008).

Predictive, *in vitro* models are essential when investigating the response of the lung to CNT exposure, and are necessary because of the diverse array of CNT forms available. In the main, immortalised cell lines are utilised, and assumed to be representative of the response of cells *in vivo*. There are several *in vitro* cell models of lung epithelial cells that have been used to assess the toxicity of CNTs, outlined in Table 10.1. Knowledge of how asbestos fibres and nanoparticles elicit toxicity has influenced the considerations made when assessing the toxicity of CNTs within epithelial cells. In addition, asbestos fibres and carbon nanoparticles are often used as controls for comparison with CNTs (e.g. do they conform to the fibre paradigm?).

10.5.1 Epithelial cells: investigating the cytotoxicity of CNTs

Mittal *et al.* (2011) recently demonstrated that a panel of CNTs interfered with many classical cytotoxicity assays, such as MTT, LDH and Neutral Red Uptake (NRU). The authors therefore stressed the need for further assay development. They were able to show cytotoxicity of CNTs *in vitro* in normal human bronchial epithelial cells, through cytometry and morphological changes. We would suggest that it is useful to use the standard viability assays in order to allow comparison with other particles and chemicals, but, before use, stringent control studies should be developed and used in order to minimise the risk for misinterpretation of data due to particle interference.

MWCNTs have been demonstrated, using such tests, to induce a dose-dependent decrease in cell viability in epithelial cell models (Srivastava *et al.*, 2011). Herzog *et al.* (2007) showed that SWCNTs (produced by HiPCO™ or arc discharge) were cytotoxic to A549 and BEAS-2B epithelial cells after a 72-hour exposure, and that they slowed down cell proliferation in lung epithelial cells after a longer incubation (of 10 days). HiPCO™ CNTs (high metal content) were found to be more toxic than the arc-discharge CNTs (low metal content, high amorphous carbon content). The authors attributed the difference in toxicity to the production method, metal content and the tendency of the HiPCO™ SWCNTs to form smaller aggregates. The metal content of MWCNTs has also been linked to their toxicity (Pulskamp *et al.*, 2007).

Casey *et al.* (2008) utilised the same particles as Herzog (Davoren *et al.*, 2007) and demonstrated the adsorption of components of cell culture medium required for maintaining cell viability (such as growth factors) to CNTs. This therefore reduced their bioavailability and triggered a cytotoxic response within A549 epithelial cells. Again, the HiPCO™ SWCNTs were demonstrated to exhibit the greatest toxicity, which was speculated to derive from their larger surface area and therefore greater capacity to adsorb medium components. Interactions between biological molecules and CNTs may therefore be important in driving their toxicity. Interestingly, double-walled carbon nanotubes (DWCNTs) are capable of binding surfactant protein A and D, in a calcium-dependent manner. The authors suggested that this 'sequestration' of surfactant proteins by CNTs may increase susceptibility to lung disease because of the important role of these proteins in lung immune defense (Salvador-Morales *et al.*, 2007).

CNTs are a very diverse population of materials, and thus direct comparisons of the toxicity exhibited by CNTs and asbestos fibres can help to reveal the properties of materials most influential in driving their toxicity. Hirano *et al.* (2010) illustrated that MWCNTs displayed a higher cytotoxic potential than crocidolite asbestos in BEAS-2B cells ($IC_{50\ MWCNT}$ 12 µg/ml, $IC_{50\ crocidolite}$ 678 µg/ml). In contrast, another study (Thurnherr *et al.*, 2011) demonstrated that MWCNTs only slightly increased the proportion of apoptotic A549 cells (30 µg/ml) while crocidolite asbestos was more cytotoxic. It is possible that differences in toxic potency between the various fibres tested can be attributed to differences in their physicochemical properties. Characterisation information is therefore integral in identifying why there are differences in the cytotoxicity exhibited by different CNT samples and what attributes of the fibres drive the responses observed.

10.5.2 Approaches to the dispersion of CNTs, and their impact on epithelial cell function

An important factor which may influence the hazard assessment of CNTs, and in fact of any nanomaterial type, is the degree of dispersion within the biological media utilised to expose cells *in vitro*. CNTs have a tendency to agglomerate within cell culture media. The most appropriate approach to dispersing CNTs is debatable, and a number of processes can be used to manipulate the dispersion of CNTs within biological media. These include physiologically relevant dispersants (e.g. serum, surfactants), solvents, mechanical processes (such as sonication) or a combination of these processes. A compromise must therefore be made between a physiologically relevant dispersion of CNTs and the use of processes to improve this dispersion within biological media, and this will depend on the aims of the study.

The lung surfactant dipalmitoylphosphatidylcholine (DPPC) has been used to disperse crocidolite asbestos fibres. However, it was observed that DPPC did not change the toxicity of fibres in A549 and NHBE cells unless they were pre-treated with the pro-inflammatory cytokine TNF-α (Herzog *et al.*, 2009). TNF-α pre-stimulation provides a model for inflamed tissue, and so this study suggests that asbestos, if dispersed with a lung lining fluid component such as DPPC, could induce enhanced toxicity if exposed to an inflamed lung, and thus that those with pre-existing disease may be more susceptible to toxicity.

Lung surfactant has also been utilised to improve the dispersion of CNTs (Wang *et al.*, 2010). Low concentrations of SWCNTs dispersed in lung lining fluid were able to stimulate BEAS-2B cell proliferation, while higher concentrations inhibited cell proliferation. However, the same SWCNTs exposed to cells in the absence of surfactant did not impact on cell viability (Wang *et al.*, 2010). This study therefore suggests that the surfactant improved the dispersion of CNTs, thereby increasing their toxicity.

In general, studies using cell lines disperse CNTs within a cell culture medium containing serum. This derives from the knowledge that serum is required to maintain the viability of cells *in vitro*, but also because serum can improve the dispersion of nanomaterial suspensions, presumably through the coating of nanomaterials with proteins in order to prevent their agglomeration. The inclusion of serum (5%) has been demonstrated to improve the dispersion of SWCNTs, which then elicited a low level of cytotoxicity within A549 cells. However, it was observed that the cytotoxicity of SWCNTs to A549 cells was increased when they were dispersed in a serum-free medium (Davoren *et al.*, 2007). This is likely to derive from the fact that serum is known to stimulate cell survival and proliferation because of its high content of growth factors, and thus the absence of serum within the SWCNT dispersion may make cells more susceptible to CNT toxicity. However, this could also be due to the protein coating of the CNTs modifying their surface properties and thence their ability to induce toxicity. Of course the relevance of using serum in a model simulating the lung surface is debatable, and so it may be that, in such studies, other dispersants should be considered.

Other methods for improving the dispersion of CNTs within biological media include the use of sonication and solvents. In order to reduce the agglomeration of MWCNTs within a cell culture medium, the impact of a combination of sonication and solvent (Tween 80®, dimethylsulfoxide (DMSO) or carboxymethyl cellulose sodium salt (CMC)) was investigated by Asakura *et al.* (2010). The duration of sonication and choice of solvent affected the dispersion of MWCNTs, with improved dispersion demonstrated to increase their cytotoxicity to CHL/IU cells. In fact, the authors hypothesised that improved dispersion of CNTs could facilitate their uptake and therefore enhance their ability to induce cytotoxicity.

10.5.3 Epithelial cells: particle uptake

The mechanism of fibre and nanotube toxicity to epithelial cells could include uptake, or an effect mediated from outside the cells. A number of studies suggest that the internalisation of fibres by epithelial cells is dictated by their length. This is exemplified by the finding that Chinese hamster lung epithelial cells incompletely internalised chrysotile fibres that were longer than 10 μm (Asakura *et al.*, 2010). However, other authors report the absence of uptake of asbestos fibres by epithelial cells, suggesting that although uptake can occur, it is not a applicable to all fibres (Wang *et al.*, 2006).

Alveolar epithelial cells are not specialised in phagocytosis. The mechanism underlying the uptake of fibres has been investigated, and it has been demonstrated that A549 cells can internalise crocidolite asbestos and that this is mediated by endocytosis (Pande *et al.*, 2006). CNTs were also internalised by A549 cells within a three-dimensional culture model (incorporating A549, human monocyte-derived macrophages and monocyte-derived dendritic cells). CNTs formed bundles that were contained in vesicles (Muller *et al.*, 2010). Uptake has been observed by A549 cells for SWCNTs (Kostarelos *et al.*, 2007) and MWCNTs (Kostarelos *et al.*, 2007; Simon-Deckers *et al.*, 2008; Hirano *et al.*, 2010; Al-Jamal *et al.*, 2011), but not for HiPCO™ SWCNTs (Davoren *et al.*, 2007). In some instances the uptake of MWCNTs was observed to be incomplete (Asakura *et al.*, 2010).

Thurnherr *et al.* (2011) conducted a chronic *in vitro* exposure of A549 cells lasting 6 months. The MWCNTs (0.5 μg/ml or 0.16 μg/cm^2) were internalised as small agglomerates or as individual CNTs. Most of the MWCNTs were surrounded by membranes, suggesting an endocytic mechanism. No sign of cytotoxicity nor oxidative stress was observed in this long-term culture at the doses studied.

Moreover, no adaptative protective mechanism against particle toxicity was developed in the long-term culture.

10.5.4 Epithelial cells: ROS and oxidative stress

In vitro studies allow for investigation of the mechanistic processes responsible for driving the toxicity of pathogenic fibres. Investigation of CNT-mediated oxidative stress and pro-inflammatory responses within cells *in vitro* is of interest because of *in vivo* observations that confirm the contribution to these processes within the lung (see above). It is well recognised that fibre-mediated ROS production can contribute to the effects mediated by fibres within lung epithelial cells, and can have many detrimental impacts on cell function, such as damage to biological molecules (such as DNA) or stimulation of signaling cascades, and can ultimately cause inflammatory, genotoxic and cytotoxic responses. A number of studies have therefore investigated the contribution of ROS to the toxicity of fibres within lung epithelial cells. Of additional interest is the source of the increased generation of ROS, as it is unclear whether the enhanced ROS production may derive from the fibre surface and/or be associated with the uptake of fibres by epithelial cells.

ROS have been shown to be involved in the cytotoxicity of SWCNTs to rat lung epithelial cells (Sharma *et al.*, 2007) and of amphibole fibres to A549 cells (Srivastava *et al.*, 2010). By studying the surface and iron content of these fibres, the authors suggested that the mobilisable iron correlated with the ability of the fibres to induce oxidative stress. However, they suggested that other physicochemical parameters such as length and shape could also be involved in the toxicity associated with fibres.

In a study by Pulskamp *et al.* (2007), exposure of A549 cells to SWCNTs or MWCNTs was associated with a dose-dependent increase in ROS production. However, when the SWCNTs were purified (removal of metal content by acidic treatment), no ROS production could be detected, suggesting that the ROS were metal-derived. Exposure of epithelial cells to MWCNTs has also been associated with ROS production, glutathione depletion and lipid peroxidation, suggesting an oxidant-mediated cytotoxic effect (Srivastava *et al.*, 2010).

Herzog *et al.* (2009) tested the toxicity of various materials to epithelial cells, and demonstrated that, depending on the cell culture medium used and the presence of dispersant, the effects observed could be drastically changed. The presence of cell culture growth factors and serum reduced the production of ROS induced by materials, while the use of DPPC increased ROS production.

There is evidence that both asbestos fibres and MWCNTs can induce apoptosis, cell signaling and pro-inflammatory events via mechanisms involving oxidative stress. For example, amosite asbestos fibres have been shown to induce oxidative stress in A549 cells (Panduri *et al.*, 2004). The same study also showed that the fibres induced caspase-9 activation and apoptosis partially via a mitochondrial dependent pathway. Exposure of C10 cells to crocidolite asbestos has been demonstrated to increase the phosphorylation of the transcription factor CREB and induce cell apoptosis (Barlow *et al.*, 2007). The authors demonstrated that CREB activation by asbestos required protein kinase A and ERK1/2 activation. MWCNT treatment of rat lung epithelial cells has been associated with disruption of mitochondrial integrity and release of cytochrome c as well as with reduced cellular ATP (Ravichandran *et al.*, 2010). Signalling studies have showed nuclear translocation of NF-κB as well as phosphorylation and reduction of its inhibitory component IκBα. Moreover, several proteins involved in the apoptotic signalling cascade, including p53, p21 and bax, were shown to be activated (Ravichandran *et al.*, 2010).

A selection of carcinogenic fibres (amosite asbestos, silicon carbide and refractory ceramic fibre) was shown to dose-dependently induce the translocation of NF-κB to the nucleus of A549 cells. By contrast, a selection of non-carcinogenic fibres (man-made vitreous fibre (MMVF10), Code 100/475 glass fibre, and RCF4) had no effect on this transcription factor. Moreover, the use of anti-oxidant reduced pathogenic fibre-induced nuclear translocation of NF-κB. These results suggest that pathogenic fibres can induce oxidative stress in lung epithelial cells, which then triggers a pro-inflammatory response (Brown *et al.*, 1999).

There may be a relationship between fibre uptake and oxidative stress, since Pande *et al.* (2006) showed that preventing endocytosis of crocidolite fibres in A549 cells reduced the intracellular glutathione depletion. The necessity of CNT uptake, and the source of ROS following the exposure of cells to asbestos fibres, requires further assessment.

Measuring the gene expression of markers associated with oxidative stress (HO-1, SOD-2), Tabet *et al.* (2009) could not demonstrate any oxidant-driven effects by MWCNTs in A549 epithelial cells. The MWCNTs were not internalised by A549 cells, nor did they affect cell permeability or induce apoptosis. In contrast, asbestos did induce apoptosis. However, the MWCNTs did decrease mitochrondrial function and cell number. The lack of oxidative stress was explained by the authors as being due to the presence of oxidisable groups at the surface of the nanotubes studied, since Fenoglio *et al.* (2006) showed that CNTs could scavenge free radicals. However, it is also possible that the endpoints chosen were not sufficiently sensitive or were too specific.

10.5.5 Epithelial cells: inflammation

An inflammatory response is associated with the exposure of the lungs to pathogenic fibres *in vivo* (Donaldson *et al.*, 1989). In addition, several studies have demonstrated that pathogenic fibres stimulate lung epithelial cell models to produce pro-inflammatory mediators which may contribute to the inflammogenic potential of CNTs in the lung. Crocidolite asbestos has been shown to strongly induce the pro-inflammatory cytokines interleukin (IL)-6 and IL-8 in A549 cells, via redox-sensitive transcription factors such as NF-κB (Luster and Simeonova, 1998). IL-8 is a major pro-inflammatory cytokine involved in the launch and continuation of pathologic chronic inflammation by stimulating the activation and recruitment of polymorphonuclear neutrophilic leukocytes (PMN), and IL-6 is involved in inflammatory and fibrotic processes in the lung. Chrysotile and crocidolite asbestos fibres have also been shown to stimulate IL-8 release from A549 and from primary human bronchial epithelial to a smaller degree (Rosenthal *et al.*, 1994). However, these fibres did not induce any release of TNF-α, IL-1β or monocyte chemoattractant protein-1 in A549 and primary bronchial epithelial cells. The authors therefore suggested that fibres can directly act on epithelial cells to modulate the inflammatory profile. Similarly, Loreto *et al.* (2009) demonstrated that A549 cells treated with four different asbestos fibres (fluoro-edenite 27, and 19, tremolite and crocidolite) released IL-1ß, IL-6 and TNF-α, with crocidolite providing the strongest pattern of secretion and tremolite the smallest.

In contrast, Herzog *et al.* (2009) found that crocidolite asbestos reduced MCP-1 release by A549 by up to 40% while it induced only a slight increase in IL-8 and IL-6 secretion at 48 hours. No significant effect of crocidolite on cytokine secretion by NHBE cells was shown. These discrepancies could be related to differences in the experimental design. Indeed, Herzog *et al.* used DPPC to disperse crocidolite, which leads to an important decrease in IL-8 and IL-6 secretion by A549 at 48 hours, while no further enhancement was observed on MCP-1 secretion.

In A549 cells, observations by Muller *et al.* (2010) suggested a relatively low capacity for SWCNTs to induce a pro-inflammatory response by epithelial cells. Similarly, at sub-lethal concentrations (below 50 µg/ml) HiPCO™ SWCNTs reduced the concentration of IL-8, IL-6 and MCP-1 in the supernatants of A549 cells, while TNF-α levels remained under the detection limit. Similar results were obtained in NHBE cells (Herzog *et al.*, 2009). These results suggest that SWCNTs do not trigger a pro-inflammatory interaction with lung epithelial cells. The authors suggested that these results corroborate *in vivo* observations showing decreased defense against bacterial infection in mice (Shvedova *et al.*, 2008a).

In contrast, Baktur *et al.* (2011) showed that HiPCO™-derived SWCNTs (0.4 and 0.8 µg/ml) stimulated IL-8 secretion by A549 cells in a time-dependent manner. Moreover, it was observed that the presence of serum enhanced these effects, while no effect on cell proliferation was observed. Interestingly, the investigators observed increased internalisation of the CNTs when serum was present in the experimental conditions that might be involved in the enhanced release of IL-8 (Baktur *et al.*,

2011). MWCNTs have also been shown to stimulate IL-6 and IL-8 production by BEAS-2B cells, which was suggested to derive from the activation of NF-κB (Hirano *et al.*, 2010).

10.5.6 Transepithelial electrical resistance and translocation of fibres across epithelial barriers

The lung epithelium acts as a barrier to external contaminants. Interference in the function of this barrier is known to be associated with disease, and may allow for the translocation of fibres from the lung to secondary target sites in the body. During the course of asbestosis and related diseases induced by asbestos exposure, lung permeability is increased. This has been replicated *in vitro*, whereby asbestos fibres (amosite and chrysotile) increased the mannitol permeability of confluent human bronchial epithelial cells (Peterson and Kirschbaum, 1998).

In an *in vitro* model using human epithelial Calu-3 cells, the effects of different CNTs on transepithelial electrical resistance (TEER) has been studied (Rotoli *et al.*, 2008). It was found that at sub-lethal concentrations (100 μg/ml) MWCNTs, and SWCNTs to a lesser extent, decreased TEER, while a mixture of SWCNTs and MWCNTs synthesised by arc discharge had no effect. Moreover, MWCNTs increased monolayer permeability to mannitol and inhibited the formation and regeneration of a high-resistance monolayer, suggesting that these CNTs inhibited the formation of cellular tight junctions. In a similar model, it was evidenced that long MWCNTs and SWCNTs reduced TEER. The authors stressed the importance of fibre shape on the effects observed; however, no modification of junction proteins occludin and ZO-1 at the RNA level were observed, so further studies are needed to understand the mechanistic drivers for this effect (Rotoli *et al.*, 2009).

10.5.7 Epithelial cells: genotoxicity

The results from studies investigating the genotoxicity of asbestos fibres to epithelial cells are often conflicting. In one study, chrysotile asbestos did not induce structural changes in chromosomes but dose-dependently increased numerical chromosome aberrations, as well as the number of bi- and multi-nucleated cells. However, chrysotile did not induce mutagenicity at the hgprt locus in these cells (Asakura *et al.*, 2010). Comparing the effects of various amphibole fibres *in vitro*, it was reported by Srivastava *et al.* (2010) for A549 cells that these fibres increased the frequency of micronuclei in binucleated cells, and that crocidolite was the most genotoxic, followed by tremolite, and finally amosite.

For SWCNTs, it was reported that these particles could induce genotoxicity by altering the mitotic process. Moreover, CNTs were observed in the nucleus of epithelial cells by confocal microscopy (Sargent *et al.*, 2010), which suggests that CNTs can directly interact with DNA to elicit genotoxic consequences.

Muller *et al.* (2008) exposed rat lung epithelial cells to ground MWCNTs (to reduce aggregation) and showed that at sub-lethal concentrations these nanotubes induced micronucleus formation in a dose-dependent manner. In the cytokinesis block micronucleus assay, MWCNTs induced genotoxicity that was correlated more to structural defects than to metal content.

In contrast, Thurnherr *et al.* (2011) demonstrated that MWCNTs did not elicit genotoxic effects in A549 cells, using both the alkaline Comet assay and the micronucleus assay. In Chinese hamster lung cells, MWCNTs were found to induce numerical chromosome aberrations but no structural effects. Moreover, although the MWCNTs increased the number of multinucleated cells, micronuclei formation was not enhanced. The authors suggested that these effects could be the result of interference by MWCNTs with the mitotic spindle or during the cytokinesis process. Moreover, MWCNTs had no effect in the hgprt mutagenicity assay (Asakura *et al.*, 2010).

In summary, many studies have been conducted in relation to the impacts of asbestos fibres and nanotubes on lung epithelial cell models *in vitro*. There seem to be similarities between these different particles, with evidence of pro-inflammatory signalling and genotoxicity. There is a clear role for oxidative stress in the asbestos-induced effects, but for nanotubes this is less clear.

10.6 Mesothelial cells

Mesothelial cells line the serosal cavities (including the pleural, peritoneal and pericardial spaces) of the body. This lining, the mesothelium, is formed from a monolayer of mesothelial cells that extends over the entire surface of these cavities (Mutsaers, 2004), with the visceral and parietal surfaces covering the organ and body-wall surfaces, respectively (Herrick and Mutsaers, 2004). The cells are predominately flattened, squamous-like cells and possess features of mesenchymal cells and epithelial cells (Mutsaers, 2004). The primary role of the mesothelium is to act as a protective barrier, although it has become apparent that mesothelial cells can carry out a multitude of important functions, such as enabling fluid and cell transport across serosal cavities, antigen presentation, inflammation and tissue repair (Herrick and Mutsaers, 2004; Mutsaers, 2004).

Malignant mesothelioma (MM) is a tumour that forms primarily within the pleural cavity of the lung (although other sites can be affected), and mesothelial cells are the progenitor cell for the tumour (Broaddus and Jaurand, 2002; see Chapter 2). Carcinogenesis is a multi-stage process that is initiated by an alteration to DNA which is passed on to daughter cells (initiation). The proliferation of these cells (promotion) and accumulation of genetic damage during the proliferation of the damaged cells (progression) ultimately lead to tumour formation (Mossman, 1994; Shukla *et al.*, 2003b). Tumour promotion and progression is typified by unregulated cell proliferation, which can occur because of resistance to cell death or because of uncontrolled cell proliferation (Timblin *et al.*, 1998). MM development is therefore driven by the unregulated stimulation of mesothelial cell proliferation.

The main risk factor for MM is exposure to asbestos fibres. The structural similarities of CNTs to asbestos fibres have prompted concerns that these materials may elicit a similar outcome following exposure. As such, it is necessary to explore the mechanisms underlying the development of MM, and specifically the processes that promote the transformation of mesothelial cells into malignant cells (*in vitro*) following exposure to asbestos fibres, and to determine if they are relevant to the response of mesothelial cells to CNTs. Both asbestos fibres and CNTs are available in a variety of forms (outlined in Table 10.2) that vary with respect to their chemical and physical properties, and as a consequence their toxic potential. A fibre-pathogenicity paradigm has been developed that enables a prediction of fibre pathogenicity using a robust structure–activity relationship. This paradigm has recognised the critical importance of the length, diameter and biopersistence of fibres to their pathogenicity (Chapter 2).

There are some uncertainties regarding the cell and molecular drivers for the pathogenicity of asbestos fibres to mesothelial cells, although it is clear that their

Table 10.2 *There are two main forms of asbestos, which vary in their composition and toxic potential (with the amphibole form being most toxic). All fibres have the commonality that they have an aspect ratio of greater than 3: 1 (length: diameter). However, the chemical and physical properties of the fibres can be vastly different, and in particular the length, biopersistence and diameter of fibres have been illustrated to be particularly important in driving their toxicity.*

Asbestos	HARN
Amphibole asbestos: crocidolite, amosite, anthophyllite, tremolite, erionite and actinolite. This form of asbestos is typified by long, thin, rod-like fibres and is the most pathogenic form of asbestos.	**SWCNTs**: composed of carbon atoms, and with a structural arrangement that resembles a graphite sheet that is rolled up into a cylindrical structure
Serpentine asbestos: chrysotile. This form has a more 'curly' structure, and is more easily broken down.	**MWCNTs**: composed of concentric layers of CNTs that can form either long, straight fibres or entangled structures

response is multifaceted. In fact, the long latency between exposure of mesothelial cells to asbestos fibres and MM development suggests that multiple, complex processes contribute to tumour development (Shukla *et al.*, 2003b). There have, to date, been limited *in vitro* investigations of mesothelial cell exposure to CNTs, despite evidence that CNTs stimulate the proliferation of mesothelial cells *in vivo* (Murphy *et al.*, 2011; Poland *et al.*, 2008). As a result, the cell and molecular mechanisms that underlie the potential carcinogenicity of CNT fibres are not well understood. This section will use knowledge of asbestos fibre-induced mesothelial cell injury observed in order to suggest future approaches for assessing mesothelial cell exposure to CNTs *in vitro*. Such tests are required to delineate the mechanisms underlying the toxicity of pathogenic fibres, and to predict the carcinogenicity potential of new fibres. In addition, greater knowledge of the cell and molecular processes underlying the transformation of mesothelial cells will allow for the identification of targets for therapeutic intervention for the treatment of MM (Heintz *et al.*, 2010).

10.6.1 Mesothelial cells: responses to asbestos fibres in vitro

There is a lack of understanding of the precise mechanisms by which asbestos fibres cause the transformation of mesothelial cells to malignant cells. However, two main processes have been hypothesised by which mesothelial injury can occur following exposure to asbestos. Specifically, there are both direct and indirect stresses on mesothelial cells that are considered to be responsible for promoting tumour development. Following inhalation, fibres can translocate to the pleural space and come into direct physical contact with mesothelial cells. Alternatively, the release of (damaging) factors (such as ROS, cytokines and growth factors) from other asbestos-exposed cells, including macrophages or epithelial cells, can impact on mesothelial cells.

The cell and molecular events underlying the response of mesothelial cells to asbestos are unclear but it is evident that their response is multifaceted, and that a number of events act in concert to stimulate cell proliferation and ultimately tumour development. Of particular importance to the mesothelial response to asbestos fibres are the uptake of fibres by mesothelial cells, the manifestation of an oxidant-driven response, the activation of cell signalling cascades, and downstream consequences of this such as modifications in gene expression. The potential contribution of each of these processes to the pathogenicity of asbestos fibres is outlined below.

10.6.2 Mesothelial cells: asbestos and oxidants

Oxidants are considered to have multiple roles in mediating the effects of asbestos fibres on mesothelial cells (Shukla *et al.*, 2003b). Regardless of whether direct or indirect stresses contribute to the toxicity mediated by asbestos, it is apparent that an oxidant-driven response is integral to the effects that manifest following exposure. ROS (including, for example, hydrogen peroxide, superoxide anion, hydroxyl anion and reactive nitrogen species) may be generated at the fibre surface, produced within mesothelial cells following exposure to asbestos, or released from other cell types in response to asbestos exposure to affect surrounding cells. There is evidence that ROS production by the asbestos fibres themselves is mediated by iron via Fenton chemistry (reviewed in Shukla *et al.*, 2003b). Similarly CNTs have been demonstrated to have metal contaminants, derived from the manufacturing process, that can contribute to their toxicity (e.g. Wick *et al.*, 2007). Accordingly, the production of oxidants at the CNT surface may contribute to effects mediated by CNTs within mesothelial cells, as observed for asbestos fibres. In order to investigate the contribution of the metal content of fibres, metal chelators are often utilised. Unfortunately metal chelators cannot completely prevent the oxidant-mediated damage inflicted by asbestos fibres (Fung *et al.*, 1997a; Swain *et al.*, 2004). Therefore, although oxidants have been observed to drive the effects of asbestos fibres within mesothelial cells, iron chelators cannot completely abolish the observed toxicity. This suggests that the iron content of asbestos is not the only source of ROS production, and that the generation of oxidants on the fibre surface is not solely accountable for the effects of asbestos. Therefore, although the role for oxidants in asbestos-related damage to mesothelial cells is well recognised, the source for these oxidants is debatable. It is likely that both direct and indirect production of oxidants contributes to the response of mesothelial cells to asbestos.

There is often an upregulation in antioxidant expression within mesothelial cells to overcome the increased oxidant presence (e.g. Fung *et al.*, 1997a). The increased expression of antioxidants suggests that cells stimulate a compensatory protective response following exposure to asbestos, thus reinforcing the involvement of oxidants in the effects mediated by asbestos in mesothelial cells.

10.6.3 Mesothelial cells: asbestos-induced cell signalling and gene expression

Although the molecular mechanisms responsible for the development of MM are unclear, it is evident that the activation of cell signalling cascades in response to asbestos exposure leads to alterations in gene expression that dictate the cell response observed, and is ultimately is responsible for stimulating unregulated mesothelial cell proliferation (Shukla *et al.*, 2003b; Swain *et al.*, 2004; Heintz *et al.*, 2010) (see Figure 10.4). The signalling events mediated by asbestos are complex, and can be stimulated by a direct interaction of asbestos with cell surface receptors (resulting in their activation) or via the generation of ROS and other factors (such as TNF-α) by mesothelial cells and other asbestos-exposed cells.

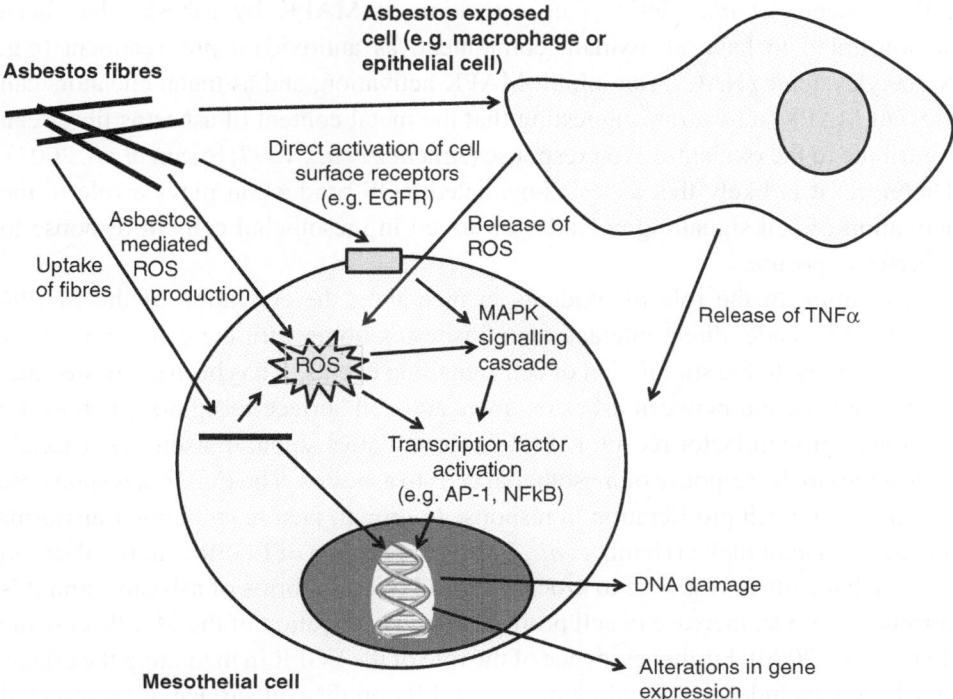

Figure 10.4 Schematic summary of the mechanism of asbestos-induced cellular effects via reactive oxygen species (ROS).

The activation of the mitogen-activated protein kinase (MAPK) signalling cascade is responsible for the initiation of a series of events that can stimulate the activation of transcription factors that control the expression of genes responsible for functions such as cell proliferation, apoptosis, inflammation and cell survival (see below). MAPKs are a family of protein kinases that phosphorylate target substrates (including, for example, other protein kinases, and transcription factors) in order to regulate these cellular activities (Johnson and Lapadat, 2002). There are three main MAPK pathways: extracellular signal-regulated kinases (ERK), c-jun NH_2-terminal kinases (JNK) or stress-activated protein kinases (SAPK), and p38 kinases. The importance of a phosphorylation cascade activated by asbestos is exemplified by the finding that activation of protein and tyrosine kinases (e.g. protein kinase C (PKC)) is involved in the response of mesothelial cells to asbestos fibres, and is responsible for downstream effects such as gene expression to control cell survival and proliferation (Fung *et al.*, 1997b). The mesothelial response to crocidolite asbestos is associated with the activation of MAPKs (ERK pathway), and this was observed to be responsible for the induction of apoptosis within the cells (Jimenez *et al.*, 1997). This activation of MAPK by asbestos has been demonstrated to have an oxidant component, as antioxidant pre-treatment (e.g. N-acetylcysteine (NAC)) can inhibit MAPK activation, and as metal chelators can prevent MAPK activation, suggesting that the metal content of asbestos fibres can contribute to the oxidant-driven response (Jimenez *et al.*, 1997; Swain *et al.*, 2004). Therefore, it is likely that asbestos-mediated ROS production plays a role in the activation of cell signalling cascades observed in mesothelial cells in response to asbestos exposure.

In addition to the role of oxidants in mediating the activation of the MAPK signalling cascade, direct interactions of asbestos fibres with the cell surface may also contribute to the stimulation of cell signalling events. It has been suggested that a direct interaction between asbestos fibres and cell surface receptors (such as the epidermal growth factor receptor (EGFR)) to initiate a signal transduction cascade contributes to the response of mesothelial cells to asbestos. The EGFR is responsible for controlling cell proliferation in response to growth factors including transforming growth factor alpha (Heintz *et al.*, 2010). Expression of EGFR is upregulated in mesothelial cells in response to crocidolite and erionite forms of asbestos, and this correlates with an increase in cell proliferation via activation of the MAPK cascade (Faux *et al.*, 2000). Further evidence of the role of the EGFR in mediating the effects of asbestos includes the accumulation of EGFRs on the cell surface of mesothelial cells, the increase in EGFRs within mesothelial cells, and the direct association of asbestos with EGFRs in stimulating the activation of the receptor, following exposure of mesothelial cells to asbestos (Pache *et al.*, 1998). The interaction of asbestos fibres with the EGFRs on mesothelial cells and the downstream consequences of this have therefore been of great interest. It has been demonstrated that EGFRs are activated by asbestos exposure, resulting in the activation of the MAPK signalling cascade (Zanella *et al.*, 1996). Furthermore, the direct interaction of crocidolite asbestos fibres with EGFRs to promote their activation has been observed, although the contribution of oxidants to the activation of the receptor was not ruled out (Zanella *et al.*, 1999). Subsequent to the activation of the EGFR, the MAPK (ERK pathway) signalling cascade is activated, which stimulates the expression of c-fos proto-oncogene to induce apoptosis (Zanella *et al.*, 1999). As such, it has been suggested that asbestos fibres can directly interact with cell surface receptors such as EGFRs to stimulate a signalling cascade that dictates the cell response. Therefore there is evidence that a combination of direct and indirect responses to asbestos stimulates the activation of the MAPK signalling cascade. The activation of MAPK signalling pathways culminates in the activation of transcription factors that control a variety of cell functions.

The activation of transcription factors (such as activator protein-1 (AP-1) and NF-κβ) is critical to the response of mesothelial cells to asbestos fibres. The activation of these transcription factors may be mediated by various means, including stimulation of the MAPK signalling cascade and directly by oxidants. The redox-sensitive transcription factor AP-1 comprises homo- or heterodimers of jun and fos protooncogenes (Heintz *et al.*, 2010). These proteins dimerise (jun:jun homodimers or jun:fos heterodimers) to form the transcription factor AP-1, to control the expression of genes associated with cell proliferation and tumour development (Shukla *et al.*, 2003b). An elevation in c-jun and c-fos expression in response to asbestos has been observed in rat pleural mesothelial cells, accompanied by AP-1 binding to DNA and increased cell proliferation (Heintz *et al.*, 1993; Janssen *et al.*, 1994, 1995; Goldberg *et al.*, 1997; Timblin *et al.*, 1998). The activation of the MAPK signalling cascade is believed to have an oxidant component as the activation of AP-1 via asbestos stimulation of MAPK is reliant on oxidant production, as the response can be inhibited by the antioxidant NAC, and GSH depletion is associated with exposure (Janssen *et al.*, 1995).

The transcription factor NF-κB has also been demonstrated to play a role in mediating the effects of asbestos in mesothelial cells. Crocidolite asbestos can increase NF-κB binding to DNA, which is indicative of NF-κB activation (Janssen *et al.*, 1995). Complementary *in vivo* studies also illustrate that NF-κB activation is critical to the response of the rat lung to asbestos exposure and tumour development.

Microarray analysis has allowed for a more detailed understanding of the subsequent modifications in gene and protein expression that are associated with the exposure of mesothelial cells to asbestos. This is exemplified by the changes in protein expression involved in cell proliferation, cell signalling, DNA repair, apoptosis and transcription activity in mesothelial cells following exposure to crocidolite asbestos (Wang *et al.*, 2011). The contribution of oxidants to this process has also been observed (Hillegass *et al.*, 2010). Specifically, genes associated with cell proliferation, signal transduction and inflammation (e.g. IL-8) were upregulated, as well as the oxidative stress markers superoxide dismutase (SOD-2)and HO-1. The involvement of ROS was further supported by an increase in protein levels and activity of SOD-2, a decrease glutathione levels and an increase in ROS production by cells following exposure to asbestos.

10.6.4 Mesothelial cell proliferation and apoptosis

A number of *in vitro* investigations have focused on the mechanisms underlying the uncontrolled proliferation of mesothelial cells required for tumorigenesis. Exposure of mesothelial cells to asbestos fibres *in vitro* has the ability to induce an apoptotic response (Goldberg *et al.*, 1997; Jimenez *et al.*, 1997; Swain *et al.*, 2004; Yang *et al.*, 2006), which is contrary to the observed and expected increase in cell proliferation (Goldberg *et al.*, 1997; Faux *et al.*, 2000) required for tumour development. It has been suggested that in response to the damaging effects of asbestos in mesothelial cells, a protective apoptotic response is stimulated and that this leads to a compensatory proliferation of cells. Therefore, the balance between cell proliferation and apoptosis is integral to understanding the response of mesothelial cells to asbestos, and to tumour development. In fact, an increase in the balance between cell proliferation and apoptosis was observed in response to asbestos exposure *in vitro* (Goldberg *et al.*, 1997). Processes that lead to an interruption in this protective apoptotic process are likely to stimulate cell proliferation and drive tumour development.

It has been demonstrated that the cytotoxic response observed in mesothelial cells *in vitro* to asbestos fibres can be reduced through pre-treatment of cells with TNF-α (Yang *et al.*, 2006). It has therefore been suggested that, on exposure of mesothelial cells to TNF-α (released from mesothelial cells or other exposed cells such as macrophages), NF-κB is activated, initiating a response that promotes cell proliferation and inhibits apoptosis (Yang *et al.*, 2006). This could promote the survival of DNA-damaged cells to encourage tumour development, and is likely to be a physiologically relevant phenomenon, as factors released from macrophages and epithelial cells (such as TNF-α) will impact on the response of mesothelial cells during the inflammatory response to the fibre exposure.

The apparent paradox that asbestos fibres *in vitro* can elicit cell death but also stimulate cell proliferation is of interest and integral to tumour development. The ability of asbestos to stimulate cell death is also likely to be associated with the concentration or form of asbestos used. In order to investigate the mechanisms underlying the ability of asbestos fibres to stimulate a proliferative response it is therefore necessary to utilise sub-lethal concentrations of asbestos.

Tumour formation is also reliant on asbestos stimulating an alteration in DNA within cells that is then inherited by daughter cells, with the DNA damage responsible for the promotion of the uncontrolled proliferation of cells. It has been suggested that asbestos fibres can elicit DNA damage via two main mechanisms: asbestos fibres can interact with DNA to elicit damage, and can stimulate the production of oxidants that cause DNA damage (Mossman, 1994). It is likely that these processes act in combination to mediate DNA damage to mesothelial cells, as tumour development is reliant on the occurrence of a number of genetic alterations. DNA damage can actually trigger apoptosis in affected cells to prevent the fixation and propagation of the genetic abnormality. For example, asbestos fibres can induce DNA damage (Chen *et al.*, 1996; Fung *et al.*, 1997a) that triggers apoptosis in mesothelial cells (Broaddus *et al.*, 1996), but pre-treatment with antioxidants or metal chelators decreases the apoptotic response. It has been suggested that oxidants elicit cell injury (DNA damage), which stimulates apoptosis as a protective response. This process enables the removal of damaged cells, and thus interruption of this pathway is likely to contribute to tumour development. However, as suggested above, this damage to mesothelial cells following exposure to asbestos can stimulate compensatory cell proliferation which may contribute to tumour development. The long latency between exposure and development of the tumour that is associated with asbestos exposure is indicative that the cell may instigate compensatory mechanisms in response to asbestos-induced cell damage (such as apoptosis and DNA damage) prior to interference of these protective responses.

Alterations in the normal cell cycle pathways often occur in response to exposure to DNA damaging agents, as failure to repair DNA means that the genomic change will be fixed and propagated (Levresse *et al.*, 1997). It has been observed that, following exposure to asbestos, inactivation of p53 (responsible for cell cycle arrest to enable DNA repair) caused the continued proliferation of cells, so there was no opportunity to repair the damaged DNA (Levresse *et al.*, 1998). Wang *et al.* (2011) investigated alterations in protein expression in response to asbestos exposure. It was demonstrated that G1/S cell cycle checkpoint pathways are suppressed by asbestos, which is indicative that DNA cannot be repaired. It was suggested that the p53 pathway is suppressed.

10.6.5 Internalisation of asbestos fibres by mesothelial cells

On exposure of mesothelial cells to fibres it is unclear whether or not fibres need to be internalised by cells to mediate adverse impacts, or whether they act via the cell surface. In addition, it is noteworthy that factors (such as ROS and TNF-α) released by other cells in response to fibre exposure also drive mesothelial responses to asbestos. In fact, it is likely that both these mechanisms act in concert to mediate the proliferation of mesothelial cells.

Broaddus *et al.* (1996) demonstrated that the internalisation of asbestos fibres by mesothelial cells was related to their ability to stimulate apoptosis. The internalisation of asbestos fibres may be a requirement for mediating their effects on mesothelial cells, and thus the way asbestos fibres are recognised and internalised by mesothelial cells is of great interest. Proteins can adsorb onto the fibre surface (e.g. vitronectin, a component of serum) to enhance their uptake by mesothelial cells through a receptor-mediated mechanism (Boylan *et al.*, 1995). Liu *et al.* (2000) expanded on this work to determine whether the enhanced internalisation of asbestos fibres could influence their behaviour. The ability of asbestos fibres to increase ROS presence within cells, elicit DNA strand breaks (via an oxidant-mediated mechanism) and stimulate apoptosis was increased by the coating of asbestos by vitronectin, and this was suggested to derive from their enhanced internalisation. Interactions between asbestos fibres and proteins is a physiologically relevant phenomenon, as on entering the body fibres will come into contact with, and be coated with, various biological media, and the interactions that occur will inevitably be dependent on the properties of the fibre under investigation. It is noteworthy that, as described for macrophages, the uptake of asbestos fibres by mesothelial cells will be reliant on their size, so that if the fibres are bigger than the cells that are attempting to ingest them, it is likely that the cell will lyse (Pache *et al.*, 1998).

Also of interest is the ability of asbestos fibres to interact with the cell surface to mediate their adverse effects on mesothelial cells. The activation of cell surface receptors (eg EGFR) has been demonstrated to be responsible for the effects of fibres (see above), which suggests that interactions of fibres with the cell surface are important.

The process of frustrated phagocytosis (described above) is known to result in the release of damaging factors which may cause mesothelial cell injury and stimulate proliferation. Adamson (1997) investigated the response of primary macrophages and mesothelial cells *in vitro* to asbestos exposure for up to 6 weeks. Macrophages were isolated from bronchoalveolar lavage fluid from the rat lung, following intratracheal exposure to crocidolite asbestos. The capacity of these macrophages to stimulate the proliferation of mesothelial cells was then investigated, and it was suggested that factors (cytokines and growth factors) released from macrophages were responsible for stimulating the observed proliferation of mesothelial cells.

10.6.6 Mesothelial cells: conclusions, gaps in research and priorities

Tumour development associated with asbestos exposure is reliant on alterations in cell function that promote cell dysfunction due to unregulated cell proliferation and resistance to apoptosis. Mesothelial cells exposed to asbestos *in vitro* exhibit a wide range of molecular and cellular alterations that are likely to be involved in the transformation of mesothelial cells to malignant cells. These processes include the release of inflammatory factors, the production of reactive oxygen species, the stimulation of cell signalling cascades, alterations in gene expression and DNA damage.

Because of the physicochemical similarities of CNTs to pathogenic fibres such as asbestos, studies are urgently needed on the potential for CNTs to induce mesothelial cells' transformation. As a result of scientific advances, methodological developments and a greater understanding of the complex signalling cascades known to be activated by fibres, a greater understanding of the molecular drivers underlying the mesothelial response to pathogenic fibres can potentially be obtained using such CNT-focused research.

There is already some evidence that mesothelial cells respond to CNT exposure in a similar manner to that observed for asbestos. For example, Pacurari *et al.* (2008) evaluated the molecular mechanisms underlying the toxicity of SWCNTs to mesothelial cells. They demonstrated that SWCNTs could stimulate ROS production in mesothelial cells. DNA damage was also observed and was partially inhibited by ROS scavengers, suggesting that there was both ROS-mediated and direct damage to DNA in response to CNT exposure. Stimulation of the MAPK signalling cascade and the subsequent activation of AP-1 were also associated with exposure to SWCNTs. This study therefore suggests that the response of mesothelial cells to CNTs could be similar to those exhibited by asbestos. Further studies are required to investigate the applicability of the response to different forms of CNTs.

Oxidants play a major role in the response of mesothelial cells to asbestos fibres, and the role of oxidants in the mesothelial cell response to CNTs warrants further study, in particular of acellular and cellular oxidant production and of the ability of antioxidants to diminish cellular responses to CNTs.

Finally, in order to replicate more accurately the interactions that are likely to occur between different cell populations in the lung in response to CNT exposure, it would be useful to utilise either co-cultures of inflammatory cells with mesothelial cells or the transfer of conditioned media from inflammatory cells to mesothelial cells.

10.7 Conclusion

The use of *in vitro* models to investigate the impact of pathogenic fibres on various cell types and the mechanisms of these impacts has been very fruitful, for example in the study of frustrated phagocytosis with macrophages, especially in relation to identifying the limitations of length on uptake. With respect to mesothelial cells, the *in vitro* models have provided a well-controlled situation in which the molecular signaling pathways controlling the development of cell survival and proliferation have been elucidated. This conundrum would have been considerably more complex if it had been limited to *in vivo* models. These models have correlated well with the *in vivo* response. We are therefore in a position to use this information in the development of nanosafety with respect to high-aspect-ratio nanomaterials. There are good protocols and models available to address many of the questions available with respect to identifying factors that drive toxicity and pathogenicity for such materials. We are not suggesting that such models will completely replace animal models, but there is an opportunity to design such studies with the priorities of refinement, replacement and reduction of animal use in mind.

References

Adamson, I. Y. (1997). Early mesothelial cell proliferation after asbestos exposure: In vivo and in vitro studies. *Environ Health Perspect* **105**(suppl. 5), 1205–1208.

Al-Jamal, K. T., Nerl, H., Muller, K. H., *et al.* (2011). Cellular uptake mechanisms of functionalised multi-walled carbon nanotubes by 3D electron tomography imaging. *Nanoscale* **3**(6), 2627–2635.

Arvizo, R. R., Miranda, O. R., Thompson, M. A., *et al.* (2010). Effect of nanoparticle surface charge at the plasma membrane and beyond. *Nano Lett* **10**(7), 2543–2548.

Asakura, M., Sasaki, T., Sugiyama, T., *et al.* (2010). Genotoxicity and cytotoxicity of multi-wall carbon nanotubes in cultured Chinese hamster lung cells in comparison with chrysotile A fibers. *J Occup Health* **52**(3), 155–166.

Baktur, R., Patel, H., and Kwon, S. (2011). Effect of exposure conditions on SWCNT-induced inflammatory response in human alveolar epithelial cells. *Toxicol In Vitro* **25**(5), 1153–1160.

Barlow, C. A., Barrett, T. F., Shukla, A., Mossman, B. T., and Lounsbury, K. M. (2007). Asbestos-mediated CREB phosphorylation is regulated by protein kinase A and extracellular signal-regulated kinases 1/2. *Am J Physiol Lung Cell Mol Physiol* **292** (6), L1361–L1369.

Boylan, A. M., Sanan, D. A., Sheppard, D., and Broaddus, V. C. (1995). Vitronectin enhances internalization of crocidolite asbestos by rabbit pleural mesothelial cells via the integrin alpha v beta 5. *J Clin Invest* **96**(4), 1987–2001.

Broaddus, V. C. and Jaurand, M. -C. (2002). Asbestos fibres and their interaction with mesothelial cells in vitro and in vivo. In Robinson, B. W. S. and Chahinian, A. P., eds., *Mesothelioma* (London: Taylor and Francis), pp. 273–294.

Broaddus, V. C., Yang, L., Scavo, L. M., Ernst, J. D., and Boylan, A. M. (1996). Asbestos induces apoptosis of human and rabbit pleural mesothelial cells via reactive oxygen species. *J Clin Invest* **98**(9), 2050–2059.

Brown, D., Donaldson, K., and Stone, V. (2004). Effects of PM10 in human peripheral blood monocytes and J774 macrophages. *Respir Res* **5**(1), 29.

Brown, D. M., Beswick, P. H., and Donaldson, K. (1999). Induction of nuclear translocation of NF-kappaB in epithelial cells by respirable mineral fibres. *J Pathol* **189**(2), 258–264.

Brown, D. M., Dickson, C., Duncan, P., Al-Attili, F., and Stone, V. (2010). Interaction between nanoparticles and cytokine proteins: Impact on protein and particle functionality. *Nanotechnology* **21**(21), 215104.

Brown, D. M., Kinloch, I. A., Bangert, U., *et al.* (2007). An in vitro study of the potential of carbon nanotubes and nanofibres to induce inflammation mediators and frustrated phagocytosis. *Carbon* **45**, 1743–1756.

Brown, D. M., Wilson, M. R., MacNee, W., Stone, V., and Donaldson, K. (2001). Size-dependent proinflammatory effects of ultrafine polystyrene particles: A role for surface area and oxidative stress in the enhanced activity of ultrafines. *Toxicol Appl Pharmacol* **175**(3), 191–199.

Casey, A., Herzog, E., Lyng, F. M., *et al.* (2008). Single walled carbon nanotubes induce indirect cytotoxicity by medium depletion in A549 lung cells. *Toxicol Lett* **179**(2), 78–84.

Chanock, S. J., el Benna, J., Smith, R. M., and Babior, B. M. (1994). The respiratory burst oxidase. *J Biol Chem* **269**(40), 24519–24522.

Chen, Q., Marsh, J., Ames, B., and Mossman, B. (1996). Detection of 8-oxo-2'-deoxyguanosine, a marker of oxidative DNA damage, in culture medium from human mesothelial cells exposed to crocidolite asbestos. *Carcinogenesis* **17**(11), 2525–2527.

Chou, C. C., Hsiao, H. Y., Hong, Q. S., *et al.* (2008). Single-walled carbon nanotubes can induce pulmonary injury in mouse model. *Nano Lett* **8**(2), 437–445.

Davoren, M., Herzog, E., Casey, A., *et al.* (2007). In vitro toxicity evaluation of single walled carbon nanotubes on human A549 lung cells. *Toxicol In Vitro* **21**(3), 438–448.

Donaldson, K., Brown, G. M., Brown, D. M., Bolton, R. E., and Davis, J. M. (1989). Inflammation generating potential of long and short fibre amosite asbestos samples. *Br J Ind Med* **46**(4), 271–276.

Donaldson, K., Murphy, F., Schinwald, A., Duffin, R., and Poland, C. A. (2011). Identifying the pulmonary hazard of high aspect ratio nanoparticles to enable their safety-by-design. *Nanomedicine (Lond)* **6**(1), 143–156.

Donaldson, K. and Stone, V. (2003). Current hypotheses on the mechanisms of toxicity of ultrafine particles. *Ann 1st Super Sanita* **39**(3), 405–410.

Dorger, M. and Krombach, F. (2000). Interaction of alveolar macrophages with inhaled mineral particulates. *J Aerosol Med* **13**(4), 369–380.

Faux, S. P., Houghton, C. E., Hubbard, A., and Patrick, G. (2000). Increased expression of epidermal growth factor receptor in rat pleural mesothelial cells correlates with carcinogenicity of mineral fibres. *Carcinogenesis* **21**(12), 2275–2280.

Fenoglio, I., Tomatis, M., Lison, D., *et al.* (2006). Reactivity of carbon nanotubes: Free radical generation or scavenging activity? *Free Radic Biol Med* **40**(7), 1227–1233.

Fiorito, S., Monthioux, M., Pierimarchi, P., *et al.* (2009). Evidence for electro-chemical interactions between multi-walled carbon nanotubes and human macrophages. *Carbon* **47**, 2789–2804.

Fiorito, S., Serafine, A., Andreola, F., and Bernier, P. (2005). Effects of fullerenes and single-walled carbon nanotubes on murine and human macrophages. *Carbon* **44**, 1100–1105.

Fung, H., Kow, Y. W., Van, H. B., and Mossman, B. T. (1997a). Patterns of 8-hydroxydeoxyguanosine formation in DNA and indications of oxidative stress in rat and human pleural mesothelial cells after exposure to crocidolite asbestos. *Carcinogenesis* **18**(4), 825–832.

Fung, H., Quinlan, T. R., Janssen, Y. M., *et al.* (1997b). Inhibition of protein kinase C prevents asbestos-induced c-fos and c-jun proto-oncogene expression in mesothelial cells. *Cancer Res* **57**(15), 3101–3105.

Goldberg, J. L., Zanella, C. L., Janssen, Y. M., *et al.* (1997). Novel cell imaging techniques show induction of apoptosis and proliferation in mesothelial cells by asbestos. *Am J Respir Cell Mol Biol* **17**(3), 265–271.

Gordon, S. and Taylor, P. R. (2005). Monocyte and macrophage heterogeneity. *Nat Rev Immunol* **5**(12), 953–964.

Grecco, A. C., Paula, R. F., Mizutani, E., *et al.* (2011). Up-regulation of T lymphocyte and antibody production by inflammatory cytokines released by macrophage exposure to multi-walled carbon nanotubes. *Nanotechnology* **22**(26), 265103.

Grimm, S., Hohn, A., and Grune, T. (2010). Oxidative protein damage and the proteasome. *Amino Acids* **42**(1), 23–38.

Guo, L., Morris, D. G., Liu, X., *et al.* (2007). Iron bioavailability and redox activity in diverse carbon nanotube samples. *Chem Mater* **19**(4), 3472–3478.

Han, S. G., Andrew, R., and Gairola, C. G. (2010). Acute pulmonary response of mice to multi-walled carbon nanotubes. *Inhal Toxicol* **22**, 340–347.

Heintz, N. H., Janssen, Y. M., and Mossman, B. T. (1993). Persistent induction of c-fos and c-jun expression by asbestos. *Proc Natl Acad Sci U S A* **90**(8), 3299–3303.

Heintz, N. H., Janssen-Heininger, Y. M., and Mossman, B. T. (2010). Asbestos, lung cancers, and mesotheliomas: From molecular approaches to targeting tumor survival pathways. *Am J Respir Cell Mol Biol* **42**(2), 133–139.

Herrick, S. E. and Mutsaers, S. E. (2004). Mesothelial progenitor cells and their potential in tissue engineering. *Int J Biochem Cell Biol* **36**(4), 621–642.

Herzog, E., Byrne, H. J., Davoren, M., *et al.* (2009). Dispersion medium modulates oxidative stress response of human lung epithelial cells upon exposure to carbon nanomaterial samples. *Toxicol Appl Pharmacol* **236**(3), 276–281.

Herzog, E., Casey, A., Lyng, F. M., *et al.* (2007). A new approach to the toxicity testing of carbon-based nanomaterials: The clonogenic assay. *Toxicol Lett* **174**(1–3), 49–60.

Hillegass, J. M., Shukla, A., MacPherson, M. B., *et al.* (2010). Mechanisms of oxidative stress and alterations in gene expression by Libby six-mix in human mesothelial cells. *Part Fibre Toxicol* **7**, 26.

Hirano, S., Fujitani, Y., Furuyama, A., and Kanno, S. (2010). Uptake and cytotoxic effects of multi-walled carbon nanotubes in human bronchial epithelial cells. *Toxicol Appl Pharmacol* **249**(1), 8–15.

Hirano, S., Kanno, S., and Furuyama, A. (2008). Multi-walled carbon nanotubes injure the plasma membrane of macrophages. *Toxicol Appl Pharmacol* **232**(2), 244–251.

Janssen, Y. M., Heintz, N. H., Marsh, J. P., Borm, P. J., and Mossman, B. T. (1994). Induction of c-fos and c-jun proto-oncogenes in target cells of the lung and pleura by carcinogenic fibers. *Am J Respir Cell Mol Biol* **11**(5), 522–530.

Janssen, Y. M., Heintz, N. H., and Mossman, B. T. (1995). Induction of c-fos and c-jun proto-oncogene expression by asbestos is ameliorated by N-acetyl-L-cysteine in mesothelial cells. *Cancer Res* **55**(10), 2085–2089.

Jia, G., Wang, H., Yan, L., *et al.* (2005). Cytotoxicity of carbon nanomaterials: Single-wall nanotube, multi-wall nanotube, and fullerene. *Environ Sci Technol* **39**(5), 1378–1383.

Jimenez, L. A., Thompson, J., Brown, D. A., *et al.* (2000). Activation of NF-kappaB by PM (10) occurs via an iron-mediated mechanism in the absence of IkappaB degradation. *Toxicol Appl Pharmacol* **166**(2), 101–110.

Jimenez, L. A., Zanella, C., Fung, H., *et al.* (1997). Role of extracellular signal-regulated protein kinases in apoptosis by asbestos and H_2O_2. *Am J Physiol* **273**(5 Pt 1), L1029–L1035.

Johnson, G. L. and Lapadat, R. (2002). Mitogen-activated protein kinase pathways mediated by ERK, JNK, and p38 protein kinases. *Science* **298**(5600), 1911–1912.

Kagan, V. E., Tyurina, Y. Y., Tyurin, V. A., *et al.* (2006). Direct and indirect effects of single walled carbon nanotubes on RAW 264.7 macrophages: Role of iron. *Toxicol Lett* **165**(1), 88–100.

Kobayashi, N., Naya, M., Ema, M., *et al.* (2010). Biological response and morphological assessment of individually dispersed multi-wall carbon nanotubes in the lung after intratracheal instillation in rats. *Toxicology* **276**(3), 143–153.

Kostarelos, K., Lacerda, L., Pastorin, G., *et al.* (2007). Cellular uptake of functionalized carbon nanotubes is independent of functional group and cell type. *Nat Nanotechnol* **2**(2), 108–113.

Lam, C. W., James, J. T., McCluskey, R., and Hunter, R. L. (2004). Pulmonary toxicity of single-wall carbon nanotubes in mice 7 and 90 days after intratracheal instillation. *Toxicol Sci* **77**(1), 126–134.

Levresse, V., Moritz, S., Renier, A., *et al.* (1998). Effect of simian virus large T antigen expression on cell cycle control and apoptosis in rat pleural mesothelial cells exposed to DNA damaging agents. *Oncogene* **16**(8), 1041–1053.

Levresse, V., Renier, A., Fleury-Feith, J., *et al.* (1997). Analysis of cell cycle disruptions in cultures of rat pleural mesothelial cells exposed to asbestos fibers. *Am J Respir Cell Mol Biol* **17**(6), 660–671.

Liu, W., Ernst, J. D., and Broaddus, V. C. (2000). Phagocytosis of crocidolite asbestos induces oxidative stress, DNA damage, and apoptosis in mesothelial cells. *Am J Respir Cell Mol Biol* **23**(3), 371–378.

Loreto, C., Carnazza, M. L., Cardile, V., *et al.* (2009). Mineral fiber-mediated activation of phosphoinositide-specific phospholipase c in human bronchoalveolar carcinoma-derived alveolar epithelial A549 cells. *Int J Oncol* **34**(2), 371–376.

Luster, M. I. and Simeonova, P. P. (1998). Asbestos induces inflammatory cytokines in the lung through redox sensitive transcription factors. *Toxicol Lett* **102**–103, 271–275.

Miserocchi, G., Sancini, G., Mantegazza, F., and Chiappino, G. (2008). Translocation pathways for inhaled asbestos fibers. *Environ Health* **7**, 4.

Mittal, S., Sharma, V., Vallabani, N. V., *et al.* (2011). Toxicity evaluation of carbon nanotubes in normal human bronchial epithelial cells. *J Biomed Nanotechnol* **7**(1), 108–109.

Mongan, L. C., Jones, T., and Patrick, G. (2000). Cytokine and free radical responses of alveolar macrophages in vitro to asbestos fibres. *Cytokine* **12**(8), 1243–1247.

Mossman, B. T. (1994). Carcinogenesis and related cell and tissue responses to asbestos: A review. *Ann Occup Hyg* **38**(4), 617–624.

Muller, J., Decordier, I., Hoet, P. H., *et al.* (2008). Clastogenic and aneugenic effects of multi-wall carbon nanotubes in epithelial cells. *Carcinogenesis* **29**(2), 427–433.

Muller, J., Huaux, F., Moreau, N., *et al.* (2005). Respiratory toxicity of multi-wall carbon nanotubes. *Toxicol Appl Pharmacol* **207**(3), 221–231.

Muller, L., Riediker, M., Wick, P., *et al.* (2010). Oxidative stress and inflammation response after nanoparticle exposure: Differences between human lung cell monocultures and an advanced three-dimensional model of the human epithelial airways. *J R Soc Interface* **7** (suppl. 1), S27–S40.

Murphy, F. A., Poland, C. A., Duffin, R., *et al.* (2011). Length-dependent retention of carbon nanotubes in the pleural space of mice initiates sustained inflammation and progressive fibrosis on the parietal pleura. *Am J Pathol* **178**(6), 2587–2600.

Mutsaers, S. E. (2004). The mesothelial cell. *Int J Biochem Cell Biol* **36**(1), 9–16.

Nel, A., Xia, T., Madler, L., and Li, N. (2006). Toxic potential of materials at the nanolevel. *Science* **311**(5761), 622–627.

Pache, J. C., Janssen, Y. M., Walsh, E. S., *et al.* (1998). Increased epidermal growth factor-receptor protein in a human mesothelial cell line in response to long asbestos fibers. *Am J Pathol* **152**(2), 333–340.

Pacurari, M., Yin, X. J., Zhao, J., *et al.* (2008). Raw single-wall carbon nanotubes induce oxidative stress and activate MAPKs, AP-1, NF-kappaB, and Akt in normal and malignant human mesothelial cells. *Environ Health Perspect* **116**(9), 1211–1217.

Pande, P., Mosleh, T. A., and Aust, A. E. (2006). Role of alphavbeta5 integrin receptor in endocytosis of crocidolite and its effect on intracellular glutathione levels in human lung epithelial (A549) cells. *Toxicol Appl Pharmacol* **210**(1–2), 70–77.

Panduri, V., Weitzman, S. A., Chandel, N. S., and Kamp, D. W. (2004). Mitochondrial-derived free radicals mediate asbestos-induced alveolar epithelial cell apoptosis. *Am J Physiol Lung Cell Mol Physiol* **286**(6), L1220–L1227.

Peterson, M. W. and Kirschbaum, J. (1998). Asbestos-induced lung epithelial permeability: Potential role of nonoxidant pathways. *Am J Physiol* **275**(2 Pt 1), L262–L268.

Poland, C. A., Duffin, R., Kinloch, I., *et al.* (2008). Carbon nanotubes introduced into the abdominal cavity of mice show asbestos-like pathogenicity in a pilot study. *Nat Nanotechnol* **3**(7), 423–428.

Porter, A. E., Gass, M., Muller, K., *et al.* (2007). Direct imaging of single-walled carbon nanotubes in cells. *Nat Nanotechnol* **2**(11), 713–717.

Pulskamp, K., Diabate, S., and Krug, H. F. (2007). Carbon nanotubes show no sign of acute toxicity but induce intracellular reactive oxygen species in dependence on contaminants. *Toxicol Lett* **168**(1), 58–74.

Ravichandran, P., Baluchamy, S., Sadanandan, B., *et al.* (2010). Multiwalled carbon nanotubes activate NF-kappaB and AP-1 signaling pathways to induce apoptosis in rat lung epithelial cells. *Apoptosis* **15**(12), 1507–1516.

Rosenthal, G. J., Germolec, D. R., Blazka, M. E., *et al.* (1994). Asbestos stimulates IL-8 production from human lung epithelial cells. *J Immunol* **153**(7), 3237–3244.

Rothen-Rutishauser, B., Brown, D. M., Piallier-Boyles, M., *et al.* (2010). Relating the physicochemical characteristics and dispersion of multiwalled carbon nanotubes in different suspension media to their oxidative reactivity in vitro and inflammation in vivo. *Nanotoxicology* **4**, 331–342.

Rotoli, B. M., Bussolati, O., Barilli, A., *et al.* (2009). Airway barrier dysfunction induced by exposure to carbon nanotubes in vitro: Which role for fiber length? *Hum Exp Toxicol* **28** (6–7), 361–368.

Rotoli, B. M., Bussolati, O., Bianchi, M. G., *et al.* (2008). Non-functionalized multi-walled carbon nanotubes alter the paracellular permeability of human airway epithelial cells. *Toxicol Lett* **178**(2), 95–102.

Salvador-Morales, C., Townsend, P., Flahaut, E., *et al.* (2007). Binding of pulmonary surfactant proteins to carbon nanotubes; potential for damage to lung immune defense mechanisms. *Carbon* **45**, 607–617.

Sargent, L. M., Reynolds, S. H., and Castranova, V. (2010). Potential pulmonary effects of engineered carbon nanotubes: In vitro genotoxic effects. *Nanotoxicology* **4**, 396–408.

Shalhoeb, J., Falck-Hansen, M. A., Davies, A. H., and Monaco, C. (2011). Innate immunity and monocyte-macrophage activation in atherosclerosis. *J Inflamm* **28**, 8–25.

Sharma, C. S., Sarkar, S., Periyakaruppan, A., *et al.* (2007). Single-walled carbon nanotubes induces oxidative stress in rat lung epithelial cells. *J Nanosci Nanotechnol* **7**(7), 2466–2472.

Shukla, A., Jung, M., Stern, M., *et al.* (2003a). Asbestos induces mitochondrial DNA damage and dysfunction linked to the development of apoptosis. *Am J Physiol Lung Cell Mol Physiol* **285**(5), L1018–L1025.

Shukla, A., Ramos-Nino, M., and Mossman, B. (2003b). Cell signaling and transcription factor activation by asbestos in lung injury and disease. *Int J Biochem Cell Biol* **35**(8), 1198–1209.

Shvedova, A. A., Fabisiak, J. P., Kisin, E. R., *et al.* (2008a). Sequential exposure to carbon nanotubes and bacteria enhances pulmonary inflammation and infectivity. *Am J Respir Cell Mol Biol* **38**(5), 579–590.

Shvedova, A. A., Kisin, E., Murray, A. R., *et al.* (2008b). Inhalation vs. aspiration of single-walled carbon nanotubes in C57BL/6 mice: Inflammation, fibrosis, oxidative stress, and mutagenesis. *Am J Physiol Lung Cell Mol Physiol* **295**(4), L552–L565.

Shvedova, A. A., Kisin, E. R., Mercer, R., *et al.* (2005). Unusual inflammatory and fibrogenic pulmonary responses to single-walled carbon nanotubes in mice. *Am J Physiol Lung Cell Mol Physiol* **289**(5), L698–L708.

Shvedova, A. A., Kisin, E. R., Murray, A. R., *et al.* (2007). Vitamin E deficiency enhances pulmonary inflammatory response and oxidative stress induced by single-walled carbon nanotubes in C57BL/6 mice. *Toxicol Appl Pharmacol* **221**(3), 339–348.

Simon-Deckers, A., Gouget, B., Mayne-L'hermite, M., *et al.* (2008). In vitro investigation of oxide nanoparticle and carbon nanotube toxicity and intracellular accumulation in A549 human pneumocytes. *Toxicology* **253**(1–3), 137–146.

Srivastava, R. K., Lohani, M., Pant, A. B., and Rahman, Q. (2010). Cyto-genotoxicity of amphibole asbestos fibers in cultured human lung epithelial cell line: Role of surface iron. *Toxicol Ind Health* **26**(9), 575–582.

Srivastava, R. K., Pant, A. B., Kashyap, M. P., *et al.* (2011). Multi-walled carbon nanotubes induce oxidative stress and apoptosis in human lung cancer cell line-A549. *Nanotoxicology* **5**, 195–207.

Stone, V. and Kinloch, I. A. (2007). Nanoparticle interactions with biological systems and subsequent activation of intracellular signaling mechanisms. In Monteiro-Riviere, N. A. and Tran, C. L., eds., *Nanotoxicology* (Boca Raton FL: CRC Press), pp. 345–362.

Swain, W. A., O'Byrne, K. J., and Faux, S. P. (2004). Activation of p38 MAP kinase by asbestos in rat mesothelial cells is mediated by oxidative stress. *Am J Physiol Lung Cell Mol Physiol* **286**(4), L859–L865.

Tabet, L., Bussy, C., Amara, N., *et al.* (2009). Adverse effects of industrial multiwalled carbon nanotubes on human pulmonary cells. *J Toxicol Environ Health A* **72**(2), 60–73.

Takahashi, N., Kozai, D., Kobayashi, R., Ebert, M., and Mori, Y. (2011). Roles of TRPM2 in oxidative stress. *Cell Calcium* **50**(3), 279–287.

Thurnherr, T., Brandenberger, C., Fischer, K., *et al.* (2011). A comparison of acute and long-term effects of industrial multiwalled carbon nanotubes on human lung and immune cells in vitro. *Toxicol Lett* **200**(3), 176–186.

Timblin, C. R., Guthrie, G. D., Janssen, Y. W., *et al.* (1998). Patterns of c-fos and c-jun proto-oncogene expression, apoptosis, and proliferation in rat pleural mesothelial cells exposed to erionite or asbestos fibers. *Toxicol Appl Pharmacol* **151**(1), 88–97.

Tomatis, M., Turci, F., Ceschino, R., *et al.* (2010). High aspect ratio materials: Role of surface chemistry vs. length in the historical 'long and short amosite asbestos fibers'. *Inhal Toxicol* **22**(12), 984–998.

Valko, M., Leibfritz, D., Moncol, J., *et al.* (2007). Free radicals and antioxidants in normal physiological functions and human disease. *Int J Biochem Cell Biol* **39**(1), 44–84.

Wang, H., Gillis, A., Zhao, C., *et al.* (2011). Crocidolite asbestos-induced signal pathway dysregulation in mesothelial cells. *Mutat Res* **723**(2), 171–176.

Wang, L., Castranova, V., Mishra, A., *et al.* (2010). Dispersion of single-walled carbon nanotubes by a natural lung surfactant for pulmonary in vitro and in vivo toxicity studies. *Part Fibre Toxicol* **7**, 31.

Wang, X., Samet, J. M., and Ghio, A. J. (2006). Asbestos-induced activation of cell signaling pathways in human bronchial epithelial cells. *Exp Lung Res* **32**(6), 229–243.

Wick, P., Manser, P., Limbach, L. K., *et al.* (2007). The degree and kind of agglomeration affect carbon nanotube cytotoxicity. *Toxicol Lett* **168**(2), 121–131.

Wu, S. and Sun, J. (2011). Vitamin D, vitamin D receptor, and macroautophagy in inflammation and infection. *Discov Med* **11**(59), 325–335.

Yang, H., Bocchetta, M., Kroczynska, B., *et al.* (2006). TNF-alpha inhibits asbestos-induced cytotoxicity via a NF-kappaB-dependent pathway, a possible mechanism for asbestos-induced oncogenesis. *Proc Natl Acad Sci U S A* **103**(27), 10397–10402.

Ye, J., Shi, X., Jones, W., *et al.* (1999). Critical role of glass fiber length in TNF-alpha production and transcription factor activation in macrophages. *Am J Physiol* **276**(3 Pt 1), L426–L434.

Zanella, C. L., Posada, J., Tritton, T. R., and Mossman, B. T. (1996). Asbestos causes stimulation of the extracellular signal-regulated kinase 1 mitogen-activated protein kinase cascade after phosphorylation of the epidermal growth factor receptor. *Cancer Res* **56**(23), 5334–5338.

Zanella, C. L., Timblin, C. R., Cummins, A., *et al.* (1999). Asbestos-induced phosphorylation of epidermal growth factor receptor is linked to c-fos and apoptosis. *Am J Physiol* **277**(4 Pt 1), L684–L693.

11

Systemic health effects of carbon nanotubes following inhalation

JACOB D. McDONALD, AMIE LUND

11.1 Introduction

Carbon nanotubes (CNTs), including single-walled carbon nanotubes (SWCNTs) and multi-walled carbon nanotubes (MWCNTs), represent an important family of emerging nanotechnologies. Their extreme strength (strongest known fiber), light weight, and electrical conductivity give promise of a wide range of applications in engineering and medicine. As documented in this book, a number of studies have demonstrated the potential for CNTs – in various forms – to cause unwanted biological effects. In general, these effects have been observed in studies that examine high doses of exposure to CNTs in forms that are most applicable to manufacturing processes. This is because CNTs typically used for toxicology studies conducted to date have not included the functionalization and modifications that are done prior to integrating the materials into products. The studies of the prefunctionalized CNTs provide a basis for occupational hazard evaluation. The potential for later release into the environment is a subject of active research.

The majority of the CNT toxicity studies to date have focused on pulmonary responses. The potential for inhalational exposure and concerns over the physico-chemical properties with parallels to toxic fibers have primarily driven these studies. As documented in previous chapters, a number of studies have shown a merit for concern over pulmonary toxicity. While the majority of the reports have shown toxicity at high doses, some studies have also shown few to no pulmonary effects. These differences suggest potentially important contrasts between CNT physico-chemical properties that likely merit further investigation.

While CNT toxicity research has primarily focused on pulmonary responses, several studies have looked at the effects of inhalational exposure to CNTs on extrapulmonary organs. As with observations of other inhaled materials, a number of systemic effects have been observed. The response mechanism may vary by organ and effect, but the overarching paradigm for systemic response to inhaled

The Toxicology of Carbon Nanotubes, ed. Ken Donaldson, Craig A. Poland, Rodger Duffin and James Bonner. Published by Cambridge University Press. © Cambridge University Press 2012.

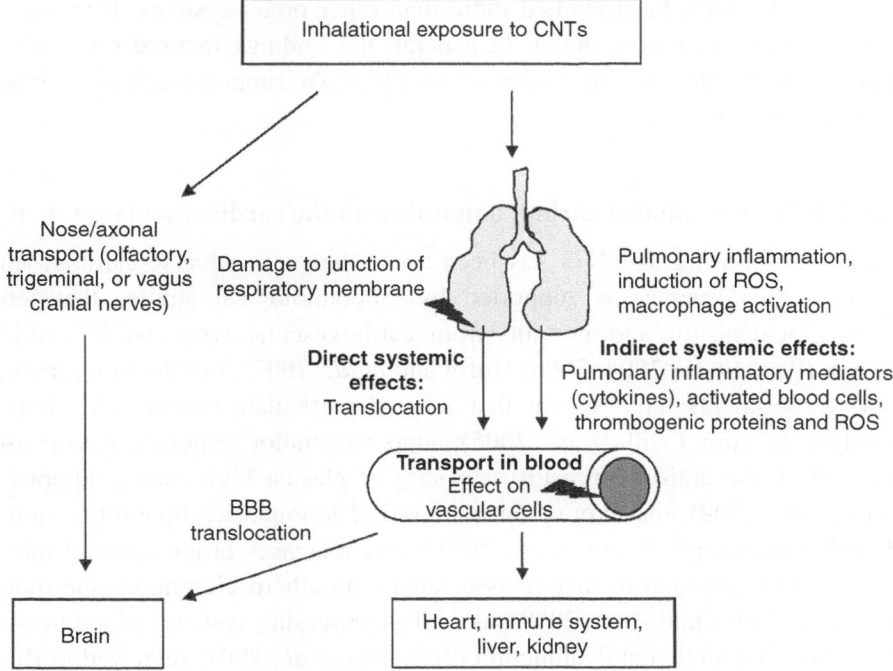

Figure 11.1 Possible routes of systemic effects from inhaled CNTs. Inhaled CNTs can elicit effects on distal organ systems through two distinct proposed mechanisms: (1) indirect pathways by induction of pulmonary inflammation, generation of ROS, and activation of macrophages, mediators of which are translocated into the circulatory system; and (2) direct pathways in which the CNTs themselves translocate across the epithelial–endothelial respiratory membrane of the lungs. Either scenario can result in toxicity in the cells of the vasculature, as well as other organ systems including (but not limited to) the liver, kidneys, spleen, and heart. Inhaled CNTs may enter the central nervous system and disrupt neuronal and glial cell structure and function, either through the axonal transport of the cranial nerves or through translocation from the circulatory system across the blood–brain barrier.

CNTs is summarized in Figure 11.1. The current hypotheses related to systemic responses include effects (1) through direct interactions of the particles, which have translocated across the alveolar epithelium of the lungs and endothelial cells of the pulmonary capillaries into the circulatory system, and into distal target organs; and (2) through indirect mechanisms which involve mediators of particle-induced inflammation, blood cell activation, and reactive oxygen species (ROS) generated in the lung but exerted as secondary effects on distal target organs.

This chapter provides an overview of the systemic effects of inhaled CNTs in studies to date. The review is limited primarily to the cardiovascular and immune systems, which have been studied more than other organ systems. However, all systemic effects were considered. In general, the findings focused on results of *in vivo* findings, unless *in vitro* findings were utilised to support mechanistic insight into the *in vivo* response.

11.2 Effects of inhaled carbon nanotubes on the cardiovascular system

Inhalational exposure to CNTs has been associated with adverse cardiovascular outcomes. This premise is supported by epidemiological studies that report increased incidence of, and mortality from, cardiovascular events on days of high pollution (Pope *et al.*, 2004, 2006; Hoffmann *et al.*, 2007). Furthermore, multiple *in vivo* experiments have shown that inhaled particulate matter (PM) impairs fibrinolytic function (Mills *et al.*, 2005); alters vasomotor responsiveness (Cascio *et al.*, 2007), the anti-inflammatory capacity of plasma high-density lipoprotein (Arujo *et al.*, 2008), the expression of oxidized low-density lipoprotein and its endothelial receptor-1 (Lund *et al.*, 2011); and increases progression of athero-sclerosis and expression of factors associated with atherosclerotic plaque rupture (Sun *et al.*, 2005; Lund *et al.*, 2009), as well as increasing systemic blood pressure and promoting endothelial dysfunction (Tornqvist *et al.*, 2007; reviewed in Brook and Rajagopalan, 2009). As nanoparticles (NPs) comprise a large percentage of combustion-derived PM, and since there is increasing evidence that the smaller particles present in the urban environment are responsible for higher levels of pulmonary deposition and resulting biological activity (Brook *et al.*, 2004, reviewed in Oberdörster *et al.*, 2005; Brook *et al.*, 2010), these studies collectively support the association of inhalational exposure to NPs with adverse effects on the cardiovas-cular system.

While the hypothesis of CNT translocation across the respiratory membrane into the systemic circulation is still controversial, NPs have previously been reported to reach the systemic circulation after both intratracheal instillation (Nemmar *et al.*, 2001; Shimida *et al.*, 2006) and inhalational methods of exposure (Nemmar *et al.*, 2002; Oberdörster *et al.*, 2002) in both animal and human models, with evidence of distribution, accumulation, and pathophysiological responses in other organ sys-tems, including the heart, brain, kidneys, and spleen (Nemmar *et al.*, 2002; Oberdörster *et al.*, 2002; Shimida *et al.*, 2006; Deng *et al.*, 2007; Mitchell *et al.*, 2007; Reddy *et al.*, 2010). Other studies report very low translocation rates for NPs into the blood circulation (Kreyling *et al.*, 2002; Mills *et al.*, 2005) and no obvious accumulation in the heart in at least two published CNT exposure studies (Ma-Hock *et al.*, 2009; Tong *et al.*, 2009); however, the hearts from CNT-exposed mice in the latter mentioned study did show significantly lower cardiac functional recovery, higher coronary flow rates, and greater infarct size, as well as exhibiting focal cardiac myofiber degeneration. Note that physical factors such as surface area dimension and aggregation, as well as the chemical nature of the NP surface, referred to as the *corona*, can alter the nanoparticle's kinetics and chemical properties, thereby dictating the way in which it interacts with a cell membrane. This phenomenon is likely both chemical (e.g. presence of transition metals) and microenvironment-dependent. For example, NPs circulating in the blood have been reported to have plasma proteins compete for binding on the NP surface. The resulting changes in surface adhesion of these proteins to the NP, or the resulting corona, can define the biological activity of the particle (Cedervall *et al.*, 2007).

While limited, studies in the current literature have examined the effects of inhaled or intratracheally instilled CNTs on the cardiovascular system both *in vitro* and *in vivo* (see Figure 11.2 for a summary of proposed CNT-mediated effects on cardiovascular cells). One of the first reported *in vivo* studies (Li *et al.*, 2007) evaluated the effects of instilled SWCNTs on alterations of aortic mitochondrial DNA (mtDNA), induction of oxidative stress, and atherosclerotic plaque formation using heme oxygenase (HO)-1 reporter transgenic mice, atherosclerotic apolipoprotein (Apo) E $-/-$ transgenic mice, and C57Bl6 control mice. HO-1 reporter transgenic mice exposed to SWCNTs displayed a significant induction of HO-1 – a marker of oxidative stress – in the lung, aorta, and heart tissues, after a single exposure (40 µg/mouse) at post-exposure day 7, but not at post-exposure days 1 or 28. The study exposed C57Bl6 mice to two doses of SWCNTs (10 and 40 µg/mouse), and also measured aortic mtDNA damage. The authors report that C57Bl6 mice exposed to SWCNTs developed mtDNA damage accompanied by changes in mitochondrial glutathione and protein carbonyl levels. Additionally, to assess potential effects of SWCNT exposure on progression of atherosclerosis, the study exposed Apo E $-/-$ mice on a high-fat diet subchronically (20 µg/mouse once every other week for 8 weeks) and histologically quantified alterations in vascular plaque area. Interestingly, while SWCNT-exposed animals did not display modified lipid profiles or alterations in vascular inflammation, they did show significant increases in plaque areas in the aortas and brachiocephalic arteries; this was associated with increased aorta mtDNA damage. The observed increase in plaque progression with SWCNT exposure may be due, in part, to increased platelet activation. Platelets represent an important linkage between inflammation, thrombosis, and the progression of atherosclerotic plaque development. Platelet-induced chronic inflammatory processes at the vascular wall result in increased leukocyte recruitment and internalization, which can lead to atherosclerotic lesions and plaque formation.

Interestingly, SWCNT exposure has been shown to increase platelet P-selectin expression, platelet aggregability, and the amount of platelet-granulocyte complexes

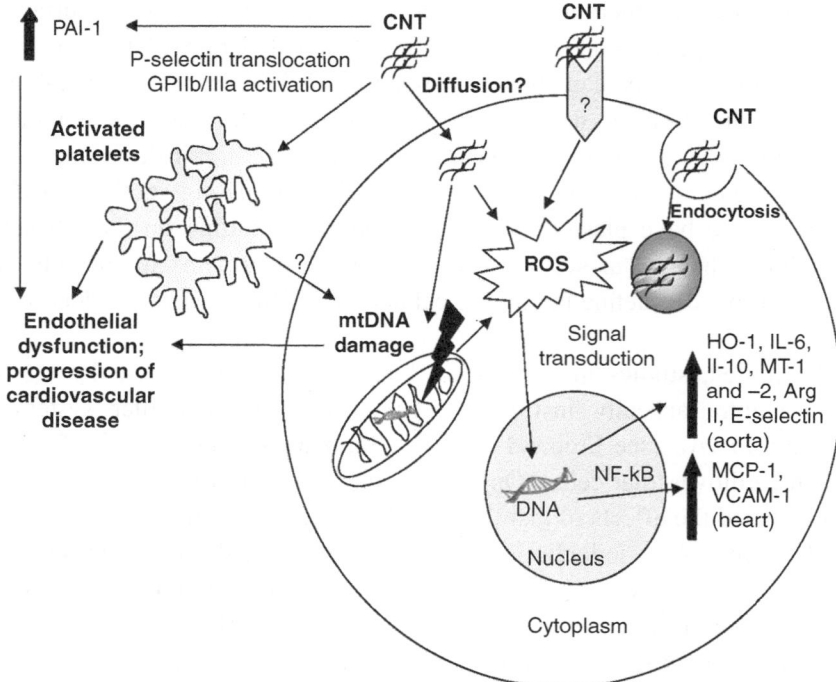

Figure 11.2 Proposed effects of CNTs on cells of the cardiovascular system. CNTs are hypothesized to be able to enter cells by diffusion, by endocytosis, or to elicit their effects through cell-membrane receptor-mediated signaling pathways. CNT exposure has been shown to increase intracellular ROS, as well as to activate transcription factors such as nuclear factor (NF)-κB, and to increase expression of genes associated with oxidative stress and lymphocyte recruitment in the vasculature, including HO-1, vascular cell adhesion molecule (VCAM)-1, monocyte chemoattractant protein (MCP-1), IL-6 and -10, MT-1 and -2, Arg II, and E-selectin. Up-regulation of these factors is associated with injury to the vessel wall, sequestration of monocytes/macrophages that can be internalized into a growing vascular plaque, and endothelial dysfunction, which collectively are the hallmarks of atherosclerosis plaque initiation and growth. CNTs have also been shown to damage mtDNA, which can result in further dysfunction of the oxidative state of the cell. Additionally, inhaled CNT exposure has been reported to increase plasma PAI-1, and activate platelets, through P-selectin translocation and glucoprotein integrin receptor (GPIIb/IIIa) activation, which can also result in progression of atherosclerotic plaque development.

through *in vitro* analysis (Bihari *et al.*, 2010). Furthermore, *in vivo* findings from this study also describe SWCNT-mediated thrombus formation in small mesenteric arteries and the microcirculation (Bihari *et al.*, 2010); this agrees with previously reported findings of SWCNT exposure promoting thrombus formation in the carotid artery of rats (Radomski *et al.*, 2005). In addition to vascular effects of CNT exposure, a few studies have examined the effects of inhaled CNTs in disruption of physiological homeostasis in the heart and vasculature. For example, a recent study reports that pulmonary SWCNT exposure (1 μg/kg body weight) in rats resulted in altered autonomic cardiovascular control regulation, as quantified by a decrease in number of baroreflex sequences (Legramante *et al.*, 2009). Altering the state of the CNT corona has been reported as leading to effects in cardiovascular function, as well. Acid functionalisation (AF) is a technique used to increase the solubility and dispersion characteristics of materials, including CNTs (Wang *et al.*, 2008). AF of SWCNTs has been reported to result in highly dispersed suspensions of particles in solution, exposure to which induced greater levels of pulmonary inflammation in mice (Saxena *et al.*, 2007). Interestingly, pulmonary exposure to AF-SWCNTs in mice (40 μg) also results in lower cardiac functional recovery after an ischemia/reperfusion challenge, and significantly higher coronary flow rates (Tong *et al.*, 2009). While no SWCNTs were detected in the heart tissues from these mice, they did exhibit signs of cardiac myofiber degeneration and alterations in systemic endpoints such as increased plasma creatine kinase levels and a decrease in red blood cell count (Tong *et al.*, 2009). SWCNT and MWCNT deposition in the lung has also been associated with alterations in aortic gene expression of mediators of inflammation, oxidative stress, remodeling, and thrombosis (including interleukin (IL)-6, IL-10, S100a8, metallothioneins (MT)-1, MT-2, HIF-3α, arginase (Arg) II, and HO-1), as well as inflammatory and prothrombotic blood proteins (plasminogen activator inhibitor (PAI-1)) and activation of E-selectin, an endothelial-specific cell adhesion molecule that signals in the recruitment of leukocytes, in the systemic vasculature (Erdely *et al.*, 2009). It is not clear from these studies whether the CNTs reach and directly act on the cells of the blood, vasculature, and/or heart or whether there is an indirect interaction through oxidative stress or inflammatory mediated pathways. However, there appears to be a clear causative association between inhaled (or instilled) CNT exposure and detrimental alterations in histological and functional endpoints that are associated with progression of cardiovascular disease.

Recently published *in vitro* studies examining the direct effect of CNTs on cells of the heart and aorta have also helped to further characterize mechanistic pathways that may be involved in the cardiovascular sequelae reported after pulmonary CNT exposure. In heart cells, SWCNT treatment (at concentrations ranging from 0.25 to 50 μg/ml) reportedly showed no increase in cardiac cell ROS, heart cell function, or alterations in myofibrillar structures (Helfenstein *et al.*, 2008). Alternatively, in studies utilizing aortic endothelial cells, purified SWCNT and MWCNT exposure (at concentrations of 1.5–4.5 μg/ml) resulted in actin filament and vascular endothelial (VE)-cadherin (endothelial barrier/junction protein) disruption, as well as reduced tubule formation, and increased cytotoxicity (Walker *et al.*, 2009). Considering that the cytoskeleton plays an active role in endothelial cell adhesion, permeability, vasculogenesis, and motility (Moreau *et al.*, 2003), and that the proximity between epithelial cell and endothelial cell membranes may play a role in translocation across the respiratory membrane (Heckel *et al.*, 2004), the finding that exposure to CNTs disrupted actin cytoskeleton arrangement and VE-cadherin in aortic endothelial cells suggests this may serve as a mechanism for CNT transloca-tion, as well as providing a route for outflow of inflammatory mediators into the systemic circulation through the capillaries at the respiratory membrane. In a separate *in vitro* study, SWCNT particles were found localized in the cytoplasm and within mitochondria inside primary aortic endothelial cells after exposure (0.8–200 μg/ml) (Zhiqing *et al.*, 2010). These findings are significant in that SWCNT exposure resulted in disruption of the mitochondrial cristae in a time-dependent matter after exposure, suggesting damage to the mitochondria and mitochondrial function. Furthermore, these authors report induced cytotoxicity, decreased membrane integrity, as well as increased intracellular ROS and expres-sion of cellular adhesion molecules (CAM), vascular cell adhesion molecule (VCAM)-1, and intercellular adhesion molecule (ICAM)-1. The expression of VCAM-1 and ICAM-1 in these cells appears to be mediated through an ROS induction of nuclear factor (NF)-κB, as co-incubation with N-acetylcysteine inhib-ited SWCNT-induced NF-κB/p65 translocation, likely through an anti-oxidative mechanism (Zhiqing *et al.*, 2010). Understanding the effects of CNTs on endothelial cells is especially important because this would be the vascular cell type with which translocated CNTs would have the most interaction if circulating in the blood. In smooth-muscle cells (SMC), the most abundant cell type found in the vessel wall, SWCNT exposure resulted in a significant dose-dependent decrease in SMC growth (Raja *et al.*, 2007). This study's reported findings also showed that NP size, as well as the agglomeration state, had significant effects on SMC growth kinetics; this is shown through comparisons of filtered SWCNT exposure (to remove aggregates) to unfiltered SWCNT exposure on SMC growth. Taken together, these *in vitro* studies indicate that the physical state of the CNTs, as well as concentration, can disrupt growth, cytoskeleton arrangement, and barrier protein expression, as well as increasing expression of intracellular ROS and expression of CAMs in the predo-minant cell types of the vasculature. Such findings suggest that CNTs can directly interact with these cell types, eliciting potentially damaging cellular effects, the implications of which suggest that further research is essential to optimize doses, structures, and routes of exposure for potential therapeutic applications of CNTs. Additionally, future studies are needed to investigate the ability of CNTs to promote, progress, or worsen systemic inflammation and ROS, whether through direct or indirect methods, as these signaling pathways are primary media-tors of initiation and progression of cardiovascular disease (Ross, 1999; Libby, 2002, 2006).

11.3 Effects of inhaled carbon nanotubes on systemic immunity

Inhaled MWCNTs have been shown to cause systemic immunosuppression and splenic oxidative stress (Mitchell *et al.*, 2007). In those studies immunosuppression was observed in C57Bl6 mice exposed for 2 weeks by inhalation at concentrations up to 1 mg/m^3 per day. Immunosuppression was evaluated in splenocytes harvested after the last day of MWCNT exposure. The studies showed an MWCNT exposure caused a decrease in response to a sheep red blood cell antigen, and that the suppression was in T cells, but not B cells. There was also a noted impact on the natural killer cells. These findings were also observed for exposure to SWCNTs (unpublished results). The mechanism of this immune suppression followed the general paradigm outlined in Figure 11.3.

The mechanism of CNT immunosuppression shown in Figure 11.3 involves the release of transformational growth factor-beta (TGF-β) from the lungs, which enters the bloodstream to signal cyclooxygenase (COX)-2-mediated increases in prostaglandin-E2 and IL-10 in the spleen, both of which play a role in suppressing T cell proliferation (Mitchell, 2009). The COX-2 pathway was deduced after an observation that the immunosuppression could be reversed by either co-treatment with a non-steroid anti-inflammatory drug (ibuprofen) that is known to block the

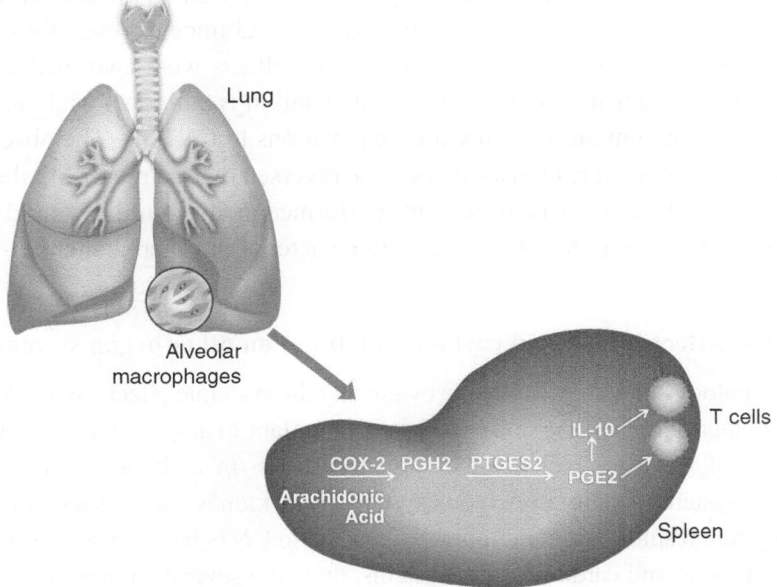

Figure 11.3 Proposed mechanism of systemic immunosuppression from inhalation exposure to CNTs. TGF-β is released from the lungs, entering the bloodstream to signal COX-2-mediated increases in prostaglandin-E2 and IL-10 in the spleen, both of which play a role in suppressing T cell proliferation.

COX pathway or by evaluating the response in COX-2 knockout mice. The response was repeated *in vitro* using protein isolated from the lavage fluid of mice exposed to CNTs.

In the paradigm defined above, the CNTs act systemically through a signal that originates in the lung and is observed in the spleen. The alternative hypothesis could have been that CNTs translocate to the spleen and cause their effects directly. The authors did not observe any CNTs in the spleens grossly (or by microscopy), but did not have a method to measure them in tissues. Few studies have attempted to measure systemic disposition of CNTs, but Deng *et al.* (2007) conducted work with ^{14}C-labeled MWCNTs that suggested limited disposition to extrapulmonary tissues. Pauluhn (2010) detected small amounts of MWCNTs in lymph nodes of rats exposed to high levels for 13 weeks. That study suggests that the potential role of CNTs acting directly may not be definitively ruled out. However, more investigations into this question are required.

Subsequent to the reports above, interesting new potential insights into the mechanism of CNTs on immune suppression have been shown. Tkach *et al.* (2011) report that immunosuppression of T cells by SWCNTs may occur through modulation of dendritic cell function, as observed by a decreased proliferation in T cells incubated with SWCNT-treated dendritic cells.

Immunosuppression as a result of exposure to inhaled materials may not be unique to CNTs. Using the same assays reported by Mitchell above, Burchiel *et al.* (2004, 2005) showed immunosuppression in AJ mice exposed for 6 months to environmental pollutants. In those studies the effects were also mediated by T cells, but the mechanisms were not studied in detail. It is unknown what the clinical significance of the immunosuppression observations to CNTs is. The observations to date show significant responses that can be reversed by treatment with ibuprofen. It is unknown if the *ex vivo* measurements performed in those studies would result in an altered immune response *in vivo* (e.g. to a bacterial or other challenge).

11.4 Effects of inhaled carbon nanotubes on other organ systems

While toxicology studies are limited in evaluating the systemic effects of inhaled CNTs in tissues other than the lung and blood, it is important to address the current field of knowledge of CNT toxicity in other organ systems. In addition to effects in the pulmonary system, toxicity in nervous system, liver, kidney, and tissues of the reproductive system resulting from pulmonary exposure to CNTs has also been reported. As with the immune and cardiovascular systems, there are several studies that report NP translocation into secondary organs, including the liver, kidney, and brain (Oberdörster *et al.*, 2002; Deng *et al.*, 2007; Kreyling *et al.*, 2009; Oberdörster, 2010; Reddy *et al.*, 2010). MWCNTs have been observed localized in Kupffer cells (hepatic macrophages) in the liver (Deng *et al.*, 2007). Furthermore, instilled MWCNTs (0.2, 1, or 5 mg/kg) have been reported to exert toxicity on the liver, resulting in dose-dependent periportal lymphocytic infiltration, congestion of sinusoids, hemorrhage, foamy degeneration of hepatocytes, focal inflammation, and necrosis in the liver (Reddy *et al.*, 2010). In these same study animals, instillation of MWCNTs resulted in significant tubular necrosis and interstitial nephritis at the highest dose administered. These histological findings were further confirmed by elevated circulating levels of biomarkers for cell injury and inflammation of the liver and kidney, serum glutamic pyruvic transaminase and serum creatinine, respectively (Reddy *et al.*, 2010).

There are also several reports that inhaled NPs have been found in the brain (Donaldson *et al.*, 2004), which can occur through translocation across the endothelial cells of the capillaries of the blood–brain barrier (BBB) or directly through the olfactory bulb/nerve (see Figure 11.1). Circulating NPs have been shown to disrupt tight junction protein expression (Chen *et al.*, 2008) and endothelial cell membrane integrity in the BBB, thereby increasing access for transport into the central nervous system (Chen *et al.*, 2008). The ability of NPs to cross the BBB appears to be dependent on size, surface modifications (e.g. coatings: surfactant, glucose, conjugations of receptor-mediated ligands such as apolipoprotein E) (Kreuter *et al.*, 1995, 2002; Calvo *et al.*, 2001; Zensi *et al.*, 2009), and their electrostatic charge (Hardebo and Kahrstrom, 1985; Sahagun *et al.*, 1990; Fenart *et al.*, 1999; Lockman *et al.*, 2004). Several *in vivo* studies have demonstrated translocation of intranasally instilled or inhaled NPs through axonal transport of the olfactory, trigeminal, and vagus cranial nerves, which innervate the nasal, face, and neck regions (Hunter and Dey, 1998; Hunter and Undem, 1999; Oberdörster *et al.*, 2004; Elder *et al.*, 2006; Yu *et al.*, 2007; Wang *et al.*, 2008). Regardless of the route of entry, the presence of NPs in the neural tissue (either neurons or neuroglia cells) has been shown to exert toxic effects, including neuronal loss, alterations in neuronal physiology, and behavioral modifications (Elder *et al.*, 2006; Wang *et al.*, 2008; Sarközi *et al.*, 2009). Collectively, these studies suggest potential detrimental effects of NPs on brain structure and function; however, it should be noted that none of these studies used CNTs as the NP for exposure. There are, however, some limited study findings using *in vitro* models, which specifically evaluated the neurotoxicity of CNTs and show that both SWCNT (Belyanskaya *et al.*, 2009) and MWCNT (Xu *et al.*, 2009) exposure results in decreased DNA content, cytotoxicity, and alterations in neuronal electrophysiology. While these studies suggest that CNTs are neurotoxic, there is insufficient evidence to determine by which mechanisms they exert their effects.

11.5 Conclusions

Inhalational exposure to CNTs in laboratory animals has been shown to cause significant biological responses in both the lungs and the extrapulmonary organs.

The effects may be due both to translocation of CNTs to a site of interest and to signaling that occurs from the lung and then causes an effect elsewhere. The studies conducted to date have demonstrated the plausibility of these effects, and have made strides in determining potential mechanisms. In many cases the responses observed after exposure to CNTs have also been reported for other inhaled materials, suggesting that the mechanisms found for CNT toxicity may extend to other materials (and vice versa). The majority of the studies have been conducted with a limited range of CNT types exposed to animals at doses that typically exceed potential human exposure. A challenge to the research community is to advance the understanding of biological effects as they relate to physicochemical form. In addition, it will be important to further understand the potential for CNT exposure to these different forms in the environment or in occupational settings, so the research results can best be placed in context.

References

Arujo, J. A., Barajas, B., Kleinman, M., *et al.* (2008). Ambient particulate pollutants in the ultrafine range promote early atherosclerosis and systemic oxidative stress. *Circ Res* **102**, 589–596.

Belyanskaya, L., Weigel, S., Hirsch, C., *et al.* (2009). Effects of carbon nanotubes on primary neurons and glial cells. *Neurotoxicology* **30**(4), 702–711.

Bihari, P., Holzer, M., Praetner, M., *et al.* (2010). Single-walled carbon nanotubes activate platelets and accelerate thrombus formation in the microcirculation. *Toxicology* **269**, 148–154.

Brook, R. D., Franklin, B., Cascio, W., *et al.* (2004). Air pollution and cardiovascular disease: A statement for healthcare professionals from the Expert Panel on Population and Prevention Science of the American Heart Association. *Circulation* **109**, 2655–2671.

Brook, R. D., Rajagopalan, S., Pope, C. A., III, *et al.* (2010). Particulate matter air pollution and cardiovascular disease: An update to the scientific statement from the American Heart Association (for American Heart Association Council on Epidemiology and Prevention, Council on the Kidney in Cardiovascular Disease, and Council on Nutrition, Physical Activity and Metabolism). *Circulation* **121**, 2331–2378.

Brook, R. D. and Rajagopalan, S. (2009). Particulate matter, air pollution, and blood pressure. *J Am Soc Hypertens* **3**, 332–350.

Burchiel, S. W., Lauer, F. T., McDonald, J. D., and Reed, M. D. (2004). Systemic immunotoxicity in A J mice following 6-month whole body inhalation exposure to diesel exhaust. *Toxicol Appl Pharmacol* **196**, 337–345.

Burchiel, S. W., Lauer, F. T., Dunaway, S. L., *et al.* (2005). Hardwood smoke alters murine splenic T-cell responses to mitogens following a 6-month whole body inhalation exposure. *Toxicol Appl Pharmacol* **202**, 229–236.

Calvo, P., Gouritin, B., Chacun, H., *et al.* (2001). Long-circulating PEGylated polycynoacrylate nanoparticles as new drug carrier for brain delivery. *PharmRes* **18**, 1157–1166.

Cascio, W. E., Cozzi, E., Hazarika, S., *et al.* (2007). Cardiac and vascular changes in mice after exposure to ultrafine particulate matter. *Inhal Toxicol* **19**, 67–73.

Cedervall, T., Lynch, I., Lindman, S., *et al.* (2007). Understanding the nanoparticle-protein corona using methods to quantify exchange rates and affinities of proteins for nanoparticles. *Proc Natl Acad Sci USA* **104**, 2050–2055.

Chen, L., Hokel, R. A., Hennig, B., and Toborek, M. (2008). Manufactured aluminum oxide nanoparticles decrease expression of tight junction proteins in brain vasculature. *J Neuroimmune Pharmacol* **3**, 286–295.

Deng, X., Jia, G., Wang, H., *et al.* (2007). Translocation and fate of multi-walled carbon nanotubes in vivo. *Carbon* **45**, 1419–1424.

Donaldson, K., Stone, V., Tran, C. L., Kreyling, W., and Borm, P. J. (2004). Nanotoxicology. *Occup Environ Med* **61**, 727–729.

Elder, A., Gelein, R., Siva, V., *et al.* (2006). Translocation of inhaled ultrafine manganese oxide particles to the central nervous system. *Environ Health Perspect* **114**, 1172–1178.

Erdely, A., Hulderman, T., Salmen, R., *et al.* (2009). Cross-talk between lung and systemic circulation during carbon nanotube respiratory exposure: Potential biomarkers. *Nano Lett* **9**(1), 36–43.

Fenart, L., Casanova, A., Dehouck, B., *et al.* (1999). Evalution of effect of charge and lipid coating on ability of 60 nm nanoparticles to cross an in vitro model of the blood-brain barrier. *J Pharmacol Exp Ther* **291**, 1017–1022.

Hardebo, J. E. and Kahrstrom, J. (1985). Endothelial negative surface charge areas and blood-brain barrier function. *Acta Physiol Scand* **152**, 495–499.

Heckel, K., Kiefmann, R., Dorger, M., Stoeckelhuber, M., and Goetz, A. E. (2004). Colloidal gold particles as a new in vivo marker of early acute lung injury. *Am J Physiol Lung Cell Mol Physiol* **287**, L867–L878.

Helfenstein, M., Miragoli, M., Rohr, S., *et al.* (2008). Effects of combustion-derived ultrafine and manufactured nanoparticles on heart cells in vitro. *Toxicology* **253**, 70–78.

Hoffmann, B., Moebus, S., Möhlenkamp, S., *et al.* (2007). Residential exposure to traffic is associated with coronary atherosclerosis (for Heinz Nixdorf Recall Study Investigative Group). *Circulation* **116**, 489–496.

Hunter, D. D. and Dey, R. D. (1998). Identification and neuropeptide content of trigeminal neurons innervating the rat nasal epithelium. *Neuroscience* **83**, 591–599.

Hunter, D. D., and Undem, B. J. (1999). Identification and substance P content of vagal afferent neurons innervating the epithelium of the guinea pig trachea. *Am J Respir Crit Care Med* **159**, 4943–1948.

Kreuter, J., Alyautdin, R. N., Kharkevich, D. A., and Ivanov, A. A. (1995). Passage of peptides through the blood-brain barrier with colloidal polymer particles (nanoparticles). *Brain Res* **674**, 171–174.

Kreuter, J., Shamenkov, D., Petrov, V., *et al.* (2002). Apolipoprotein-mediated transport of nanoparticle-bound drugs across the blood-brain barrier. *J Drug Target* **10**(4)317–325.

Kreyling, W. G., Semmler, M., Erbe, F., *et al.* (2002). Translocation of ultrafine insoluble iridium particles from lung epithelium to extrapulmonary organs is size dependent but very low. *J Toxicol Environ Health A* **65**(20), 1513–1530.

Kreyling, W. G., Semmler-Behnke, M., Seitz, J., *et al.* (2009). Size dependence of the translocation of inhaled iridium and carbon nanoparticle aggregates from the lung of rats to the blood and secondary organs. *Inhal Toxicol* **21**, 55–60.

Legramante, J. M., Valentini, F., Magrini, A., *et al.* (2009). Cardiac autonomic regulation after lung exposure to carbon nanotubes. *Hum Exp Toxicol* **28**, 369–375.

Li, Z., Hulderman, T., Salmen, R., *et al.* (2007). Cardiovascular effects of pulmonary exposure to single-wall carbon nanotubes. *Environ Health Perspect* **115**, 377–382.

Libby, P. (2002). Inflammation and atherosclerosis. *Nature* **420**, 868–874.

Libby, P. (2006). Atherosclerosis: Disease biology affecting the coronary vasculature. *Am J Cardiol* **98**, 3Q–9Q.

Lockman, P. R., Koziara, J. M., Mumper, R. J., and Allen, D. D. (2004). Nanoparticle surface charges alter blood-brain barrier integrity and permeability. *J Drug Target* **12**, 635–641.

Lund, A. K., Lucero, J., Lucas, S., *et al.* (2009). Vehicular emissions induce vascular MMP-9 expression and activity associated with endothelin-1 mediated pathways. *Arterioscler Thromb Vasc Biol* **29**, 511–517.

Lund, A. K., Lucero, J., Harman, M., *et al.* (2011) The oxidized low density lipoprotein receptor mediates vascular effects of inhaled vehicle emissions. *Am J Respir Crit Care Med* **184**, 82–91.

Ma-Hock, L., Treumann, S., Strauss, V., *et al.* (2009). Inhalation toxicity of multi-wall carbon nanotubes in rats exposed for 3 months. *Toxicol Sci* **112**, 468–481.

Mills, N. L., Tornqvist, H., Robinson, S. D., *et al.* (2005). Diesel exhaust inhalation causes vascular dysfunction and impaired endogenous fibrinolysis. *Circulation* **112**, 3930–3936.

Mitchell, L. A., Gao, J., Wal, R. V., *et al.* (2007). Pulmonary and systemic immune response to inhaled multiwalled carbon nanotubes. *Toxicol Sci* **100**, 203–214.

Mitchell, L. A., Lauer, F. T., Burchiel, S. W., and McDonald, J. D. (2009). Mechanisms for how inhaled multiwalled carbon nanotubes suppress systemic immune function in mice. *Nat Nanotechnol* **4**(7), 451–456.

Moreau, V., Tatin, F., Varon, C., and Genot, E. (2003). Actin can reorganize into podosomes in aortic endothelial cells, a process controlled by Cdc42 and RhoA. *Mol Cell Biol* **23**, 6809–6822.

Nemmar, A., Vanbilloen, H., Hoylaerts, M. E., *et al.* (2001). Passage of intratracheally instilled ultrafine particles from the lung into the systemic circulation in hamster. *Am J Respir Crit Care Med* **164**, 1665–1668.

Nemmar, A., Hoet, P. H., Vanquickenbuorne, B., *et al.* (2002). Passage of inhaled particles into the blood circulation in humans. *Circulation* **105**, 411–414.

Oberdörster, G., Sharp, Z., Atudorei, V., *et al.* (2002). Extrapulmonary translocation of ultrafine carbon particles following whole body inhalation exposure in rats. *J Toxicol Envir Health A* **65**, 1531–1543.

Oberdörster, G., Sharp, Z., Atudorei, V., *et al.* (2004). Translocation of inhaled ultrafine particles to the brain. *Inhal Toxicol* **16**, 437–445.

Oberdörster, G., Maynard, A., Donaldson, K., *et al.* (2005). Principles for characterizing the potential human health effects from exposure to nanomaterials: Elements of a screening strategy (ILSI Research Foundation/Risk Science Institute Nanomaterial Toxicity Screening Working Group). *Part Fibre Toxicol* **1**(2), 8.

Oberdörster, G. (2010). Safety assessment for nanotoxicology and nanomedicine: Concepts of nanotoxicology. *J Intern Med* **267**, 89–105.

Pauluhn, J. (2010). Subchronic 13-week inhalation exposure of rats to multiwalled carbon nanotubes: Toxic effects are determined by density of agglomerate structures, not fibrillar structures. *Toxicol Sci* **113**(1), 226–242.

Pope, C. A., III, Burnett, R. T., Thurston, G. D., *et al.* (2004). Cardiovascular mortality and long-term exposure to particulate air pollution: Epidemiological evidence of general pathophysiological pathways of disease. *Circulation* **109**, 71–77.

Pope, C. A., III, Muhlestein, J. B., May, H. T., *et al.* (2006). Ischemic heart disease events triggered by short-term exposure to fine particulate air pollution. *Circulation* **114**, 2443–2448.

Radomski, A., Jurasz, P., Alonso-Escolano, D., *et al.* (2005). Nanoparticle-induced aggregation and vascular thrombosis. *Br J Pharmacol* **146**, 882–893.

Raja, P. M. V., Connolley, J., Ganesan, G. P., *et al.* (2007). Impact of carbon nanotubes exposure, dosage and aggregation of smooth muscle cells. *Toxicol Lett* **169**, 51–63.

Reddy, A. R. N., Krishna, D. R., Reddy, Y. N., and Himabindu, V. (2010). Translocation and extrapulmonary toxicities of multi wall carbon nanotubes in rats. *Toxicol Mech Methods* **20**, 267–272.

Ross, R. (1999). Atherosclerosis is an inflammatory disease. *Am Heart J* **138**, S419–S420.

Sahagun, G., Moore, S. A., and Hart, M. N. (1990). Permeability of neutral vs. anionic dextrans in culture brain microvascular endothelium. *Am J Physiol* **259**, H162–H166.

Sarközi, L., Horvath, E., Konya, Z., *et al.* (2009). Subactue intratracheal exposure of rates to manganese nanoparticles: Behavioral, electrophysiological, and general toxicological effects. *Inhal Toxicol* **21**, 83–91.

Saxena, R. K., Williams, W., McGee, J. K., *et al.* (2007). Enhanced in vitro and in vivo toxicology of poly-dispersed acid-functionalized single wall carbon nanotubes. *Nanotoxicology* **1**, 291–300.

Shimida, A., Kawamura, N., Okajima, M., *et al.* (2006). Translocation pathway of the intratracheally instilled ultrafine particles from the lung into blood circulation in the mouse. *Toxicol Pathol* **34**, 949–957.

Sun, Q., Wang, A., Jin, X., *et al.* (2005). Long-term air pollution exposure and acceleration of atherosclerosis and vascular inflammation in an animal model. *JAMA* **21**, 3003–3010.

Tkach, A. V., Shurin, G. V., Shurin, M. R., *et al.* (2011). Direct effects of carbon nanotubes on dendritic cells induce immune suppression upon pulmonary exposure. *ACS Nano* **5**(7), 5755–5762.

Tong, H., McGee, J. K., Saxena, R. K., *et al.* (2009). Influence of acid functionalization on the cardiopulmonary toxicity of carbon nanotubes and carbon black particles in mice. *Toxicol Appl Pharmacol* **239**, 224–232.

Tornqvist, H., Mills, N. L., Gonzales, M., *et al.* (2007). Persistent endothelial dysfunction in humans after diesel exhaust inhalation. *Am J Respir Crit Care Med* **176**, 395–400.

Walker, V. G., Li, Z., Hudlerman, T., *et al.* (2009). Potential in vitro effects of carbon nanotubes on human aortic endothelial cells. *Toxicol Appl Pharmacol* **236**, 319–328.

Wang, J., Liu, Y., Jiao, F., *et al.* (2008). Time-dependent translocation and potential impairment on central nervous system by intranasally instilled TiO(2) nanoparticles. *Toxicology* **254**, 82–90.

Xu, H., Bai, J., Meng, J., *et al.* (2009). Multi-walled carbon nanotubes suppress potassium channel activities in PC12 cells. *Nanotechnology* **20**, 285102–285111.

Yu, L. E., Lanry Yung, L. Y., Ong, C. N., *et al.* (2007). Translocation and effects of gold nanoparticles after inhalation exposure in rats. *Nanotoxicology* **1**, 235–242.

Zensi, A., Begley, D., Pontikis, C., *et al.* (2009). Albumin nanoparticles targeted with Apo E enter the CNS by transcytosis and are delivered to neurons. *J Control Release* **137**, 78–86.

Zhiqing, L., Zhuge, X., Fuhuan, C., *et al.* (2010). ICAM-1 and VCAM-1 expression in rat aortic endothelial cells after single-walled carbon nanotubes exposure. *J Nanosci Nanotechnol* **10**, 8562–8574.

12

Dosimetry and metrology of carbon nanotubes

LANG TRAN, LAURA MACCALMAN, ROB AITKEN

To assess the potential health risk of multi-walled carbon nanotubes (MWCNT) it is essential to quantify the dose of MWCNT in a meaningful way for the assessment of both exposure and toxicology. Traditionally, the preferred measure of dose is in units of mass. For particles of low solubility and low toxicity, it is the mass (or volume) of the particles which drive the toxicity – in this case, inflammation. This is known as the Particle Overload Phenomenon. Recently, it has been shown that particle surface area is a better descriptor of the particle dose than mass or volume. This particle surface area metric is applicable for particle-like materials and is different from the 'fibre paradigm' explored elsewhere in this book. In this chapter, we demonstrate that this metric is also applicable for some forms of MWCNT and we discuss the implications for exposure assessment and calculation of no-adverse effect level of concentration for MWCNT.

12.1 The volumetric-overload phenomenon

In a ground-breaking paper, Morrow (1988) hypothesised that the observed impairment of the alveolar macrophage (AM)-mediated clearance is due to dust *overloading* of individual AMs which affects their motility. Overload was hypothesised to be initiated when the particle volume exceeded an average of 6% of the macrophage volume, and was complete, with a virtual cessation of clearance, when the particle volume exceeded an average of 60% of the macrophage volume. This was based on three estimates or assumptions:

- the assumption of a uniform distribution of the entire particle burden over the AM population ($\sim 2.5 \times 10^7$ cells in the Fischer 344 rat)
- an average rat macrophage volume of 1000 μm^3, estimated by Dethloff and Lehnert (1988)
- an estimate of the lung burden level from which overload becomes apparent as being 1 mg for unit-density particles (Morrow, 1988).

The Toxicology of Carbon Nanotubes, ed. Ken Donaldson, Craig A. Poland, Rodger Duffin and James Bonner. Published by Cambridge University Press. © Cambridge University Press 2012.

Thus, overload was hypothesised to be initiated when the particle volume exceeded ~60 μm^3 per AM and was complete when the particle volume exceeded ~600 μm^3 per AM. The theoretical volumetric burden of 600 μm^3 suggested by Morrow is equivalent to 42 particles of 3 μm diameter, or one particle of 10.5 μm diameter. The hypothesised effect of this volumetric burden in individual AM was supported by Oberdörster *et al.* (1992), who showed that there was no detectable clearance of iron particles of 10 μm diameter at 200 days after the particles had been instilled into rat lungs.

For non-unit-density particles, the mass burden corresponding to the volumetric overload level is scaled according to the density. For example, with TiO_2 this '6% level' of lung burden will be approximately 4.25 mg (Figure 12.1).

Clearance is known to become slower as lung burden increases from the '6% level' to the '60% level'. Studies involving measurement of clearance rates with tracer particles have indicated that clearance rates decrease quite rapidly as lung burden rises above the '6% level' (Yu *et al.*, 1988). A severe reduction in clearance rate is suggested by the marked change in the rate of accumulation of lung burden during exposure to TiO_2 as shown in Figure 12.2 at approximately the '6% level' of lung burden (4.25 mg).

The dependence of overload on particle volume *only* (hence the name 'volumetric overload') implies that particles with different densities overload pulmonary clearance at *different masses*. This overload dependence has been elegantly modelled mathematically by Stöber *et al.* (1989). However, in recent studies, Oberdörster *et al.* (1994, 1996) have shown that ultrafine dusts can cause impairment of clearance without overloading individual AMs, as judged by the Morrow criterion. They also showed that the inflammatory response appeared to be related to the surface area of the particulate dose. Driscoll *et al.* (1996) have provided further

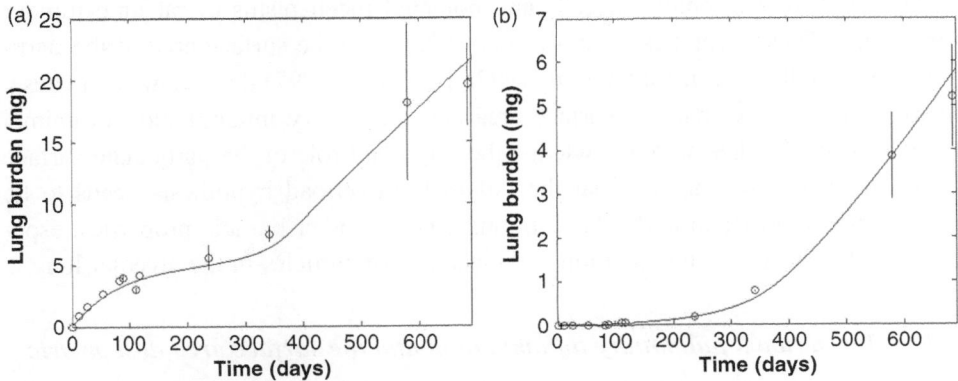

Figure 12.1 (a) Lung burden of rats exposed to titanium dioxide (TiO_2) at 10 mg/m^3. (b) Lymph node burden data from the same experiment (Jones *et al.*, 1988).

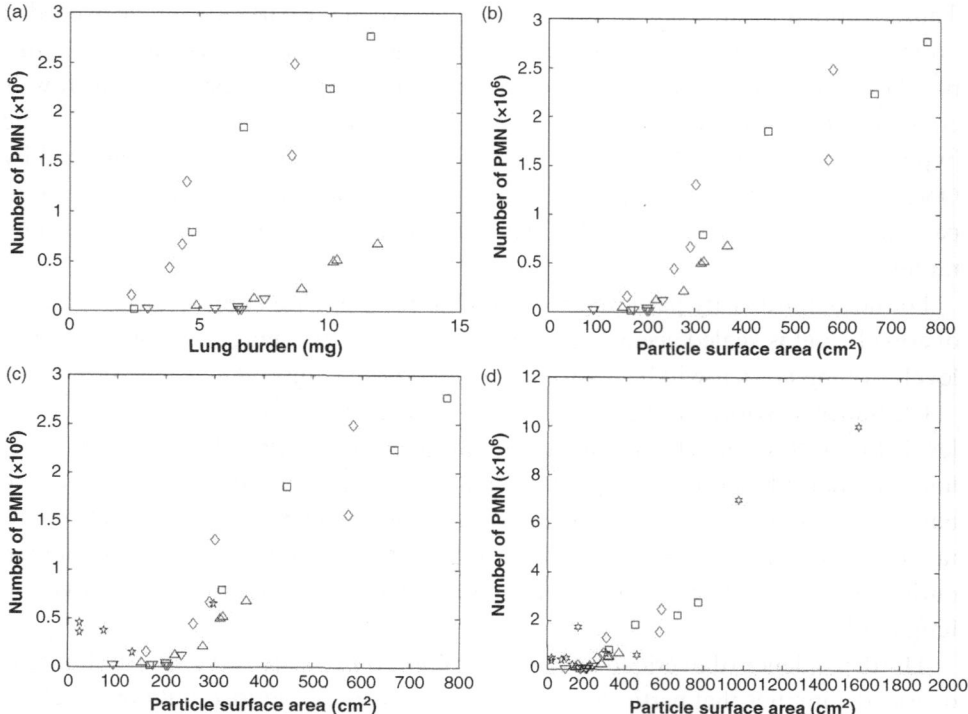

Figure 12.2 Inflammation as a function of lung burden. Top row: (a) burden expressed as mass; (b) burden expressed as surface area; symbols: TiO$_2$ high (□) and low (◊) exposure experiments; BaSO$_4$ 'high' (△) and 'low' (▽) exposure experiments. Bottom row: graphs include data from Oberdörster *et al.* (1994) for (c) fine rutile TiO$_2$ (□) and (d) ultrafine TiO$_2$ (■); note the extended range of axes for the ultrafine TiO$_2$ data (d).

hypotheses regarding the chain of events from particle-induced inflammation to mutagenic effects in epithelial cells and potential mechanisms of rat lung tumour formations. These hypotheses involve dependence on the surface area of the particulate dose. Others (e.g. Lison *et al.* 1997; Tran *et al.* 1997) demonstrated a causal relationship between particle surface area and pulmonary inflammation in animal models. These studies all acknowledge the important role of the particulate surface area. These findings suggest that the volumetric-overload hypothesis needs to be re-evaluated in order to understand the influence of other particle properties, especially surface area, on the retention and clearance of particles in the alveolar lung.

12.1.1 Overload, pulmonary inflammation and the surface area dose metric

In their inhalation studies with TiO$_2$ and BaSO$_4$, two poorly soluble particles, Tran *et al.* (1997) have demonstrated a dose–response relationship between lung burden,

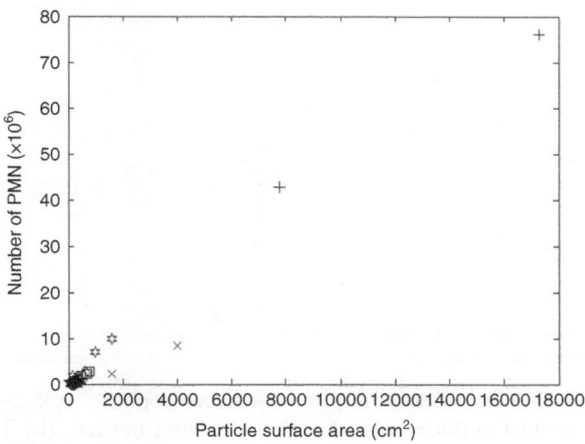

Figure 12.3 Number of polymorphonuclear neutrophilic leukocytes (PMN) in BAL plotted against surface area of particulate lung burden (from Figure 12.2(d)) with the inclusion of carbon black data from Driscoll *et al.* (1996). Carbon black data are from experiments at 1 mg/m^3 (o), 7 mg/m^3 (x) and 50 mg/m^3 (+).

described in units of surface area, and pulmonary inflammation. The lung burdens for TiO$_2$ and BaSO$_4$ were obtained under high, overload-inducing, exposure regimens (Tran *et al.*, 1997). Figure 12.2 describes the dose–response relationship for the TiO$_2$ and BaSO$_4$ from Donaldson *et al.* (2008) as well as fine and ultrafine TiO$_2$ from Oberdörster *et al.* (1994).

For higher surface area burdens and inflammation, data for carbon black from Driscoll *et al.* (1996) are combined with existing data in Figure 12.2 and shown in Figure 12.3. Altogether, the data demonstrate a consistent dose–response of inflammation and surface area.

12.1.2 *Baytubes*® *and overload*

The study of Pauluhn (2010) investigated the inhalation exposure of rats to a particle-like form of MWCNT, known as Baytubes®. The study exposed the rats, through inhalation, for 13 weeks at a range of doses (0.1, 0.4, 1.5 and 6 mg/m^3), with a post-exposure period of 200 days. The mass median aerodynamic diameter (MMAD) of the particles was 2.84 μm, with a surface area (as measured using Brunauer-Emmett-Teller (BET)) of 257 m^2/gm. Less than 1% of the MWCNT mass was cobalt (0.117%).

The results of the experiments were used to calculate the burden of carbon nanotubes and converted to surface area (Figure 12.4).

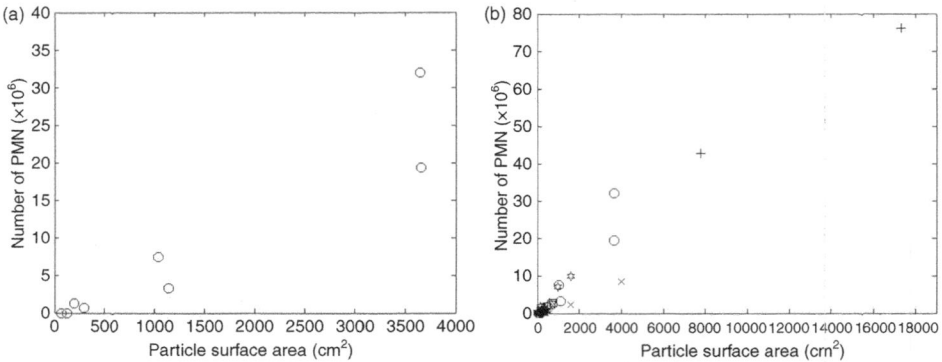

Figure 12.4 (a) Number of polymorphonuclear neutrophilic leukocytes (PMN) in BAL plotted against surface area of particulate lung burden. (b) The same plot including the data of Figure 12.3.

12.1.3 Surface or volumetric overload

Since the results so far have shown a high correlation between particle surface area and pulmonary inflammation in rats, a hypothesis regarding the underlying mechanism of its toxicity is needed. We offer a tentative hypothesis regarding the mechanism of toxicity of particle surface area. We hypothesise that there are two important alveolar cell types involved in particle overload. They are the epithelial cells, which largely constitute the alveolar surface, and the alveolar macrophages that protect them. As particles are deposited in the alveolar region, the first cell type they come into contact with are epithelial cells (Brody *et al.*, 1981). The contact between particles and cells is facilitated by the particle surface area. For some dusts, this contact triggers local chemotactic signals such as the C5A complement which attracts alveolar macrophages to the site of deposition (Warheit *et al.*, 1988). The larger the particle surface area, the larger the contact area and consequently the more activated the epithelium is likely to become.

A normal chemotactic gradient would act to attract AMs to the mucociliary escalator with their particle load; this is effective clearance. In a low-exposure situation (i.e. lung burden with low particulate surface area), this activation may not be strong enough to change the chemotactic gradient.

In a high-exposure situation (i.e. lung burden with high particulate surface area), the alveolar epithelium becomes increasingly stressed as a result of increasing contact between particles and cells, releasing chemokines such as IL-8, leading to a greater influx of neutrophils into the alveolar region. Pulmonary inflammation is associated with (i) the abnormal abundance of type II epithelial cell-derived products (e.g. phospholipids) in the alveoli, products which have been observed to impair particle clearance (Ferin, 1981); or (ii) the continual alteration of the permeability of the epithelial layer in a manner which allows the passage of constituents in the lung capillaries into the alveolar space. The presence of these factors in the alveolar space may facilitate AM aggregation and prevent them from migrating to the mucociliary escalator (Lehnert, 1993) regardless of their particle loading; this is impairment of clearance.

Once they are held back by the chemotactic signals in the alveolar region, these AMs may still be locally mobile and able to phagocytose particles. However, their phagocytic ability either may become compromised because of eventual overload (this is impairment of motility) or may be relatively less effective due to rapid interstitialisation of particles. This appears likely in the case of ultrafine particles. The former case, the impairment of AM motility, leads to an undefended epithelial surface and hence a higher rate of epithelial stimulation and damage, resulting in the movement of particles across the epithelial barrier into the interstitial space (this is known as the interstitialisation of particles). So either case produces higher interstitialisation. Higher interstitialisation is evident from an increase in dust burdens in the lymph nodes.

As demonstrated by Oberdörster *et al.* (1992) with 10 μm particles, impairment of motility of AMs leads to impairment of clearance. However, under the current hypothesis, impairment of clearance is due to a change in chemotactic gradient (triggered by particulate surface area) which keeps AMs from migrating out of the lung. Under this hypothesis, impairment of clearance is not necessarily associated with impairment of (local) macrophage motility.

12.2 Extrapolation to humans

As demonstrated above, the inflammatory reaction and the cessation of clearance are due to overload. In this section, we describe a method for calculating the human equivalent of a no-effect exposure level observed in animal studies.

Prior to overload, the retention of particles can be described in first-order kinetics. Let L be the retained dose in the lung; then

$$\frac{dL}{dt} = (\text{Deposited dose}) - kL, \tag{12.1}$$

where

$$(\text{Deposited dose}) = (\text{Concentration}) \times (\text{Ventilation}) \times (\text{Depostion fraction}). \tag{12.2}$$

Concentration (airborne concentration)	mg/m^3
Ventilation	m^3/day
Deposition fraction (for a particulate; dependent on size, density etc.)	unitless
k (alveolar macrophage-mediated clearance)	day^{-1}

At steady state, $dL/dt = 0$. Hence, the steady-state lung retained dose is

$$L_i = \frac{(\text{Deposited dose})_i}{k_i}, \tag{12.3}$$

where i represents the species (e.g. rat, human, mouse).

For a rat, we can normalise this retained dose with respect to any of the following metrics (P_i):

(i) rat lung surface area
(ii) rat lung weight
(iii) rat AM volume.

Let P represent one of the above; then the normalised retained dose for the rat, $L_{\text{rat, normalised}}$, is

$$L_{\text{rat,normalised}} = \frac{L_{\text{rat}}}{P_{\text{rat}}}. \tag{12.4}$$

The normalised human retained dose $L_{\text{human, normalised}}$ equivalent to the rat dose is

$$\begin{aligned} L_{\text{human,normalised}} &= \frac{(\text{Deposited dose})_{\text{human}}}{k_{\text{human}}} \\ &= \frac{(\text{Deposited dose})_{\text{rat}}}{k_{\text{rat}}} \times \frac{P_{\text{human}}}{P_{\text{rat}}}. \end{aligned} \tag{12.5}$$

Using the definition of the Deposited dose in Equation 12.2, we derive the unadjusted human airborne concentration equivalent to the rat as

$$\text{Conc}_{\text{human}} = \text{Conc}_{\text{rat}} \times \frac{\text{AF}_{\text{kinetics}}}{\text{AF}_{\text{deposited dose}}} \times \text{AF}_P, \tag{12.6}$$

where

$$\text{AF}_{\text{kinetics}} = \frac{k_{\text{human}}}{k_{\text{rat}}} \tag{12.7}$$

$$\text{AF}_{\text{deposited dose}} = \frac{(\text{Ventilation})_{\text{human}}}{(\text{Ventiltion})_{\text{rat}}} \times \frac{(\text{Deposition fraction})_{\text{human}}}{(\text{Deposition fraction})_{\text{rat}}} \tag{12.8}$$

$$\mathrm{AF}_P = \frac{P_{\mathrm{human}}}{P_{\mathrm{rat}}}. \tag{12.9}$$

P can be any of the normalising metrics above; Pauluhn uses the AM volume. The other difference between this approach and Pauluhn is in the clearance kinetics. Pauluhn used the half-time, $t_{1/2}$, (1 year in humans, 60 days in rats) while this approach uses the clearance rate k. The relation between k and $t_{1/2}$ is

$$k = \frac{\ln 2}{t_{1/2}}. \tag{12.10}$$

In doing the calculations above, we have introduced the following extrapolations:

(i) short-term to chronic
(ii) rat to human
(iii) average individual (i.e. no intra-species variation).

Therefore the derived concentration must be adjusted to take account of the uncertainty factors (UF) associated with these extrapolations. In short, the derived unadjusted human airborne concentration (Equation 12.6) must be divided by $\mathrm{UF}_1 \times \mathrm{UF}_2 \times \mathrm{UF}_3$.

We are now ready to compute the value of $\mathrm{Conc}_{\mathrm{human}}$. Tables 12.1 and 12.2 summarise the values used in the calculation.

Based on a concentration of $0.1 \, \mathrm{mg/m}^3$, the deposited dose in the rat was estimated to be $0.058 \, \mathrm{mg/day}$ and the steady-state retained lung dose was calculated as $4.85 \, \mathrm{mg}$.

Table 12.3 shows the results of normalising this retained dose (based on a number of different normalising factors) for humans and rats as well as the equivalent human concentration (derived from Equation 12.6).

Dividing the resulting human concentration by the uncertainty factors, the equivalent human airborne concentration is $0.1 \, \mu\mathrm{g/m}^3$ (when normalising based on lung weight or surface area) or $0.6 \, \mu\mathrm{g/m}^3$ (when normalising based on AM volume).

Table 12.1 *The parameters used to derive the human airborne concentration equivalent.*

Parameter	Unit	Rat	Human
Ventilation	$\mathrm{m}^3/\mathrm{day}$	0.102	9.8
Deposition fraction	unitless	0.057	0.118
Half-time clearance	day^{-1}	0.012	0.0019
AM volume per lung	$\mu\mathrm{m}^3$	3.0×10^{10}	3.5×10^{13}
Lung weight	kg	0.00175	0.53
Lung surface area	m^2	0.65	200

Table 12.2 *The adjustment and uncertainty factors used to derive the human airborne concentration equivalent. The values of UF are taken from REACH Guidance R.8.*[a]

Factor	Value
$AF_{kinetics}$	0.16
$AF_{deposited\ dose}$	200
AF_P	
Lung weight	302
Lung surface area	308
Lung AM volume	1167
$UF_{1,2,3}$	
Sub-acute to chronic	6
Intra-species	10
Inter-species	2.5

[a] ECHA, 2008

Table 12.3 *The normalised rat and human retained lung doses and the equivalent human airborne concentration, normalised according to the various metrics (P) considered here.*

Metric (P)	$L_{rat,\ normalised}$ (mg)	$L_{human,\ normalised}$ (mg)	$Conc_{human}$ (mg/m^3)
Lung surface area	36.2	1 494	0.02
Lung weight	2 771	1 465	0.02
AM volume	1.62×10^{-10}	5 660	0.09

12.3 Exposure assessment

Realistic interpretation of dose–response relationships is facilitated by knowledge of plausible exposure levels and duration. As a result, critical questions relating to exposure are how much, how long and how many people are exposed. Thus, exposure is usually assessed in terms of its intensity (concentration) and duration (or frequency). Control of exposure to zero effectively removes the risks from the toxic agent.

12.3.1 Historical methods for measuring particles and fibres

In early studies investigating the health effects of inhaled particles, dust samples were collected by drawing air through a filter which collected the material suspended in the air. These samples were subsequently analysed off-line to produce estimates of exposure, expressed as a concentration in air. In early studies of the coal industry, the samples were analysed by counting particles collected on the filter under a light microscope (Walton and Vincent, 1998). This resulted in an estimate of exposure in terms of particle number concentration, expressed as number of particles per cm^3 or per m^3 of air. Later epidemiology studies showed a good correlation between pneumoconiosis, the primary health effect of concern, and the mass concentration, expressed typically as mg/m^3. This subsequently became the preferred approach since it was less demanding and more accurate than manual counting. Since then, Occupational Exposure Limits (OELs), based on mass concentrations, have become the norm for measuring or regulating exposure for most hazardous substances (HSE, 2005).

The one class of aerosols that is treated differently is fibres. Exposure to fibres is characterised by the number (concentration) of fibres in the air with a specific shape and composition, rather than by mass concentration.

To assess exposure, fibres are collected on a filter and counted by optical microscopy according to a method published by the World Health Organization (WHO, 1997). Only fibres with diameters less than 3 μm with a length greater than 5 μm and an aspect ratio of at least 3 : 1 are counted. These are defined as 'respirable' fibres, i.e. fibres which can enter the alveolar region of the lung. The method relies on a set of counting rules governing size (as above), number of areas (fields of view) scanned, number of fibres scanned, number density of fibres on the collection substrate, and how to deal with 'bundled' or overlapping fibres.

Optical microscopy is very insensitive for very fine fibres (of diameters less than 300 nm) and is likely to underestimate the total number of fibres collected. The fibre count is therefore only an index of the total number of fibres which may be present, rather than an absolute measure of exposure.

The scope of application of the WHO method is broad, as indicated in the following statement: 'The method . . . is applicable to the assessment of concentrations of airborne fibres in workplace atmospheres, most commonly personal exposures – for all natural and synthetic fibres, including the asbestos varieties, other naturally occurring mineral fibres and man-made mineral fibres' (WHO, 1997). There is therefore no principled reason why this should not be applied to the measurement of MWCNT.

12.3.2 Possible approaches for measuring MWCNT

Multi-walled carbon nanotubes (MWCNT) could be considered to fall within the scope of the WHO method, and it has been suggested that fibre counting could be an appropriate method to assess exposure to MWCNT and other types of

(a)   (b)

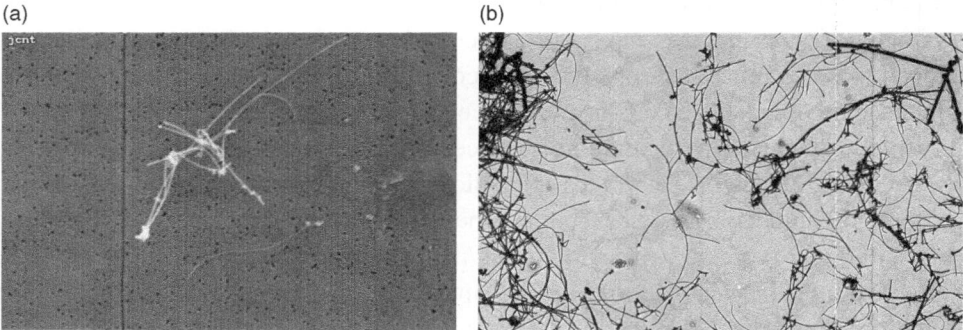

Figure 12.5 Images of a sample type of carbon nanotube by (a) SEM and (b) TEM. Images courtesy of Craig A. Poland, Institute of Occupational Medicine and SafeNano.

high-aspect-ratio nanoparticles (HARN) (BSI 6699–2:2007; BSI, 2007). However, there are a number of problems with this approach. Since the diameters of MWCNT are substantially less than 300 nm, optical microscopy would not detect individual MWCNT. It could, however, detect bundles, ropes or other aggregations of MWCNT.

To overcome this limitation, the option suggested by the BSI (2007) is to use scanning electron microscopy (SEM) or transmission electron microscopy (TEM). Figure 12.5 shows that at least some types of carbon nanotubes are visible using these methods.

However, the necessary higher magnification that these methods could provide would mean that the size of the individual fields of view being assessed would be much smaller. This would have two important effects. First, in order to cover the same area of the sample, many more individual fields would need to be counted. Second, WHO rules state that a fibre is only counted if at least one of its ends is within the field of view (if only one end is in this field it is counted as half a fibre; if both ends are in this field it is counted as a whole fibre). The smaller field of view would mean a greater chance that the fibre ends would not lie within the field and thus not be counted. Both of these effects would substantially increase the time required to count the samples.

A further complexity is that the higher magnifications employed would (as intended) detect very fine fibres that would not be observed by optical microscopy. This would lead to difficulties in making comparisons with limit values for fibres set using optical microscopy. Finally, from published images of MWCNT, it is clear that many of the forms observed are highly entangled. Strict application of the WHO counting rules would require that touching fibres should not be counted.

Only a few published studies have attempted to measure particle release and/or occupational exposure to carbon nanotubes (CNT) in workplace settings. Maynard *et al.* (2004) looked at the exposures associated with production of SWCNT and a simulation of a mixing operation set up in a laboratory. Methner *et al.* (2007) considered handling processes associated with the use of carbon nanofibres and potential release from processing of composite material containing these nanofibres. Han *et al.* (2008) measured the release of MWCNT during CNT manufacture and in processing (at a laboratory scale) where the CNT were being mixed in the production of a composite material. Bello *et al.* (2009) reported exposure to CNT during machining of advanced composites containing CNT. Johnson *et al.* (2010) measured the release of fullerenes (C60), MWCNT and carbon black (CB) during laboratory tasks, weighing and sonication. Tsai *et al.* (2009) reported measurements during CNT production by chemical vapour deposition (CVD).

Only Han *et al.* (2008) used an approach based on the WHO method, in which they collected samples onto a filter using a cowelled sampler and analysed them using TEM to report fibre concentrations. Bello *et al.* (2009) also collected samples from the air onto a filter for electron microscopy analysis, but no fibres were identified during the analysis. Johnson *et al.* (2010) observed clumped MWCNT on filter samples taken close to a laboratory sonication process but did not quantify these. Other investigators used condensation particle counters (CPC), optical particle counters (OPC) or scanning mobility particle sizers (SMPS) to try to detect the release of CNT. However, these devices provide no morphological information and cannot distinguish fibres from other types of particles.

In the Han *et al.* study it is clear that not all of the formal WHO counting rules were applied, but no statement of compliance with these rules is reported. For example, many of the images shown in the paper indicate clumped or overlapping fibres but it is not stated whether these were counted. Fibres counted were short; the reported maximum length of fibres observed was 1500 nm, too small to be considered a fibre under the WHO criteria, where the minimum length of a fibre is 5000 nm.

The fibres were found in only one scenario – blending of CNT into the composite mix – and were found on both personal and area samples. The reported number concentration was very high (193.6 and 172.9 fibre/ml). This compares with the typical asbestos workplace exposure limits of 0.1 fibre/ml but, as stated above, these were short and would not be considered fibres under WHO criteria. Enclosure of the blending activity reduced the fibre count by four orders of magnitude, indicating effective control of fibre release from the process.

This study indicates that it is possible for 'fibre-type' aerosols, in relatively high concentrations, to be generated by certain activities if appropriate control measures are not implemented. It also seems to indicate that there is potential for applying fibre-counting measurement approaches, using TEM or perhaps even SEM in order to more effectively quantify the level of exposure.

Despite the many difficulties, no alternative approaches for counting fibre number concentration for MWCNT have been proposed or developed, and it is difficult to see how quantification of fibre number concentration could be achieved if not by a method derived from the WHO approach, suitably modified as described above.

In the earlier part of this chapter we identified the usefulness of measuring dose in terms of mass. Assessment of release and/or exposure in terms of mass is feasible and has been reported in the literature. Mass-based exposure measurements are recommended in the recent NIOSH Current Intelligence Bulletin, *Occupational Exposure to Carbon Nanotubes and Nanofibers* (NIOSH, 2010). The basis for the NIOSH approach is NIOSH Method 5040 for Diesel Particulate Matter (as Elemental Carbon) (NIOSH, 2003). This method is based on a thermal-optical analysis technique for organic carbon (OC) and elemental carbon (EC), and quantifies total carbon (TC) in a sample as the sum of OC and EC. Although the method was developed for measurement of diesel particulate matter (DPM) in occupational settings, it is applicable to other types of carbonaceous aerosols and is widely used for environmental and occupational monitoring. NIOSH claims an upper limit of quantitation (LOQ) of NIOSH Method 5040 of 7 $\mu g/m^3$ for MWCNT. At the present time, few studies have reported mass measurements for MWCNT and none have independently reported using the NIOSH method.

Han *et al.* (2008) measured total carbon using a portable aethalometer. A recent review of options for CNT detection and analysis (SWA, 2010) concluded that the ELPI spectrometer may have some utility in this respect. Various off-line measurement approaches were reviewed by Tantra and Cumpson (2007), who concluded that none were immediately appropriate for measurement of occupational exposure. Currently there is no consensus on the most appropriate approach. While mass measurements are clearly possible, much more work is needed in order to routinely detect and quantify mass-based exposure to MWCNT with these approaches.

12.4 Conclusions

In estimating the human exposure level equivalent to the rat no-observed-adverse-effect-level (NOAEL) for MWCNT, we have normalised the steady-state lung dose by lung weight, surface area and alveolar macrophage (AM) volume. One can argue that, at non-overload level, the AM volume is irrelevant. In this case, normalising according to lung weight or to surface area gives similar values of 0.1 $\mu g/m^3$.

We assigned a value of 6 to the uncertainty factor associated with the extrapolation from sub-chronic to chronic exposure. The clearance half-time is short for both rats and humans (3 and 12 months, respectively) in comparison to the respective lifetime exposure, resulting in a steady-state lung burden within the sub-chronic exposure time-scale. Therefore, if we do not account for the uncertainty due to extrapolation from sub-chronic to chronic exposure (i.e. not including UF_1 in the definition of UF) we obtain a value of 25 for UF and 0.8 µg/m^3 for the estimated human exposure to MWCNT, which is below the value of 7 µg/m^3 suggested by NIOSH (NIOSH, 2010). Regardless of the argument for UF_1, the calculation above has yielded values less than the control limit suggested by NIOSH, confirming that the value suggested by NIOSH is a reasonable conservative choice of control limit for MWCNT.

The challenge is of course to control the occupational exposure level of MWCNT at the level suggested by NIOSH. As pointed out above, mass measurement is possible, but much more work is needed in implementing the current methods for routine detection and quantification of MWCNT exposure.

References

Bello, D., Wardle, B., Yamamoto, N., *et al.* (2009). Exposure to nanoscale particles and fibers during machining of hybrid advanced composites containing carbon nanotubes. *J. Nanopart. Res.* **11**(1), 231–249.

BSI (2007). Nanotechnologies, Part 2: Guide to Safe Handling and Disposal of Manufactured Nanomaterials. British Standards Institution. PD 6699–2:2007.

Brody, A. R., Hill, L. H., Adkins, B., O'Connor, R. W. (1981). Chrysotile asbestos inhalation in rats: Deposition pattern and reaction of alveolar epithelium and pulmonary macrophages. *Am. Rev. Resp. Dis.* **123**, 670–679.

Dethloff, L. A. and Lehnert, B. E. (1988). Pulmonary interstitial macrophages: Isolation and flow cytometric comparisons with alveolar macrophages and blood monocytes. *J. Leuk. Biol.* **43**, 80–90.

Donaldson, K., Borm, P. J., Oberdorster, G., *et al.* (2008). Concordance between in vitro and in vivo dosimetry in the proinflammatory effects of low-toxicity, low-solubility particles: The key role of the proximal alveolar region. *Inhal. Toxicol.* **20**, 53–62.

Driscoll, K. E., Carter, J. M., Howard, B. W., *et al.* (1996). Pulmonary inflammatory chemokine and mutagenic responses in rats after subchronic inhalation of carbon black. *Toxicol. Appl. Pharmacol.* **136**(2), 372–380.

ECHA (2008). Characterisation of dose [concentration]-response for human health. Guidance on information requirements and chemical safety assessment, chapter R.8. (Helsinki: European Chemicals Agency).

Ferin, J. (1981). Alveolar macrophage mediated pulmonary clearance suppressed by drug-induced phospholipidosis. *Exp. Lung Res.* **4**, 1–10.

Han, J. H., Lee, E. J., Lee, J. H., *et al.* (2008). Monitoring multiwalled carbon nanotube exposure in carbon nanotube research facility, *Inhal. Toxicol.* **20**(8), 741–749.

HSE (2005). Workplace Exposure Limits. EH40/2005, Health and Safety Executive. (London: HMSO).

Johnson, D. R., Methner, M. M., Kennedy, A. J., and Steevens, J. A. (2010). Potential for occupational exposure to engineered carbon-based nanomaterials in environmental laboratory studies. *Environ. Health Perspect.* **118**(1), 49–54.

Jones, A. D., McMillan, C., Johnston, A. M., *et al.* (1988). *Animal Studies to Investigate the Deposition and Clearance of Inhaled Mineral Dusts.* Final report on CEC

Contract 7248/33/026, IOM Report TM/88/05 (Edinburgh: Institute of Occupational Medicine).

Lehnert, B. E. (1993). Defence mechanisms against inhaled particles and associated particle-cell interactions. In Guthrie G. D., Jr. and Mossman, B. T., eds., Health Effects of Mineral Dusts. *Reviews in Mineralogy, vol. 28* (Washington DC: Mineralogical Society of America), pp. 427–469.

Lison, D., Lardot, C., Huaux, F., Zanetti, G., Fubini, B. (1997). Influence of particle surface area on the toxicity of insoluble manganese dioxide dusts. *Arch. Toxicol.* **71**, 725–729.

Maynard, A. D., Baron, P. A., Foley, M., *et al.* (2004). Exposure to carbon nanotube material: Aerosol release during the handling of unrefined single-walled carbon nanotube materia. *J. Toxicol. Environ. Health A*, **67**, 87–107.

Methner, M. M., Birch, M. E., Evans, D. E., *et al.* (2007). Identification and characterization of potential sources of worker exposure to carbon nanofibers during polymer composite laboratory operations. *J. Occup. Environ. Hyg.* **4**(12), D125–D130.

Morrow, P. E. (1988). Possible mechanisms to explain dust overloading of the lungs. *Fundam. Appl. Toxicol.* **10**, 369–384.

NIOSH (2003). *Method 5040 for Diesel Particulate Matter (as Elemental Carbon).* NIOSH Manual of Analytical Methods, 4th edn, issue 3 (Washington DC: National Institute for Occupational Safety and Health). http://www.cdc.gov/niosh/docs/2003-154/pdfs/5040.pdf

NIOSH (2010). *Occupational Exposure to Carbon Nanotubes and Nanofibers.* Current Intelligence Bulletin (Washington DC: National Institute for Occupational Safety and Health). http://www.cdc.gov/niosh/docket/review/docket161A/pdfs/carbonNanotube CIB_PublicReviewOfDraft.pdf

Oberdörster, G., Ferin, J., Morrow, P. E. (1992). Volumetric loading of alveolar macrophages (AM): A possible basis for diminished AM-mediated particle clearance. *Exp. Lung Res.* **18**, 87–104.

Oberdörster, G., Ferin, J., Soderholm, S., *et al.* (1994). Increased pulmonary toxicity of inhaled ultrafine particles: Due to lung overload alone? *Ann. Occup. Hyg.* **38**(suppl.1), 295–302.

Oberdörster, G., Finkelstein, J., Ferin, J., *et al.* (1996). Ultrafine particles as a potential environmental health hazard. *Chest* **109**(suppl. 3), 68s–69s.

Pauluhn, J. (2010). Subchronic 13-week inhalation exposure of rats to multiwalled carbon nanotubes: Toxic effects are determined by density of agglomerate structures, not fibrillar structures. *Toxicol. Sci.* **113**(1), 226–242.

SWA (2010). *Developing Workplace Detection and Measurement Techniques for Carbon Nanotubes* (Canberra: Safe Work Australia). http://www.safeworkaustralia.gov.au/About SafeWorkAustralia/WhatWeDo/Publications/Documents/375/DevelopingWorkplaceDete ctionandMeasurementTechniquesforCarbonNanotubes.pdf

Stöber, W., Morrow, P. E., Hoover, M. D. (1989). Compartmental modeling of the long-term retention of insoluble particles deposited in the alveolar region of the lung. *Fundam. Appl. Toxicol.* **13**, 823–842.

Tantra, R. and Cumpson, P. (2007). The detection of airborne carbon nanotubes in relation to toxicology and workplace safety. *Nanotoxicology*, **1**(4), 251–265.

Tsai, S. J., Hofmann, M., Hallock, M., *et al.* (2009). Characterization and evaluation of nanoparticle release during the synthesis of single-walled and multiwalled carbon nanotubes by chemical vapor deposition. *Environ. Sci. Technol.* **43**(15), 6017–6023.

Tran, C. L., Jones, A. D, Donaldson, K. (1997). Overloading of particles and fibres. *Ann. Occup. Hyg.* **41**(suppl. 1), 237–243.

Walton, W. H. and Vincent, J. H. (1998). Aerosol instrumentation in occupational hygiene: An historical perspective. *Aerosol Sci. Technol.* **28**, 417.

Warheit, D. B., Overby, L. H., George, G., Brody, A. R. (1988). Pulmonary macrophages attracted to inhaled particles through complement activation. *Exp. Lung Res.* **14**, 51–66.

WHO (1997). *Determination of Airborne Fibre Number Concentrations: A Recommended Method, by Phase-Contrast Optical Microscopy (Membrane Filter Method)* (Geneva: World Health Organization).

Yu, C. P. Morrow, P. E. Chan, T. L. Strom, K. A. Yoon, K. J. (1988). A non-linear model of alveolar clearance of insoluble particles from the lung. *Inhal. Toxicol.* **1**, 97–107.

Index